VOLUME FIFTY SEVEN

DEVELOPMENTS IN PETROLEUM SCIENCE

PRESSURE TRANSIENT FORMATION AND WELL TESTING

DEVELOPMENTS IN PETROLEUM SCIENCE 57

Volumes 1–7, 9–18, 19b, 20–29, 31, 34, 35, 37–39 are out of print.

8	Fundamentals of Reservoir Engineering
19a	Surface Operations in Petroleum Production, I
30	Carbonate Reservoir Characterization: A Geologic-Engineering Analysis, Part I
32	Fluid Mechanics for Petroleum Engineers
33	Petroleum Related Rock Mechanics
36	The Practice of Reservoir Engineering (Revised Edition)
40a	Asphaltenes and Asphalts, I
40b	Asphaltenes and Asphalts, II
41	Subsidence due to Fluid Withdrawal
42	Casing Design Theory and Practice
43	Tracers in the Oil Field
44	Carbonate Reservoir Characterization: A Geologic-Engineering Analysis, Part II
45	Thermal Modeling of Petroleum Generation: Theory and Applications
46	Hydrocarbon Exploration and Production
47	PVT and Phase Behaviour of Petroleum Reservoir Fluids
48	Applied Geothermics for Petroleum Engineers
49	Integrated Flow Modeling
50	Origin and Prediction of Abnormal Formation Pressures
51	Soft Computing and Intelligent Data Analysis in Oil Exploration
52	Geology and Geochemistry of Oil and Gas
53	Petroleum Related Rock Mechanics
55	Hydrocarbon Exploration and Production
56	Well Completion Design

VOLUME FIFTY SEVEN

DEVELOPMENTS IN PETROLEUM SCIENCE

PRESSURE TRANSIENT FORMATION AND WELL TESTING

Convolution, Deconvolution and Nonlinear Estimation

By

Fikri J. Kuchuk

Mustafa Onur

Florian Hollaender

ELSEVIER

Amsterdam • Boston • Heidelberg • London • New York • Oxford
Paris • San Diego • San Francisco • Singapore • Sydney • Tokyo

Elsevier
Radarweg 29, PO Box 211, 1000 AE Amsterdam, The Netherlands
Linacre House, Jordan Hill, Oxford OX2 8DP, UK

First edition 2010

British Library Cataloguing in Publication Data
A catalogue record for this book is available from the British Library

Library of Congress Cataloging-in-Publication Data
A catalog record for this book is available from the Library of Congress

ISBN: 978-0-444-52953-4
ISSN: 0376-7361

For information on all Elsevier publications
visit our web site at books.elsevier.com

Printed and bound by CPI Group (UK) Ltd, Croydon, CR0 4YY

Transferred to Digital Print 2012

This book is dedicated to Fikri Kuchuk's daughters Michelle and Eliza; Mustafa Onur's wife and partner in life Ayşegül, and daughters Berfin and Selin; and Florian Hollaender's wife Maud and daughter Anouck and son Alexis

Contents

Preface

This book is devoted to three main topics of pressure transient formation and well testing: namely convolution, deconvolution, and nonlinear parameter estimation. It also presents introduction to testing hardware, test types, flow regimes, and interpretation. These topics are of interest not only from a theoretical pressure transient testing point of view, but also their practical applications for evaluation and characterization of gas, oil, water bearing formations and reservoirs. Since its introduction in the 1920s, pressure transient testing has advanced substantially for better system identification and parameter estimation to include geological complexities, a wide variety of sophisticated reservoir and well models, and measurement uncertainties. For well testing, reservoir models normally include a few wells but not the entire field, unless it is very small. For wireline formation testing, reservoir or formation models (both will be used interchangeably) typically include a one hundred-ft formation around the wellbore with a high degree of vertical resolution. Both formation and reservoir models include storage, skin, fracture, etc. effects.

Pressure transient testing techniques, tools, and gauges have improved significantly during the last four decades. The pressure gauge resolution is now better than 0.01 psi with a few psi absolute accuracy. Most pressure transient tests are now interpreted by using commercial or noncommercial software via powerful computers. Testing and data acquisition systems with wirelines, slicklines, wireless, etc. and interpretation software allow production and reservoir engineers to monitor wellbore pressure remotely, and interpret in real-time at the wellsite or in any office around the globe. This allows completion, production, and reservoir management to better optimize production and recovery throughout the life of the reservoir from early exploration drilling to secondary recovery phases.

Because of these significant advances in hardware, testing, and interpretation techniques, the Society of Petroleum Engineers selected well testing to be the topic of its first monograph. The SPE first monograph *Pressure Buildup and Flow Tests in Wells* by Matthews and Russell (1967) was published in 1967. The second SPE testing monograph *Advances in Well Test Analysis* by Earlougher (1977) was published in 1977. Since then more than six books have been published on pressure transient testing. The third SPE testing monograph titled *Transient Well Testing* (Kamal et al., 2009) was published recently in 2009.

This book is not an undergraduate or graduate level text book, nor a comprehensive treatise on pressure diffusion in porous media; it is rather a practical book, sharing the accumulated knowledge and experience of the authors on interpretation of pressure transient formation and well tests. We have treated each topic by presenting the essential mathematics rigorously without any proof and theorem. We have also treated each topic comprehensively, so the book can be a useful reference for undergraduate and graduate students, and researchers. In addition, we have given many test examples to facilitate the understanding of each topic. Hence, the book will help the readers to develop a better understanding when using commercial or noncommercial pressure transient formation and well test interpretation software.

Chapter 1 presents a brief introduction of pressure transient formation and well test hardware and test types. Chapter 2 presents basic pressure transient formulas for interpretation and flow regime identification. It also introduces skin and wellbore storage effects.

Chapter 3 presents the convolution integral, or Duhamel's superposition theorem, for dealing with simultaneously measured pressure and flow rate data sets in the wellbore or in any other spatial location in the system. These data sets (pressure and rate; pressure, rate, pressure; or pressure and pressure) can also be acquired at different discrete times and spatial locations during the same or different tests. Additionally, these data sets could be acquired at any location in the wellbore from the bottomhole to the wellhead, at any spatial location in the formation, and/or among wells; not necessarily at the same location, e.g., surface flow rate measurements at the production well and downhole pressure measurements at observation wells.

Chapter 4 presents deconvolution techniques for pressure-rate and pressure-pressure data sets. Deconvolution is basically used to solve the convolution integral to obtain the influence (impulse) function of the system (reservoir and well). The deconvolved solution corresponds to the solution of the pressure-diffusion equation for a time-independent or constant-rate boundary condition. Deconvolution is simply the inverse of the convolution process.

Chapter 5 presents nonlinear parameter estimation methods, namely: (1) Least-squares estimation (LSE) and (2) Maximum likelihood estimation (MLE). The LSE is the most widely used estimation procedure in pressure transient testing because it can be applied in an ad hoc manner directly to a deterministic model. The MLE treats observations as random variables with certain probability distributions and thus is more suitable for statistical inference about the match.

In Chapter 6, we briefly present an interpretation methodology for pressure transient formation and well tests. Although each chapter presents synthetic and field examples to illustrate the application of each technique, the interpretation methodology will be applied to two examples in the final Chapter 6.

Throughout this book all equations and definitions are given in any consistent unit system such as SI or CGS. However, when we use the oilfield

(API American Petroleum Institute) units, we will state them explicitly. Furthermore, the flow rate q is assumed to be at the downhole conditions. Therefore, the formation volume factor B is omitted in all equations. The oilfield pressure unit will be psi rather than psia.

We thank following organizations for giving permissions to use a number of figures in the book: American Petroleum Institute for Figure 1, Oilfield Review of Schlumberger for Figures 1.7 and 1.12, SEG for Figure 6.2, and Society of Petroleum Engineers for Figures 1.8, 3.12, and 6.1. We also thank TES Technical Editing Services, particularly Dominic Haughton for re-drawing a number of figures and illustrations.

We are grateful to Schlumberger and the Technical University of Istanbul for allowing us the time and the use of their facilities to write this book. In this book we used copious material from the papers written with our colleagues. In particular, we are indebted to Bilgin Altundas, Cosan Ayan, Ihsan M. Gok, Peter Goode, Tarek Habashy, Peter S. Hegeman, Evgeny Pimonov, T. S. Ramakrishnan, David Wilkinson, and Murat Zeybek.

INTRODUCTION

A pressure transient test is a field experiment that is, like any experiment, only partially controlled. It cannot be repeated under the same conditions, but can be rerun using the results from earlier tests (experiments). There are many ways to interpret pressure transient test data; there are many models with a set of parameters that may match the observed data, but there is only one correct and more than a few probable answers.

The primary objective of pressure transient formation and well testing is to obtain the productivity of a well and properties of the formation from downhole and/or surface pressure and flow-rate measurements. The formation and reservoir information obtained from pressure transient measurements are essential because they reflect the *in situ* dynamic properties of the reservoir under realistic production conditions. When pressure transient test data are incorporated with geoscience data such as geophysical, geological, core, log, etc., it considerably improves reservoir characterization. Particularly, when long-term production data are not available for undeveloped reservoirs, it is necessary to complement the volumetric estimate of oil or gas in-place with long-duration well tests to estimate well productivity and reservoir size before the optimization of the field development.

The rate change at the surface or subsurface creates pressure diffusion (transient) in porous but permeable formations. The pressure diffuses away from the wellbore deep into the formation and brings information about the properties and characteristics of the reservoir. This process is traditionally called *pressure transient well testing*. Pressure transient tests are also conducted with Wireline Formation Testers (WFT). Such tests are called *formation pressure transient tests.*

Pressure transient formation and well testing (*reservoir testing*) for oil, gas, and/or water exploration, and production and injection wells are two of the most powerful tools for determining well and reservoir parameters under dynamic conditions. Because they are dynamic and direct, pressure measurements provide essential information for well productivity and dynamic reservoir description and hold critical importance for exploration as well as production and reservoir engineering. Introduced in the 1920s, pressure transient well testing was first used for taking fluid samples and obtaining average reservoir pressure. Gradually, in addition to pressure and samples, formation permeability and skin (wellbore damage or stimulation) have been also obtained from transient pressure measurements. Innovation and refinements in testing hardware have made it possible to measure pressure accurately across the sandface (downhole). Although measuring downhole pressure remains one of the fundamental functions of reservoir testing, today it is possible to measure downhole flow rate, fluid density, and temperature simultaneously with pressure as a function of time and depth in the wellbore as well as taking fluid samples.

Acquiring accurate downhole pressure data is the most critical part of pressure transient testing for the interpretation. The first downhole well-test system was introduced by the Johnston brothers in the 1920s and was called the *formation tester*. This system was basically a packer system that temporary isolated the zone to be tested from the well hydrostatic pressure. After the packer setting, the downhole valve was opened to produce the formation fluids through the drillstring. In this system, both flow rate and pressure were measured at the surface, and bottomhole pressure was obtained from the hydrostatic pressure of the fluid in the drillstring and surface pressure measurements. O'Neill (1934) reported that the Johnston formation tester was used the first time in the Mid-Continent, Texas in 1926. In parallel, Geophysical Research Corporation of Amerada introduced the first bottomhole pressure gauge in 1929, and it was called the *Amerada gauge* or *bomb*. In 1930, Millikan and Sidwell (1931) reported that the Amerada gauge was used in several wells in Oklahoma in 1930.

Since its introduction in the 1920s, pressure transient testing has held a great promise for drilling, production, and reservoir engineers. It offers a potential to assess well condition, and to obtain formation transmissibility, reservoir pressure, and inhomogeneities, such as faults and fractures, and heterogeneities. Circular reservoirs with a constant pressure or no-flow boundary condition have been well studied since the beginning of the industry. In fact unsteady-state (transient) solutions for both constant pressure and no-flow boundary circular reservoirs as a function of time and outer radius were presented by Moore et al. (1933) and Hurst (1934), based on earlier works on heat conduction.

Furthermore, Moore et al. (1933) also presented a history match, as shown in Figure 1, of a pressure transient test to their infinite-acting 1D radial solution to estimate formation permeability. The Moore et al. (1933) pressure transient test consisted of a drawdown test, during which both pressure and flow rate were measured simultaneously, and a subsequent buildup test. Moore et al. (1933) described that the flow rate was measured from the changing annulus liquid level by using a sonic tool. Perhaps this was the first downhole flow rate measurements obtained with the downhole transient pressure. Furthermore, they attributed the change in the downhole flow rate to the wellbore storage effect during the production period. As can be seen from Figure 1, the match is excellent [digitized from a 2-by-2.5-inch graph given by Figure 2 of Moore et al. (1933)]. It should be pointed out that the match was obtained manually by trial-and-error. It should be also noticed that this is a very short test, about 2 hr, and has a few measured data points (about 10). Their infinite-acting 1D radial solution did not include skin and wellbore storage effects because van Everdingen and Hurst (1949) formulated the wellbore storage and van Everdingen (1953)

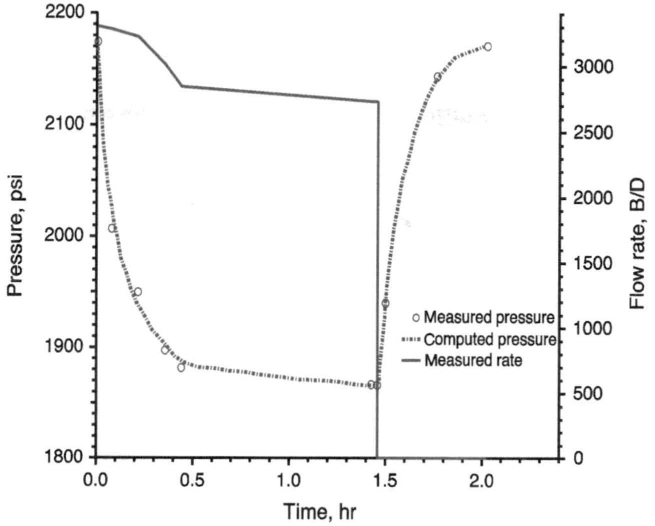

Figure 1 The history match of drawdown and buildup tests given by Moore et al. (1933).

and Hurst (1953) introduced the concept of damage skin about 20 years later. The five important contributions of the work of Moore et al. (1933) are:

1. The first transient (unsteady-state) solution of pressure diffusion in 1D radial porous media,
2. The type-curves for dimensionless transient pressure versus dimensionless time for infinite acting, and both constant-pressure and no-flow boundary circular reservoirs, and also as a function of outer reservoir radius,
3. The first downhole flow rate measurements and their usage in well test interpretation,
4. The first history matching for parameter estimation, and
5. The first realization of wellbore storage effects (Ramey, 1976b).

In his classic book on unsteady-state flow problems, Muskat (1937a) presented many analytical solutions to both incompressible and compressible single-phase fluid flow in porous media and the relationship between the flow rate (input) and pressure (output) as a convolution integral (Duhamel's principle). Furthermore, Muskat (1937b) presented a trial-and-error procedure to determine both reservoir pressure and formation permeability from downhole pressure buildup data.

Many of the modern developments in pressure transient test interpretation and the understanding of the theoretical reservoir and well behaviors have been made by applications of Laplace transforms and Green's functions to fluid flow problems. In reservoir engineering, van Everdingen and Hurst (1949) were the first to apply Laplace transforms to solve compressible single-phase fluid flow problems in 1D radial infinite and bounded reservoirs with both constant-pressure and constant-rate inner boundary conditions on a finite-radius cylindrical wellbore and no-flow and constant-pressure outer boundary conditions.

They also presented an equation describing the wellbore storage phenomenon. Muskat (1937a) used Green's functions to solve a few steady-state and transient flow problems.

Horner (1951) applied the superposition (Duhamel's) principle to the constant-rate line-source solution to obtain a pressure buildup equation similar to the one described by Theis (1937). He presented an interpretation technique (now called the *Horner method*) to estimate both the formation permeability, the distance to a sealing fault, and to obtain the static reservoir pressure (extrapolated) from pressure buildup test data. At around the same time, Miller et al. (1950) presented a different semilog interpretation technique (now called the *MDH method*), where the shut-in pressure was plotted as a function of the logarithm of the shut-in time for buildup tests. The concept of damage skin was not known to Horner (1951), but fortunately, both the extrapolated reservoir pressure and the permeability obtained from the Horner and MDH methods are independent of skin factor for buildup tests, as is now well known.

The decade between 1950 and 1960 were very productive years for the understanding of pressure behavior of reservoirs and wells. New phenomena such as damage skin (Hawkins, 1956; Hurst, 1953; van Everdingen, 1953), geometric skin (Brons & Marting, 1961; Hantush, 1957; Nisle, 1958), and wellbore storage (Moore et al., 1933; van Everdingen & Hurst, 1949), and new interpretation techniques such as determination of average pressure (Matthews et al., 1954), slug tests for estimating formation transmissibility (Ferris & Knowles, 1954), isochronal tests for gas wells (Cullender, 1955), the effect of multiphase flow on pressure buildup tests (Perrine, 1956), and wireline formation testing (Lebourg et al., 1957) were introduced.

In the 1960s, many papers were published on layered, fractured, and heterogenous reservoirs, and a few papers on fractured and partially penetrated wells in homogenous systems. Further investigations were performed on wellbore storage and skin effects on pressure transient tests, interference tests for characterization of inter-well properties, and characterization of faults and hydraulic fractures, etc. In addition to the radial and linear flow regimes identified by Muskat (1937a), spherical, unit-slope, and semi-radial flow regimes were identified, and better understanding of the linear flow due to hydraulic fractures was provided. The SPE first monograph *Pressure Buildup and Flow Tests in Wells* by Matthews and Russell (1967) was published in 1967 on the fundamentals of well testing.

In the 1970s, a new direction in pressure transient testing was taken to interpret difficult pressure data such as tests that are too short and affected by wellbore storage and skin. Actually, the analysis was found to be simple, and important results began to appear by the early 1970s. For instance, log-log type curves presented by Ramey (1970) and McKinley (1971) were very useful for interpreting pressure transient data dominated by skin and wellbore storage effects. Often, these type curves frequently matched field data for the entire duration of a test. Ramey (1976b) presented a detailed study on the well testing work done from 1930 to 1975.

The new results were so remarkable that SPE published its second monograph *Advances in Well Test Analysis* by Earlougher (1977) on well testing in 1977. For the identification of certain characteristics of reservoir-well systems, terms like semilog-straight line, unit slope, half slope, and semi-radial were extensively used for the analysis of transient tests. Furthermore, the time for the end of wellbore storage and skin effects was given quantitatively (Ramey, 1976b).

Particularly for exploration wells, a new generation of wireline-conveyed formation testers was introduced in the early 1970s to obtain formation fluid samples and pressure gradients. Although they had first been introduced in the 1950s, those tools had no repeat capability, while the new tools were able to measure formation pressure along the wellbore at many different locations, from which pressure gradient and fluid contacts (Schultz et al., 1975) could be obtained.

Since the 1970s well testing has become more sophisticated, with developments in downhole electronics, pressure sensors, computer technology, better reservoir models, and applied mathematics. Pressure measurements have been radically improved by the introduction of quartz crystal pressure gauges. Better and faster methods (Stehfest, 1970) were introduced for inverting the transient solutions from the Laplace transform domain to the time domain. The Laplace transform technique has been one of the useful methods for solving advanced mathematical models for the pressure diffusion equation. Type curves and analysis methods based on pressure derivatives (Bourdet et al., 1983; Tiab & Crichlow, 1979; Tiab & Kumar, 1980) were introduced in the petroleum engineering literature. Introduction of these type curves and analysis methods further enhanced the identification of flow regimes, the likelihood of obtaining a unique type-curve match and a consistent analysis. Finally, type-curve matching has been executed on computers, and nonlinear estimation methods have been introduced.

From 1930 to 1980, analytical solutions have been presented for single-phase fluid flow for many complicated well and boundary conditions. Analytical methods are applied to fractured, layered, and laterally and radially composite systems. For more complicated and heterogeneous systems, finite difference and finite element methods have been used for the simulation of single- and multi-phase fluid flow in porous media. Other numerical methods, such as the boundary integral equation method, have also been used for single- and multi-phase fluid flow simulation. In general, single-phase numerical models are called the *well simulator* or *well model* and are mainly used for well test interpretation. The modeling capabilities for pressure transient testing have been remarkable. Pressure transient formation and well testing developments from 1980 to 2008 in terms of analytical and numerical solutions, interpretation, software, and tools were presented in great detail by Kamal et al. (2009), and Gringarten (2008) presented a detailed summary of the evolution of well test interpretation *From Straight Lines to Deconvolution*. Therefore we will not cover the period from 1980 to 2008. Moreover, the third SPE testing monograph (Kamal et al., 2009) has for the first time incorporated wireline formation testing into pressure transient testing.

Nomenclature

Abbreviations

BU	buildup
DD	drawdown
DST	drillstem test
FRFR	first radial flow regime
IPPT	interval pressure transient test
L-M	Levenberg-Marquardt
LWD	logging while drilling
MAP	maximum *a posteriori* estimate
ML	maximum likelihood
MLE	maximum likelihood estimation
QA	quality assurance
QC	quality control
pdf	probability density function
p-p	pressure-pressure
p-r	pressure-rate
PLT	production logging tool
r-p	rate-pressure
RMS	root-mean-square error
SPE	Society of Petroleum Engineers
TRFR	third radial flow regime
UWLSE	unweighted least-squares
WFT	wireline formation tester
WLS	weighted least-squares

Symbols

a	model parameter or a generic constant for a probability density function
b	model parameter or a generic constant for a probability density function
c	compressibility

C	wellbore or tool storage coefficient
$\mathbf{C}$	matrix-valued function of the response coefficients z
C_A	Dietz shape factor
$\mathbf{C}_D$	covariance matrix for pressure measurement errors
$\mathbf{C}_M$	*a priori* covariance matrix of model parameters
$\mathbf{C}_{MP}$	*a posteriori* covariance matrix of model parameters
d_w	distance between the observation and active well
cov	covariance
$\mathbf{D}$	matrix of curvature constraints
det	determinant of a square matrix
$\mathrm{E_i}, \mathrm{E_k}, \mathrm{E_1}$	exponential integral functions
erfc	complimentary error function
exp	exponential function
$\mathbf{f}$	vector of model response functions
$\mathbf{F}$	Dawson's integral
g	unit-impulse response of the system
g_f	unit-impulse response of the formation without skin
g_w	unit-impulse response at the well including only skin
g_{wf}	unit-impulse response at the well including skin and wellbore storage
$\mathbf{g}$	vector of unit-impulse responses
G	Green's function
$\mathbf{G}$	sensitivity coefficient matrix
h	thickness
h_w	thickness of the producing interval
H	Heaviside unit-step function
$\mathbf{H}$	Hessian matrix
$\mathbf{i}$	unit coordinate vector in the x-direction
$\mathbf{I}$	identity matrix
$\mathbf{j}$	unit coordinate vector in the y-direction
J_w	rate-normalized pressure change
J_0	Bessel's function of the first kind
k	permeability
k_s	spherical permeability
$\mathbf{k}$	unit coordinate vector in the z-direction
$\mathrm{K_0}, \mathrm{K_1}$	modified Bessel's functions of the second kind
l_w	half length of packer interval
L	likelihood function
L_x	length of the reservoir in the x-direction

L_y	length of the reservoir in the y-direction
L_w	horizontal half well length
m	slope on a log-log plot or model parameter
m_r	radial slope (slope of a straight line on a semilog plot of pressure vs. time)
$\mathbf{m}$	vector of model parameters
$\mathbf{m}_{\text{prior}}$	vector of prior means of model parameters
$\mathbf{m}_{\infty}$	maximum *a posteriori* estimate after conditioning to IPTT data
M	number of model parameters
N_g	number of nodes used to generate deconvolved unit-rate responses
N_d	number of observed data points
N_p	number of measured pressures
N_q	number of flow rates
N_s	number of observed data sets
p	pressure
$\Delta \boldsymbol{p}_m$	vector of pressure change measurements
q	flow rate
q_N	normalized flow rate
q_{wf}	rate impulse response due to constant-pressure wellbore boundary condition
$\mathbf{q}_c$	vector of flow rates
Q	amount of fluid volume produced instantaneously
r	radial distance
r_w	wellbore radius
r_e	external radius of the reservoir
$\mathbf{r}$	vector of residuals or position vector
s	Laplace transform variable
S	skin factor or some of the squares of the residuals
O	objective function
t	time
t_H	Horner time ratio
t_p	producing time
$\mathbf{t}$	vector of times at which observed data are recorded
X_a	x-coordinate of the active well from the lower left corner of the rectangle
X_o	x-coordinate of the observation well from the lower left corner of the rectangle
y_c	model response function

y_m	observed (measured) response
$\mathbf{y}_c$	vector of model response functions
$\mathbf{y}_m$	vector of observed (measured) responses
Y_a	y-coordinate of the active well from the lower left corner of the rectangle
Y_o	y-coordinate of the observation well rom the lower left corner of the rectangle
V	volume
V_p	reservoir pore volume
w	weight factors
$\mathbf{W}$	matrix of weight factors
z	response coefficients defined as $\ln(tg)$
$\boldsymbol{z}$	vector of response coefficients
z_o	vertical distance from center of packer interval to the vertical probe
z_w	vertical distance for the bottom of the source layer to the center of packer interval or horizontal well
ε	stochastic error term
ψ	convolution function
$\boldsymbol{\psi}$	distribution parameter
θ	binary operator or well inclination angle
δ	Dirac delta function
$\boldsymbol{\vartheta}$	Levenberg-Marquardt regularization parameter
ν	relative error weight
λ	mobility, i.e., permeability divided by viscosity or regularization parameter
κ	convolution kernel
ϕ	porosity
μ	viscosity of the fluid
∇	gradient operator
$\mathcal{G}$	impulse function for the pressure-pressure convolution/deconvolution
Ω	steady-state shape factor
ρ	correlation coefficient
σ	standard deviation
σ^2	variance
φ	storativity defined as the product of porosity and total isothermal compressibility
$\mathcal{F}$	Fourier transform operator
$\mathcal{L}$	Laplace transform operator

$\mathcal{P}$	integral of the convolution kernel κ
$\|\|\ \|\|_F$	Frobenius norm of a matrix
$\|\|\ \|\|_2$	l_2 norm of a vector

Subscripts

c	computed or curvature constraint
d	deconvolved
D	dimensionless
e	external
h	horizontal or horizontal probe
i, j	index
int	intersection time
k	iteration index
m	measured
nD	turbulence and intertial effects (non-Darcy)
o	initial or observation well
p	packer or pressure
pss	pseudo-steady state
q	rate
r	radial direction
s	sink probe
sf	sandface
sp	spherical
t	total
u	unit-rate
v	vertical or vertical probe
w	wellbore
z	vertical direction
θ	angular direction

Superscripts

T	transpose
-1	inverse
$'$	first derivative
$''$	second derivative
$-$	Laplace transformation
$\equiv$	Fourier transformation

CHAPTER 1

Formation and Well Testing Hardware and Test Types

Contents

1.1. Testing Hardware

Pressure transient testing hardware are divided into three basic categories according to their conveyance systems:

1. Wireline,
2. Pipe, tubing, coil tubing, slickline, and
3. Permanent.

Moreover, wireline units are sometimes combined with pipes or coil tubing. There are also other possible combinations. In principle, formation testing is usually conducted with a wireline unit and well testing is usually conducted with pipe, tubing, coil tubing, slickline, and/or permanent systems.

1.1.1. Well testing hardware

Well testing hardware basically has to conduct pressure transient tests to acquire pressure data and fluid samples in the wellbore and/or at the surface and perform a wide variety of other tasks related to safety and efficiency, and sometimes including perforating. Both Kikani (2009) and Schlumberger (2006) have given good details for both formation and well testing hardware, and other operation related tasks.

Developments in Petroleum Science, Volume 57
ISSN 0376-7361, DOI: 10.1016/S0376-7361(10)05707-9

During well testing, a vast amount of data has to be collected. For a basic test, the downhole pressure has to be acquired as a function of time, and the corresponding flow rate has to be measured downhole and/or at the surface. During transient tests conducted with production logging tools (PLT), the downhole flow rate, density, and temperature of the fluid are measured simultaneously with pressure as a function of time and vertical depth.

A wide variety of bottomhole and surface tools (hardware) are available from service and oil companies for performing well tests. Many improvements in downhole and surface testing equipment and technology during the last two decades have satisfied the safety requirements and environmental concerns with better efficiency. Nevertheless, a proper test design, correct handling of surface effluents, high performance gauges, flexible downhole tools and perforating systems, wellsite validation, and comprehensive real-time interpretation are keys to successful well testing.

1.1.1.1. Drillstem test (DST)

Because of their wide usage and crucial importance for the exploration and appraisal phases of reservoirs, DSTs are briefly described here. The first DST downhole well test system was introduced by the Johnston brothers in the 1920s and was called the *formation tester* (also referred to *Johnston formation tester*), as shown in Figure 1.1. As stated above, this system was basically a packer system that temporary isolated the zone to be tested for productivity from the well hydrostatic pressure. After packer setting, the downhole valve was opened to produce the formation fluids through the drillstring. In this system, both flow rate and pressure were measured at the surface, and bottomhole pressure was obtained from the hydrostatic pressure of the fluid in the drillstring and surface pressure measurements.

DST hardware has gone through many changes and modifications since its introduction in the 1920s. For more details on modern DST hardware, the readers should refer to Earlougher (1977), Streltsova (1988), and Murray (2009). A openhole DST string, as shown in Figure 1.2, is a temporary downhole completion with packer elements that are designed to provide a perfect hydraulic seal between the formation to be tested and the wellbore, which may contain drilling and/or completion fluids at a pressure usually higher than the formation pressure. With packers, an opening and closing flow control valve system for production and buildup, sample chambers, pressure and temperature gauge carriers (presently a few pressure gauges), and operation control system are assembled together as a DST string and run on drillpipes or special tubing strings. Another important element is a multiflow evaluation system that makes possible repeated flow and shut-in cycles rather than the single flow and buildup periods offered by earlier tools (Vella, Veneruso, Lefoll, McEvoy, & Reiss, 1992). Casedhole DST strings

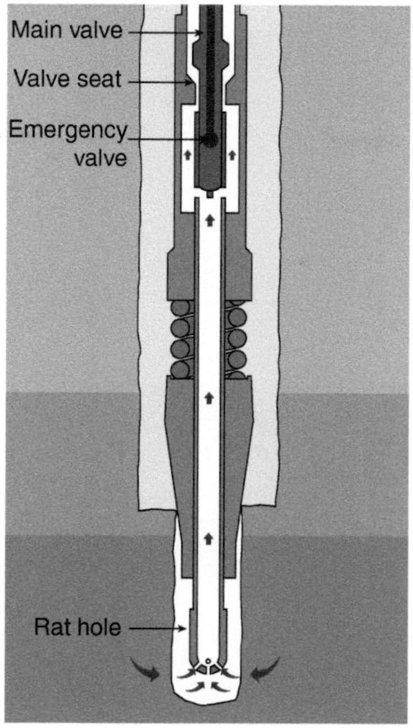

Figure 1.1 Johnston formation tester setup from the Johnston brochure (1927).

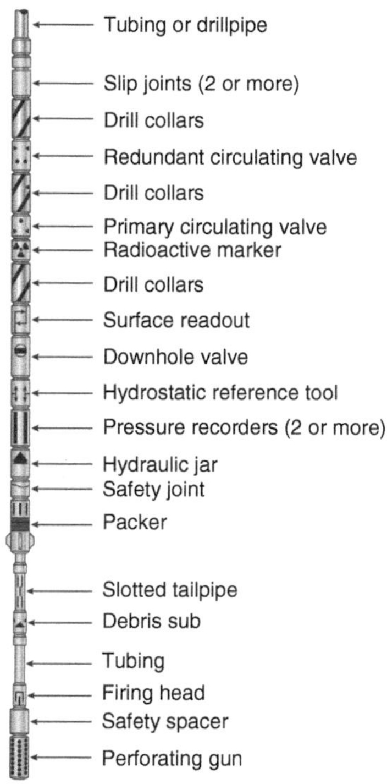

Figure 1.2 A basic schematic of a DST setup, after Schlumberger (2006).

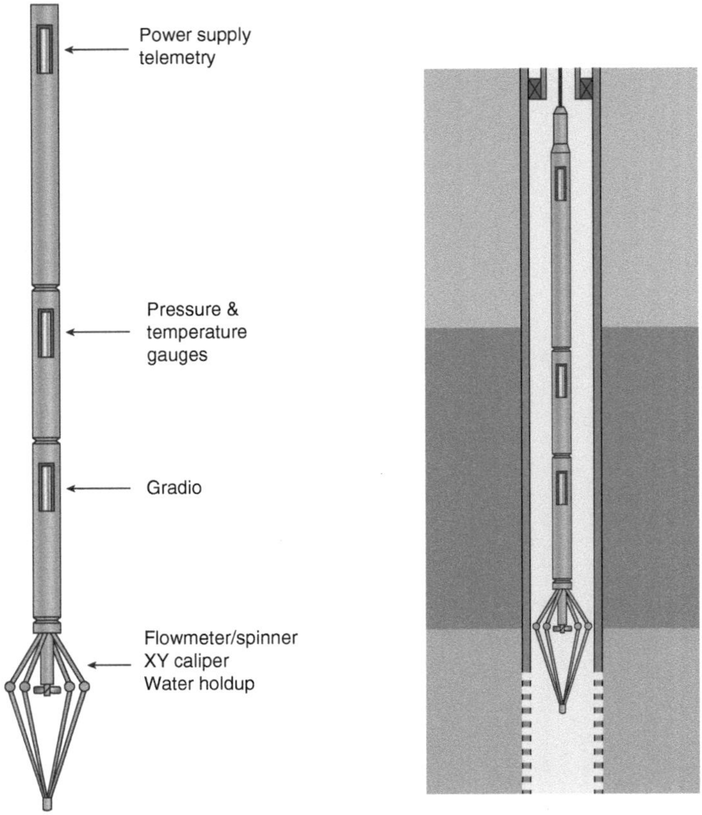

Figure 1.3 Production logging tool string and setup in a wellbore, after Lenn et al. (1990).

are slightly different from openhole DSTs, but essentially provide the same functionalities, and can be run on slickline, wireline, coil tubing, or pipe.

1.1.1.2. Testing with production logging tools (PLT)

There are many types of flow metering devices used in production logging tools to measure flow rate, holdup, pressure, temperature, and density of the wellbore fluid. Figure 1.3 presents a typical string of a production logging tool that measures pressure and flow rate simultaneously. The same figure also presents a PLT testing downhole setup. Several different tool combinations may be required to perform correct rate measurements under multiphase conditions: spinner, caliper, density, temperature, and fluid hold-up sensors constitute a typical production logging tool string. In the SPE Transient Testing Monograph, Kikani (2009) has given details of pressure and flow rate tools.

PLTs have increased the scope of transient well testing by providing new measurements. Drawdown tests, for which it has often been difficult to keep the flow rate constant, can today provide interpretable pressure transient data simultaneously measured with flow rate data both acquired by production logging tools. Thus, the possibility of obtaining reliable information about the reservoir features by using characteristic features of both transient tests (drawdown and buildup) has increased considerably. In addition, PLTs provide a production profile as a function of depth that is related to the thickness of the producing zone or open interval that is necessary for well test interpretation. Furthermore, PLTs give oil, water, and gas flow rates as a function of depth that are essential to conduct multilayer tests.

A propeller-type spinner (rotor) is most commonly used in PLTs to measure flow rate. There are three main basic types of spinner flowmeter: (1) Continuous, (2) Fullbore, and (3) Packer types. Venturi Flowmeters are commonly used at the surface and in permanent downhole systems to measure flow rates. There are many other flow rate measurement devices such as acoustic, tracer, etc. but they are used infrequently.

Testing with PLTs has opened up new possibilities for testing of low-energy wells that otherwise would not start to flow if the well is shut in, injection wells that may go on vacuum during falloff tests, layered reservoirs, and high-flow-rate wells where fluid momentum in the wellbore hinders pressure measurements. Moreover, measuring downhole flow rate and its profile along the wellbore offers a simple way of obtaining the effective thickness of the producing interval, where fluids enter into the wellbore, for permeability and skin estimation, as well as determining perforation efficiency.

1.1.1.3. Testing with downhole shut-in tool

When the fluid is moving upward in the production string in naturally flowing oil wells, as the pressure drops, the oil expands. At some point in the production string, the wellbore pressure becomes lower than the bubble-point pressure, and then the solution gas in the crude oil comes out continuously until it reaches the separator or the test tank. The gas evolution process makes the compressibility of the fluid in the production string vary considerably, which in turn produces a variable wellbore storage condition in the wellbore. Furthermore, when a well with a multiphase fluid in the wellbore is shut in, gas, oil, and water, if any, segregate and it is called *phase redistribution* (Fair, 1981). With and without phase redistribution, wellbore storage dominates the pressure behavior of a well for a significantly long time, particularly during the early time period. For horizontal wells, the wellbore storage effect continues for a long time. For instance, as shown in Figure 1.4, the buildup pressure and its derivative of the horizontal well JX-2 in the Prudhoe Bay field (Kuchuk, Goode, Brice,

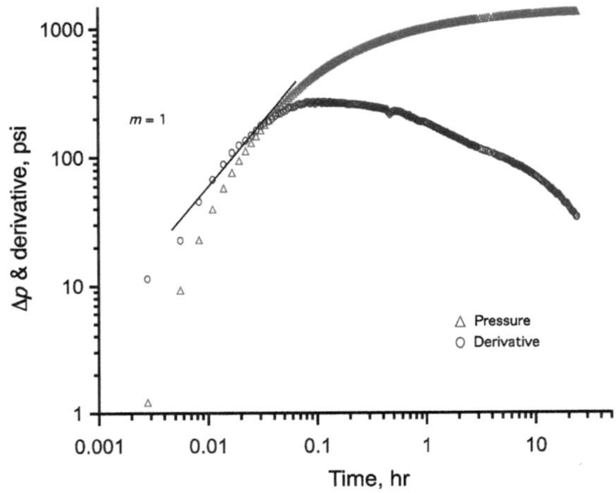

Figure 1.4 The buildup pressure change and its derivative for the horizontal well JX-2 in the Prudhoe Bay field.

Sherrard, & Thambynayagam, 1990b; Rosenzweig, Korpics, & Crawford, 1990) is dominated by the changing wellbore storage (see also Figure 2.10 as another example). Therefore, downhole shutting is crucial for most wells to minimize the wellbore storage effect in order to obtain information about skin, permeability, reservoir characteristics, etc. of the system as well as to conduct the test for a manageable time period. Furthermore, minimizing the wellbore storage effects by downhole shutting is important for system identification. Figure 1.5 presents a schematic representation of a downhole shutting system. Pressure and temperature in a wellbore below the downhole shut-in valve are normally transmitted to the surface in real-time.

Figure 1.6 presents the derivatives for two buildup tests from the horizontal well X-184 in a carbonate reservoir (Kuchuk, 1997). For the first buildup test, Well X-184 was shut in at the surface for almost 3 days, after producing for approximately 190 days following the completion of the well. For the second buildup test, Well X-184 was shut in at the downhole for 4.5 days in order to assess the rapid production decline after one year of production (Kuchuk, 1997). As can be seen in Figure 1.6, the early time slope of the first buildup derivative is much greater than the unit slope $m = 1$, while after this initial constant wellbore storage period, the derivative of the second buildup exhibits a unit-slope $m = 1$ due to the wellbore-storage dominated flow period. After this period, the derivative starts declining rapidly and becomes almost flat at about 0.06 hr, clearly indicating the first radial flow regime before the effect of the bottom no-flow boundary, which is the nearest boundary to the wellbore. In any case, these two derivatives, less than a year apart, from the same well have almost no similarity. The first buildup derivative was mainly the result of events in the wellbore.

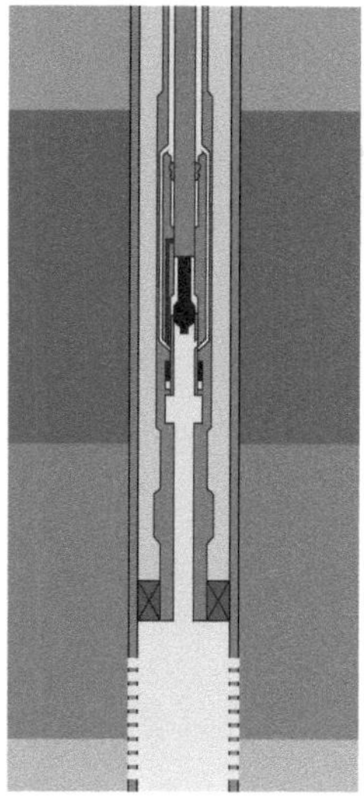

Figure 1.5 A basic schematic of a downhole shut-in setup, after Ayestaran (1987).

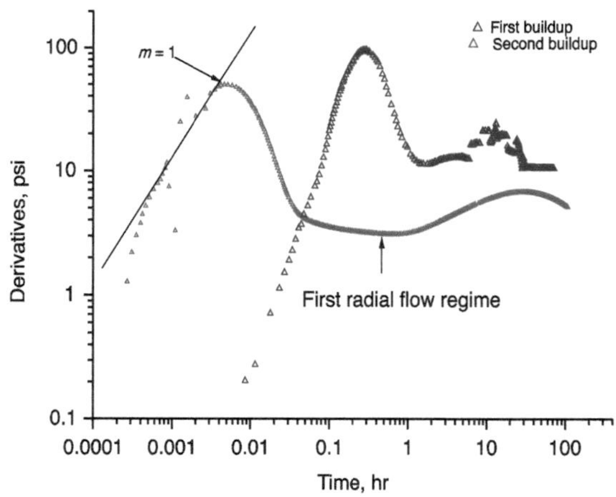

Figure 1.6 Derivatives of two buildup tests from the horizontal well X-184.

1.1.1.4. Permanent downhole testing systems

The first semi-permanent downhole system was described by Riordan (1951) as *Surface Indicating Pressure, Temperature and Flow Equipment* for evaluating flow characteristics of wells, particularly to locate gas entries. Bannies (1957) described another semi-permanent downhole system to continuously measure downhole pressure and to regulate the fluid level in oil wells, although these two first semi-permanent downhole systems were not for transient well testing per se. Lozano and Harthorn (1959) described a

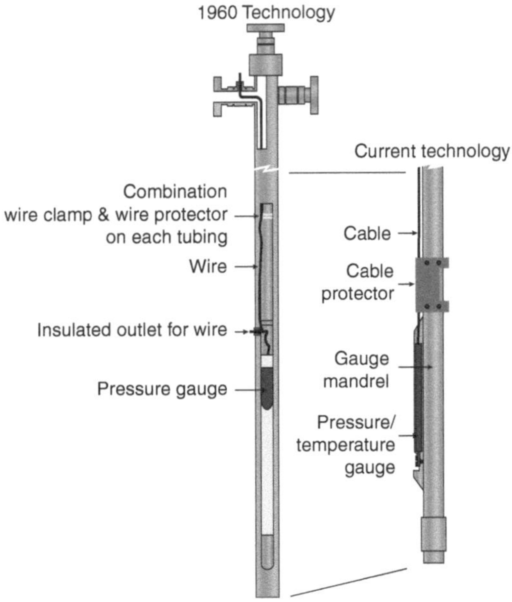

Figure 1.7 Schematic of permanent downhole systems from two different time periods, after (Baker et al., 1995).

permanent downhole pressure-temperature system with surface monitoring using Maihak bottomhole pressure gauges that were field tested in three wells in the San Joaquin Valley, California. The pressure measurements from this system were compared with downhole buildup pressure measurements from Maihak gauges with wireline. The difference between two gauges was from 0.7% to 19.4% in different wells. Maihak gauges and transmitters, and surface data acquisition systems were improved significantly in the 1960s (Engel, 1963; Kolb, 1960). Figure 1.7 on the left side presents a typical permanent monitoring system from the 1960s. As shown in this figure, the downhole pressure gauge is mounted on the bottom of the tubing, and the communication with the gauge from the surface is accomplished via a cable mounted on the outside of the tubing. This type of permanent monitoring system normally creates flow restrictions and also prevents access to the wellbore below the gauge (Baker et al., 1995).

Kolb (1960) presented a remarkable reservoir limit test example, as shown in Figure 1.8, acquired by a permanent downhole system. As can be seen from this figure, the linear plot of flowing wellbore pressure vs. time exhibits a perfect pseudosteady-state flow regime.

Permanent monitoring systems were not frequently used until the 1990s; however, many components of the system have continuously improved since the 1960s. In the North Sea, particularly in the deep sea areas, new permanent monitoring systems were installed with completion in the 1990s

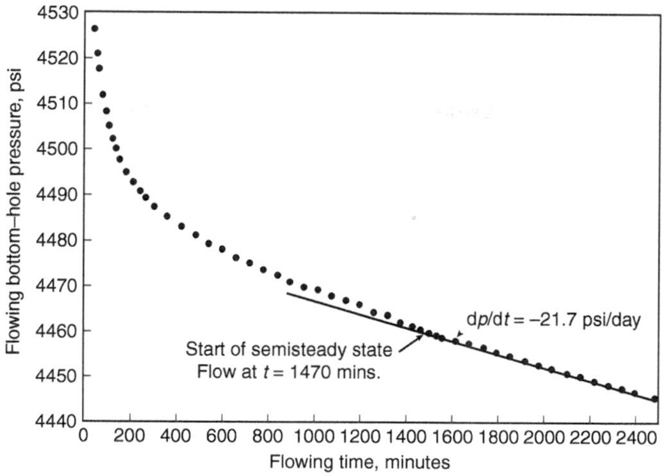

Figure 1.8 A reservoir limit test from a permanent downhole pressure gauge, after Kolb (1960).

(Bezerra, Da Silva, & Theuveny, 1992; Carter & Morel, 1990; Shepherd, Neve, & Wilson, 1991), and the number of installations has increased rapidly due to the reliability, longevity, and accuracy of pressure gauges (Baker et al., 1995).

Most permanent monitoring systems as shown in Figure 1.7 are installed during the completion of the well and provide uninterrupted downhole pressure and temperature data, access to the data being almost instantaneously available anywhere in the world. In some installations they can also acquire production rates and/or fluid densities, particularly if free gas is not present in the fluid. For downhole pressure, flow rate, temperature, etc. monitoring, reliability is an essential requirement for permanent systems because downhole well intervention, repair, and replacement are very costly and sometimes not even feasible.

Modern systems use downhole pressure/temperature sensors that are designed for permanent installations. These can be used to monitor tubing and/or annulus pressure as well as downhole two-phase liquid flow by a venturi flowmeter configuration. Various recording devices may be used at the wellhead to record and transmit the relevant data. The downhole and surface monitoring systems and communication equipment integrate downhole permanent sensors, surface sensors, surface controls, a remote terminal unit, remote control (via radio, telephone line, leased lined, private network, etc.) and application software to control well safety and optimize production and recovery.

Since the 1990s, reservoir permanent monitoring systems have been installed in numerous reservoirs around the world and provide essential data for production and reservoir management. Here are a few applications of permanent monitoring systems (Baker et al., 1995):

Reservoir Pressure Monitoring. Continuous monitoring of reservoir pressure is very important to minimize subsidence, sand production, early gas and/or water breakthroughs, and to avoid producing below the bubble point in the reservoir, keeping a balanced production throughout the field, particularly in compartmentalized reservoirs, etc.

Well Production Optimization. Nodal and gas lift analyses can be performed continuously to optimize well performance.

Reservoir Performance and Prediction. Continuous pressure and temperature, and in some cases production and density data will reduce uncertainties in reservoir description and history matching, thus future predictions, estimated reserves and recovery will be more accurate. Furthermore, with accurate models, both gas and water conformance can be improved to increase sweep efficiency and to avoid early gas and water key breakthroughs.

Pressure Transient Well Tests. Buildup and interference tests data can be automatically acquired when wells are intentionally or unintentionally shut in, and drawdown data when the well is put on line. With derivatives and deconvolution (covered in Chapter 4), permanent downhole pressure and flow rate data become critical for system identification and reservoir characterization in addition to conventional analyses. Because deconvolution derivatives may extend the time for system identification and may be longer than the duration of the buildup test, therefore permanent downhole pressure and flow rate data may increase the scope of investigation (Gringarten, 2008).

In the recent *SPE Transient Well Testing* monograph, Kikani (2009) has given a brief introduction of permanent downhole systems, and Houze, Kikani, and Horne (2009) have given details for handing and interpreting permanent downhole pressure and production data, particularly data storage and filtering. Figure 1.9 presents permanent downhole pressure and surface production data given by Unneland, Manin, and Kuchuk (1998). Notice that the pressure data are very noisy because this is the flowing well pressure. Therefore, it is difficult to interpret this data using derivative and deconvolution techniques, on the other hand using the new deconvolution algorithms of (Levitan, 2005; Pimonov, Ayan, Onur, & Kuchuk, 2009a; Pimonov, Onur, & Kuchuk, 2009b; von Schroeter, Hollaender, & Gringarten, 2004), these data could have been very valuable with buildup test data. However, for this case the reservoir model was known from an earlier two-day buildup test, Unneland et al. (1998) presented an interpretation of this data using the variable-rate decline curve analysis technique. It was stated that the interpretation of the 75-day permanent downhole pressure data had overcome the ambiguity of the reservoir model obtained from the two-day buildup test.

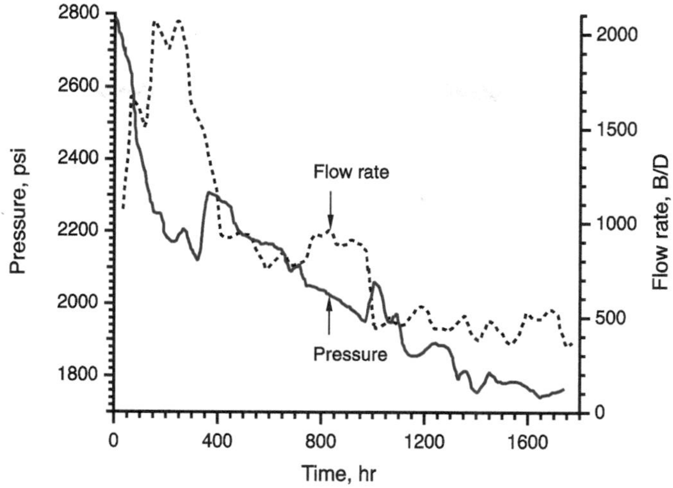

Figure 1.9 Permanent downhole pressure and surface rate history for Field Case 1, after Unneland et al. (1998).

1.1.2. Formation testing hardware

Lebourg, Fields, and Doh (1957) reported that the first commercial *Formation Tester Tool*, as shown in Figure 1.10, was run on a logging cable in 1955 in the Gulf Coast of Louisiana and Texas. Lebourg et al. (1957) stated that Formation Testers offer a method of rapidly testing possible producing formations in openhole wells. Furthermore, they stated the following applications of the tool:

1. Determination of productive fluid—particularly differentiating between gas and oil,
2. Determination of reservoir pressure as an aid to:
   - Safety in completion technique,
   - Reduction in drilling costs, and
   - Reservoir analysis.
3. Determination of minimum gas-oil ratios,
4. Location of gas-oil and/or oil-water contacts, and
5. Obtaining representative fluid samples for examination and analysis: oil (density); gas (hydrocarbon content); water (salinity).

Additional applications envisioned for the future are also given:

1. Subsurface exploration based on accurate pressure measurements,
2. Recovery of fluid samples from the undisturbed reservoir for PVT work, and
3. Possible quantitative study of permeabilities or productivity index.

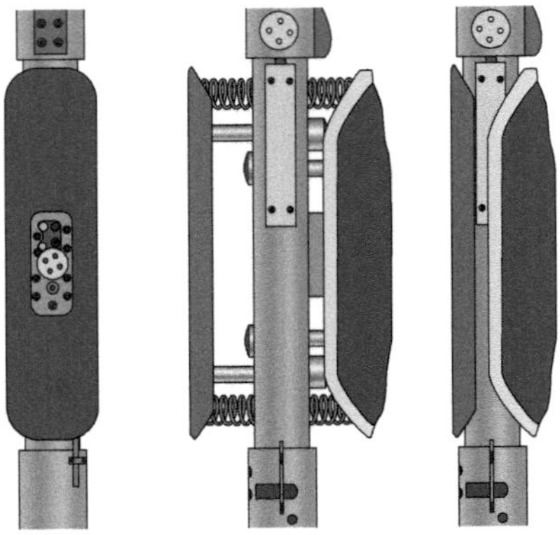

Figure 1.10 The first Formation Tester Tool, after Lebourg et al. (1957).

In the first tool, opening of the single sample chamber was initiated from the surface and the shutting in of the valve was performed either when the sample chamber was full or when the allocated time has elapsed. The first generation of formation testers had several limitations:

- Only one test could be performed during each downhole trip,
- Pressure gauges had a poor resolution and accuracy,
- There was no surface readout for monitoring operations, acquired data, etc.
- The overall test run time was significant.

In 1975, repeat formation testers, as shown in Figure 1.11, were presented by Schultz, Bell, and Urbanosky (1975). As can be seen from this figure, the tool essentially consists of a packer and probe, pressure sensors, and two pretest and sample chambers. The tool setting and testing operations are controlled hydraulically by an electrically powered oil-pump. The packer is forced against the borehole to ensure hydraulic communication between the formation and probe. In addition to the applied force, the mud hydrostatic pressure also pushes the packer against the mud cake in a normally overbalanced pressure condition in the wellbore. The probe, which is a shape edge tube connected to a piston and located at the center of the packer, as shown in Figure 1.12, can extend into the formation by cutting through the mud cake. The probe has a filter to restrict particle and sand transport into the tool and to minimize plugging of the probe due to sand and solids. When the tool is retracted, the piston pulls the probe back into the tool. All these process are hydraulically controlled from the surface.

Unlike formation testers, with one downhole trip per test, the repeat formation testers can successively test a number of zones in a single downhole trip without bringing the tool to surface. The second important addition is the possibility to perform a pretest in order to

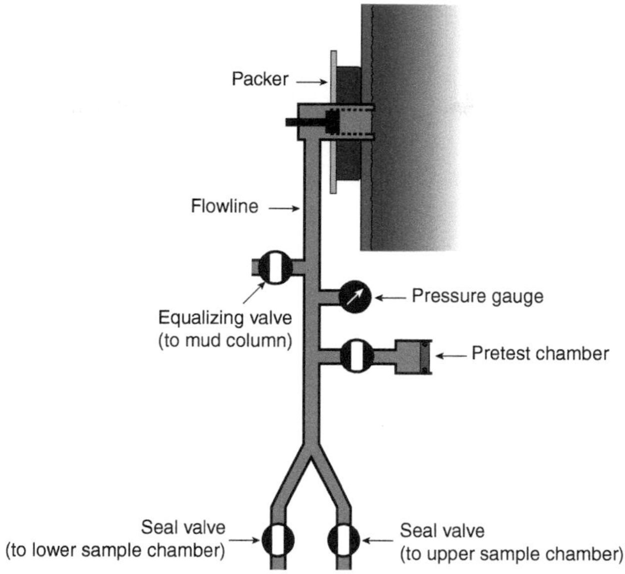

Figure 1.11 Schematic of a repeat formation tester, after Schultz et al. (1975).

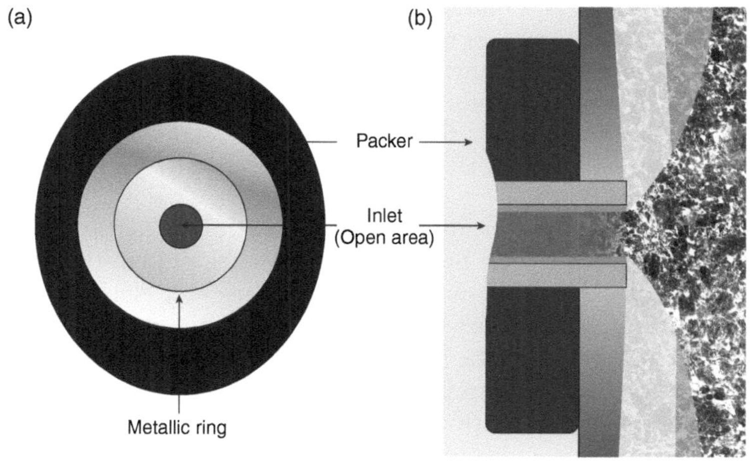

Figure 1.12 The front view of a single probe and packer element (a) and in a set position against the formation (b), after Akkurt et al. (2007).

ensure communication with the formation. If any problem is detected (no significant production from the formation or inadequate seal from the packer), the tool can then be retracted to be set at the same or another location. The other important improvement is the addition of a second sample chamber in order to test at two different rates to obtain a better second sample, more representative of formation fluids. The pretest chamber had a fixed volume of 15 cc, while the main sample chambers could have different sizes: 1, 2 3/4, 6 or 12 gallons. Additional features are the reduction in the size of the probe, thus reducing the probability of tool sticking to the formation as well as improvements on gauge accuracy. Because the pretest has a known volume, the time needed to fill it up was a good indicator of the time required for sampling in the main chambers. The overall impact of repeat formation testers compared to the formation testers was a reduction in test time, an increase in success rate to 90% and a significant improvements in terms of pressure measurement quality and quantity that led to a development of pressure transient interpretation techniques for wireline formation tester data.

After the probe setting, the piston of the pretest chamber withdraws fluid at a constant speed while the flowing pressure is acquired as a function of time. This flowing period, although short, is a conventional drawdown test, called the *first pretest*. Normally the flowing formation pressure may reach a spherical steady-state condition in the formation if it is not very tight because of a very small production and flowing time. If the mud filtrate clean up continues the steady-state condition may not be reached. When the first piston reaches the end of the first chamber, then it stops. At the same time the second chamber piston begins to withdraw fluid from the formation while the flowing pressure is acquired. The second drawdown test is called the *second pretest*. The flow rate for the second pretest is approximately twice the first one. These two pretests are exactly the same as the conventional two-rate flow test. The flowing pressure for the second pretest usually reaches a steady-state condition because the formation clean up mostly occurs during the first pretest. For more details on repeat formation testers, one can refer to Schlumberger (1981).

When the second piston stops, the flow rate from the formation drops rapidly and the pressure start building up against the fluid in the flow lines (afterflow effects). This last test is called the *buildup test*. Figure 1.13 shows typical repeat formation tester events. After the transient tests, the pretest chambers are emptied by the resetting of the pistons and the system is ready for another test. If required, a fluid sample may be taken after the second pretest. Basically, with repeat formation testers, pressure transient formation test interpretation commenced in obtaining reservoir pressure and permeability or mobility (the ratio of permeability and viscosity) along the wellbore.

A new generation formation testing tool was introduced by Zimmerman, MacInnis, Hoppe, Pop, and Long (1990), called the *Modular Formation Dynamics Tester*, in 1990. This was a significant improvement over formation and repeat formation testers. In this book, we call it the *Wireline Formation Tester* (WFT) or *Multiprobe Formation Tester*. The WFT is composed of several modules that can be assembled in numerous ways, providing the necessary functionalities in order to meet specific objectives. The main functions are

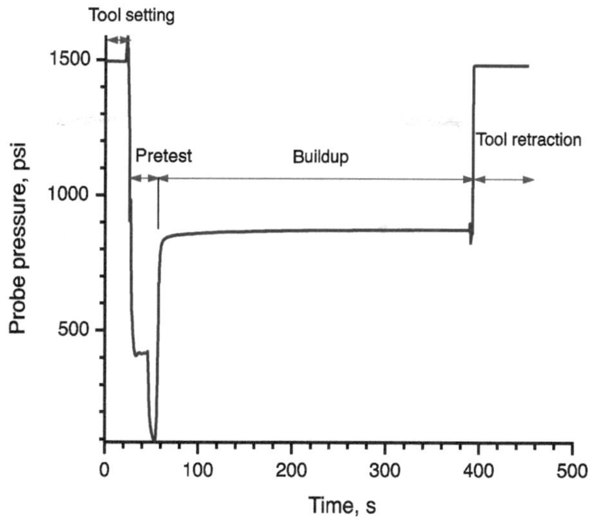

Figure 1.13 Tool events and pressure transient tests from a repeat formation tester.

essentially the same as those of the repeat formation tester: collecting fluid samples, conducting pretests, and longer drawdown and buildup tests. Apart from this modularity, the main additions are:

- Controlled volume withdrawal during pretests,
- Measuring flow rate simultaneously with pressure during pretests and pressure transient tests,
- Better pressure gauge specifications,
- Versatility of sampling options, better fluid samples in terms of mud filtrate contamination, and preserving samples at the reservoir conditions,
- Fluid type recognition and providing a limited number of fluid properties downhole, such as bubble point pressure, viscosity, etc.,
- Ability to produce for hours with the pumpout module,
- Stress testing to obtain formation parting pressure,
- Multiprobe module, and
- Dual-packer module.

Figure 1.14 illustrates several possible combinations of probes and packer:

1. a—Two single-probe with sample chambers,
2. b—A dual-probe module with two single-probes (vertical observation probes) and sample chambers, and
3. c—A dual-packer module with a single-probe (vertical observation) and sample chambers.

The selection of modules for a given test objective depends essentially on the objectives of the test and on practical constraints: well condition,

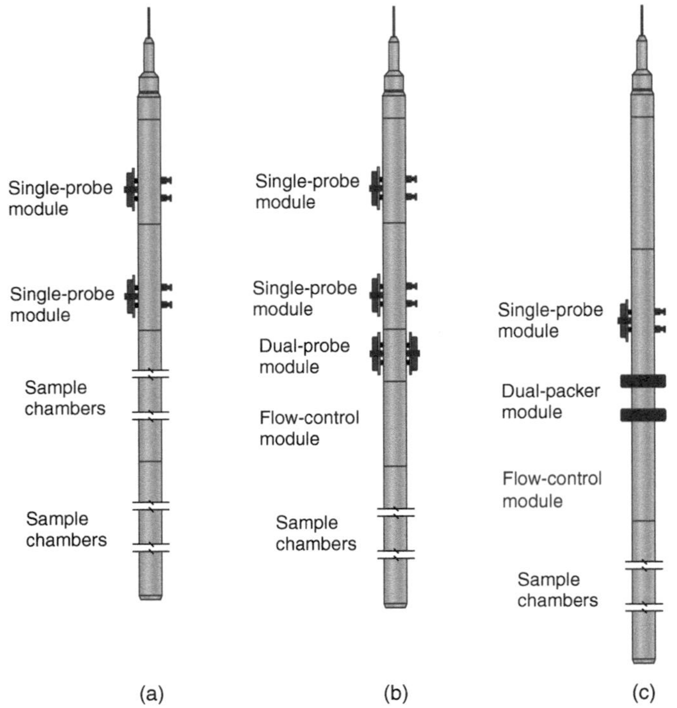

Figure 1.14 Different tool configurations and modules for new generation wireline formation testers, after Schlumberger (2006).

hydrostatic mud pressure, formation type, etc. The selection of a specific probe arrangement is crucial for pressure transient testing and sampling. Further details of new generation WFT are given by Kikani (2009), Hegeman (2009), Schlumberger (2006), and Schlumberger (1996).

Figure 1.15 presents a typical pressure transient test with a WFT packer module to illustrate the sequence of the events: tool setting, pretest, drawdown, buildup, and finally tool retraction. After tool setting, the pretest establishes communication with the formation by withdrawing up to 1000 cc through the packer. The same figure also presents the flow rate that was measured by the pumpout module during the packer pretest and pressure transient test. Notice that the flow rate is more or less constant during the entire drawdown period. As can be seen in this figure, the drawdown pressure increases slightly, indicating a degree of cleanup during the production period.

1.1.3. Pressure gauges and their metrology

Acquiring accurate downhole pressure data is the most critical part of pressure transient testing for interpretation. As we stated in the introduction

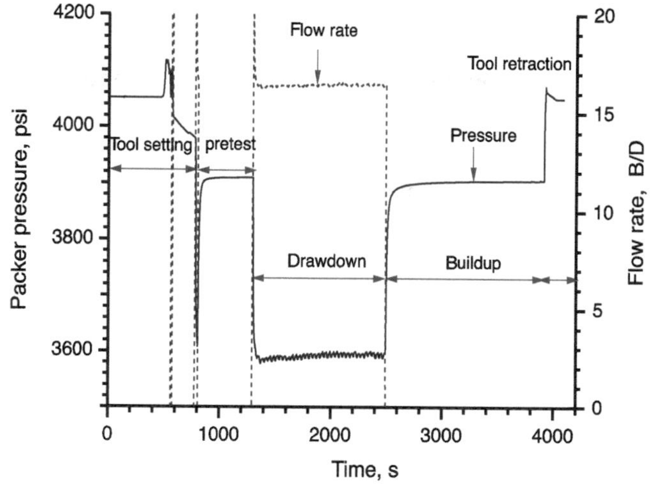

Figure 1.15 Packer pressure and flow rate for a WFT pressure transient test.

section, for DSTs conducted with a Johnston formation tester, both flow rate and pressure were measured at the surface, and bottomhole pressure was obtained from the hydrostatic pressure of the fluid in the drillstring and surface pressure measurements.

From the 1930s to the 1960s, mechanical pressure gauges were mostly used to acquire downhole pressure measurements for pressure transient testing with surface flow rate measurements. In the 1940s, the mechanical pressure gauge resolution was about 3 to 5 psi and has continuously been improved since then. In principle, a mechanical pressure transducer converts pressure into a mechanical displacement or deformation that is recorded by a stylus on a sensitive cylindrical surface, which is rotated by a mechanical clock. Thus, this system provides a pressure versus time chart to be digitized manually.

In the 1970s, electronic gauges were introduced, where the mechanical displacement or deformation of the sensing element is converted into electrical signals, then into pressure measurements. This digital form of pressure measurements has expanded the usage and applications of pressure transient tests and is easily stored in computers. Initially all electronic gauges used strain sensors for most testing applications. However, the accuracy, dynamic response, stability, and resolution of these electronic gauges drastically improved, and about 0.1 psi resolution was achieved.

In the early 1970s Hewlett-Packard introduced a new electronic downhole pressure gauge utilizing a quartz transducer. Although other pressure gauges were introduced over the years, the Hewlett-Packard vibrating quartz crystal pressure sensor was a step change in downhole and surface pressure measurements in terms of the accuracy, dynamic response, stability, and resolution if the temperature variations were negligible during the test. The resolution of quartz sensors was about 0.01 psi, with high stability and without hysteresis. Their pressure range was 0-10,200 psi with an accuracy of 0.025% in reading pressures greater than 1000 psi (Miller, Seeds, & Shira, 1972). However, the early time pressure data for both formation and well testing were inaccurate, particularly for buildup tests during which temperature may change more than a few degrees Celsius, although it was a perfect gauge for interference tests, where the downhole temperature remains constant during the test.

The combinable quartz gauge has a single quartz crystal with two vibration modes (Kikani, 2009; Vella et al., 1992). One mode is particularly sensitive to temperature, and is used as a thermometer to correct the frequency-temperature behavior of the other more pressure-sensitive mode. Because this happens on the same piece of quartz, there is no possibility of a temperature lag or discrepancy. Combinable quartz gauges have an accuracy about 1 psi over a range from atmospheric pressure to 15,000 psi at temperatures 35 °C to 175 °C [95 °F to 350 °F]. The response to changes in temperature is extremely fast—less than a few seconds—rather than the 30 minutes or so taken by previous quartz gauges (Vella et al., 1992). The resolution of the combinable quartz gauges is about 0.003 psi ((@ 15,000 psi; 1-sec sampling).

It should be pointed out that most formation and well test setups now have one high resolution quartz and strain pressure sensor. In addition to pressure sensor characteristics, the setup, gauge carriers, and all associated electronics affect measurement metrology. As stated above, downhole pressure is the most critical measurement acquired during formation and well tests. The validity of the reservoir and well parameters obtained from the interpretation of pressure measurements depends on the meteorological characteristics of pressure gauges. Meteorological characteristics fall into two distinct categories:

1. Static characteristics which describe the performance of the gauge under stabilized temperature conditions with respect to an ideal response, and
2. Dynamic characteristics which describe the performance of the gauge when it is submitted to varying pressure and temperature conditions.

Static parameters describing the transducer performance at static conditions are:

- Accuracy,
- Resolution,
- Stability, and
- Sensitivity.

Accuracy is the difference between a measured value and the true value generated by a reference standard DWT (dead weight tester). Manufacturers generally refer to static accuracy. This is the algebraic sum of all errors influencing pressure measurements.

Resolution is the minimum pressure change that can be detected by the sensor. When referring to a gauge resolution, it is important to take into account the associated electronics because the two are always used together. It is also important to measure the resolution with respect to a specific sampling rate. The gauge resolution is equal to the sum of three factors:

1. Pressure sensor resolution,
2. Digitizer resolution, and
3. Electronic noise induced by the amplification chain.

In the case of tools equipped with strain gauge transducers, mechanically induced noise may be an additional factor that limits gauge resolution because some gauges behave as microphones or accelerometers. This may be an important consideration during tests where there is fluid movement or tool movement downhole.

Stability is the ability of a sensor to retain its performance characteristics over a relatively long period of time. The stability of the sensor is the mean drift in psi per day obtained at a given pressure and temperature. Three levels of stability can be defined:

- Short term stability for the first day of a test,
- Medium term stability for the following six days, and
- Long term stability for a minimum of one month.

Sensitivity is the ratio of the transducer output variation induced by a change of pressure to this change of pressure. In other words, the sensitivity represents the slope of the transducer output versus the pressure curve.

Dynamic parameters describing the transducer performances in dynamic conditions are:

1. *Transient response during temperature variation.* The sensor response is monitored under dynamic temperature conditions whilst the applied pressure is kept constant. The peak error represents the maximum discrepancy between the applied pressure and the stabilized sensor output. By general consensus, the stabilization time represents the time needed to be within 1 psi of the stabilized pressure. The offset represents the difference between the initial and final pressure. This parameter provides, for a given temperature variation, the time required to get a reliable pressure measurement.
2. *Transient response during pressure variation.* The sensor response is recorded before and after a pressure variation whilst the temperature is kept constant. Peak error and stabilization time is measured as previously described for a temperature variation.

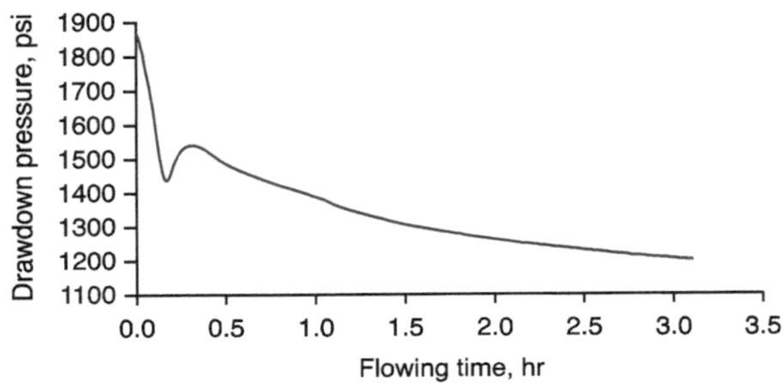

Figure 1.16 Wellbore pressure during the drawdown test.

Accuracy, drift, and resolution are the most important pressure gauge characteristics from the pressure transient testing point of view, provided that all other meteorological characteristics are in order. Furthermore, the apparent resolution, which is defined as the gauge resolution that is stated by the manufacturer (service companies, vendors, etc.) plus the natural background noise (Kuchuk, 2009a), will affect the pressure transient tests most. In addition to the natural background noise, background transients from nearby wells and/or tidal effects also contribute to noise. The apparent resolution can also be defined as the resolution at which a pressure gauge starts measuring an observable pressure change due to a flow rate pulse at a point in the reservoir or in the wellbore, where the pressure is constant or stable everywhere and the natural background noise does not contain high-frequency components. For instance, pressure measurements have a significant level of high-frequency noise in a producing wellbore. Except during the initial period of shut-in, high frequency noise usually disappears during the buildup period.

For instance, it is theoretically possible to conduct a very short drawdown test and a very long subsequent buildup test to investigate reservoir characteristics away from the well, say at a spatial location $\mathbf{r}$. However, if the pressure change at the spatial location $\mathbf{r}$ is smaller than the apparent gauge resolution at any time during the buildup test, then the buildup pressure or its derivative will not resolve reservoir characteristics beyond any spatial location greater than $\mathbf{r}$, where the pressure change is less than the apparent gauge resolution. Next, we will look at some field data to examine the pressure gauge and apparent resolution.

Figure 1.16 presents a field drawdown test, and Figure 1.17 presents a subsequent buildup test. This was a producing well, but it was shut in for about 20 minutes to allow the lowering of a production logging tool (PLT) into the well. After this short shut-in period, the well was put back in production at a constant surface flow rate. In order to obtain a stabilized downhole flow rate, the well produced for a little more than 3 hr, as shown in Figure 1.16. After downhole rate stabilization, the production profile was

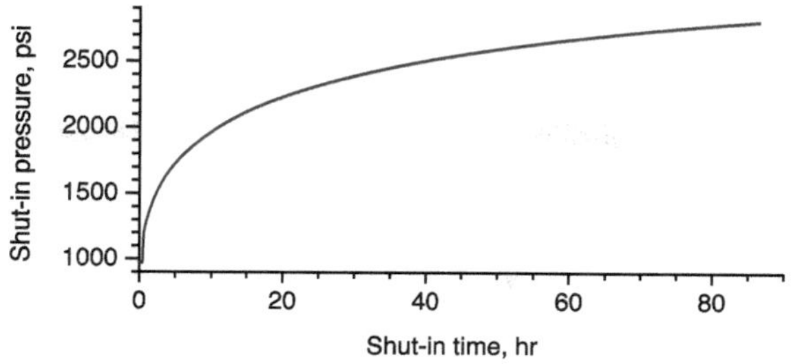

Figure 1.17 Wellbore shut-in pressure during the buildup test.

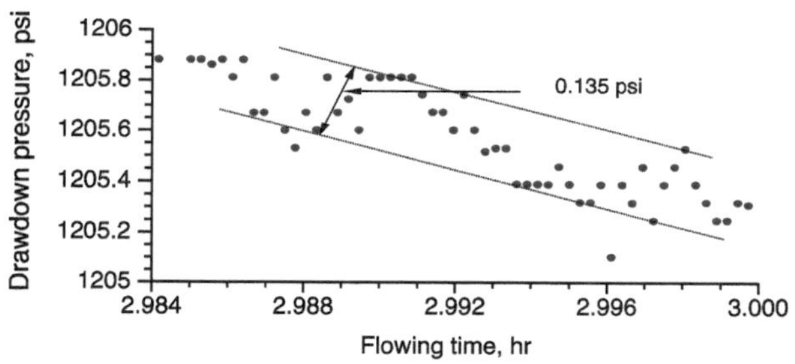

Figure 1.18 A time interval showing the wellbore pressure for the drawdown test in Figure 1.16.

obtained with the PLT at two different flow rates (the second drawdown during the second rate is not shown here). It should also be pointed out that during the production profiling the flowing wellbore pressure was not measured at a stationary point in the wellbore. After a total of 6-hr production, the PLT was stationed at the same location as the first drawdown and acquired wellbore pressure about 30 minutes before the start of the buildup. The last wellbore drawdown pressure, just before the buildup, was 240 psi lower than the pressure at the end of the 3-hr drawdown, as can be seen in Figures 1.16 and 1.17, due to the rate change during the second drawdown. Both drawdown and buildup pressure data look very smooth, except for the initial hump in the drawdown pressure, as can be seen in Figure 1.16.

Figures 1.18 and 1.19 present two intervals at very extended pressure scales towards the end of both drawdown and buildup tests given in Figures 1.16 and 1.17. As shown in these figure, the spread of pressure data in these test intervals is about 0.135 psi for drawdown and 0.06 psi for the buildup. The resolution for this pressure gauge was stated to be 0.01 psi, although it could have changed with time for some of the sensors. The difference in spread of pressure measurements for the drawdown test and the subsequent buildup must be due to external noise. It is impossible

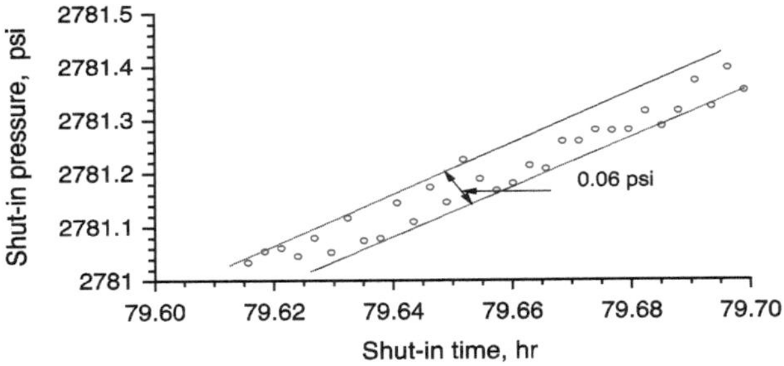

Figure 1.19 A time interval showing the wellbore pressure for the buildup test given in Figure 1.17.

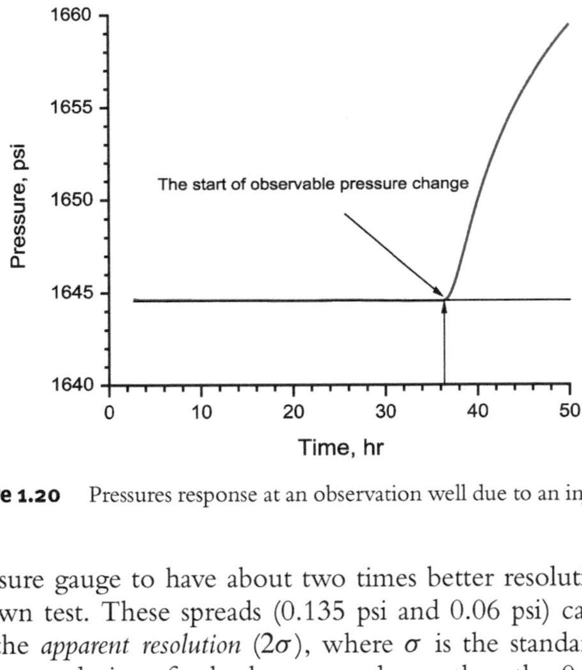

Figure 1.20 Pressures response at an observation well due to an injector.

for the pressure gauge to have about two times better resolution just after the drawdown test. These spreads (0.135 psi and 0.06 psi) can be loosely defined as the *apparent resolution* (2σ), where σ is the standard deviation. The apparent resolutions for both tests are larger than the 0.01 psi stated gauge resolution. As in this example, even for drawdown tests, it is possible to attain a 0.1 psi apparent resolution, as shown in Figure 1.18. For buildup and interference tests, it is possible to achieve an apparent resolution between 0.01 and 0.05 psi with modern quartz pressure gauges, which typically have resolutions ranging from 0.002 to 0.01 psi.

Next, another example will be given for gauge resolution. Figure 1.20 presents the pressure response in an observation well due to a water injector. After about 36.25 hr injection period at the injector, the pressure diffusion (change) becomes observable at the observation well; i.e. the change in pressure becomes observable from the "static" reservoir pressure. At the scale

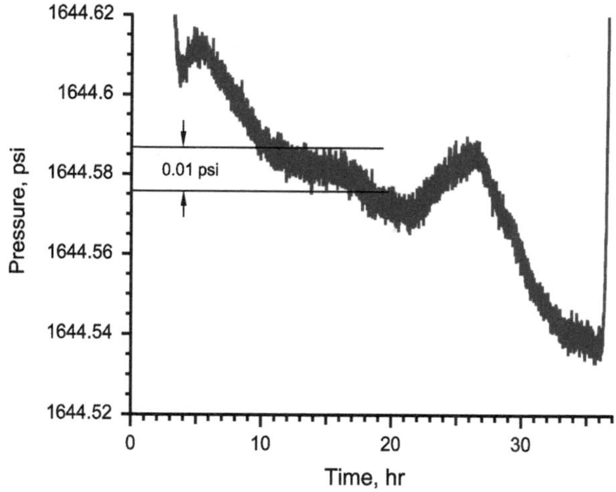

Figure 1.21 The expanded plot of the pressure at an observation well due to an injector given in Figure 1.20.

shown in Figure 1.20, which is about 1 psi per division, the "static" pressure appears perfectly smooth.

Figure 1.21 presents a portion of the same data where the plot is expanded to 0.01 psi per division from the start of the test to the start of the observable pressure change. The pressure looks noisy, and almost randomly varies within 0.01 psi. The observed pressure decreased about 0.1 psi during the 36 hr observation period. If this change was due to injection then the observed pressure should have increased. Therefore, this variation could be due to many different phenomena: background transients from other wells, tidal effects, or perhaps due to a slight gauge drift.

Figure 1.22 presents a short interval of the data given in Figure 1.20, where the plot is expanded to 0.001 psi per division around the start of the observable pressure change. Again, the pressure looks noisy and almost randomly varies within 0.01 psi from 35 to 36 hr. Figure 1.22 presents the average pressure and the standard deviation, σ, of the pressure measurements for this stable 1-hr period, where $\sigma = 0.00183$ psi. Thus, the apparent resolution for this test is $2\sigma = 0.00366$ psi. If we want to be conservative, then $\sigma_{max} = 0.0103$ psi, as shown in the figure can be used as an apparent resolution.

1.2. Pressure Transient Test Types

When a rate or pressure pulse is applied at a point in the reservoir and/or in the wellbore, the resulting transient pressure (diffusion) and/or rate data may be acquired at the source, for instance in a wellbore, and/or at

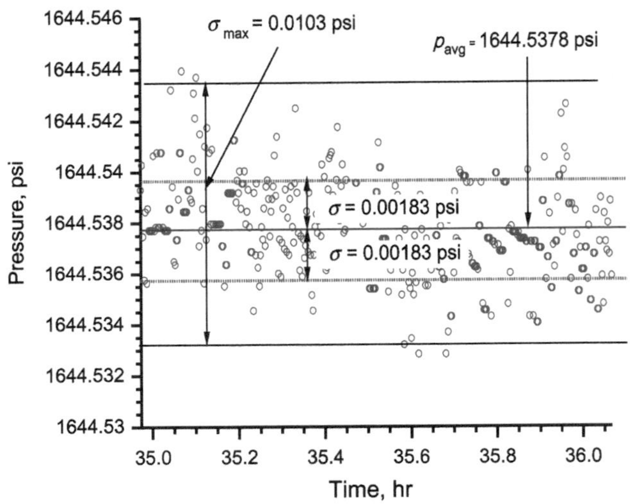

Figure 1.22 The apparent gauge resolution for the interference test data given in Figure 1.20.

different spatial locations in the reservoir and/or in an observation well or in the formation along the wellbore. Therefore, at the most basic level, pressure transient tests can be categorized spatially as (1) Production/injection tests and (2) Interference tests at observation points in the reservoir or along the sandface. In general, it is common to use flow rate as a pulse (input) and to measure pressure as an output of the system in pressure transient testing. Therefore, when the production or injection rate is changed from a nonzero value to a different value, including from zero to another value or from a nonzero value to zero at a well, it creates pressure diffusion in the reservoir; monitoring and measuring the diffusion at any point in the system is called *pressure transient testing*.

Thus, depending on the disturbance, sink or source type, well tests are also divided as: (1) Production test and (2) Injection test. If the production or injection rate is changed to zero from a nonzero value, the production test is called a buildup test and the injection test is called falloff. If the production or injection rate is changed to a nonzero from zero, the production test is called a drawdown test and the name injection test remains the same. If the production or injection rate is changed several times with the possible inclusion of zero rate periods, the entire test is called a multirate test. If the measurements are made at an observation well during the multirate test, then this type of test is called a pulse test: normally with equal pulse duration and magnitude. Well tests are also categorized according to their objectives, such as reservoir limit test, multilayer test, vertical interference test, deliverability test, isochronal gas well test, pumping well test, horizontal well test, impulse test, closed chamber test, repeat formation test, interval

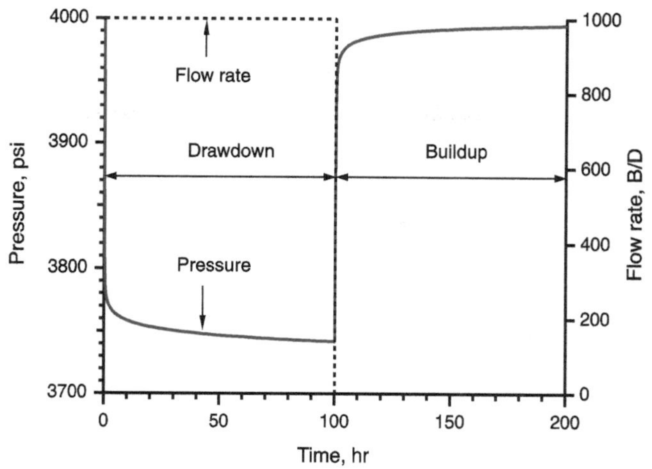

Figure 1.23 Drawdown test and a subsequent buildup test for a constant-rate production period.

pressure transient test (IPTT), etc. Therefore, a basic pressure transient test consists of a production/injection rate change, during which the wellbore pressure is measured in general, the downhole and production/injection rate is monitored (measured directly or indirectly) either at the wellbore or surface, a subsequent shut-in/falloff period during which the wellbore pressure is usually measured downhole.

A pressure transient test during production is also called a *drawdown test* and is called a *buildup test* during the shut-in period, as shown in Figure 1.23. Understanding of these two basic test types and the constant-rate drawdown behavior (pressure, pressure change and its logarithmic derivative) is fundamental to pressure transient test interpretation.

1.2.1. Drawdown tests

The principal purpose of the drawdown or buildup tests shown in Figure 1.23 is to estimate the reservoir pressure, permeability, skin factor, etc. These tests are often also used to obtain reservoir characteristics, such as the boundaries of the reservoir or identification of the reservoir type. A drawdown test consists of measuring the downhole and/or surface pressure and flow rate as a function of time in the wellbore. However, for conventional drawdown tests, the sandface flow rate is normally not measured continuously, but the surface flow rate is usually regulated by a choke or other mechanical systems. In some cases, it is measured sporadically as a function of time in the gathering or test tank after the separation of gas from crude oil. Nevertheless, the theoretical drawdown response provides the basis for pressure transient testing.

1.2.2. Pressure buildup tests

Compared to other transient tests, buildup/falloff tests have been widely used for production and injection wells since the 1930s. Buildup tests are performed by closing the wellhead valve (shut in) or using a downhole shut-in device to cut off the well production for a finite time period. Before a buildup test, the well should produce at a constant or somewhat stabilized rate at the sandface for a specified time period, or the sandface rate has to be measured. In general, the wellbore storage, wellbore geometry, and near-wellbore complexities considerably affect the wellbore pressure of buildup tests. Figure 1.23 shows a typical buildup test after a drawdown test.

CHAPTER 2

Mathematical Preliminaries and Flow Regimes

Contents

2.1. Introduction

In this chapter, we review the basic principles of pressure transient flow regimes. In order to achieve this, first we present the well-known wellbore pressure solutions and unit impulse responses for basic formation and well testing flow geometries. Unit impulse responses will often be used in the succeeding chapters.

In this book, we often use the words sink or source. For the purpose of clarification, it is important to note that the source refers to a point in the system where the fluid is to be added (injected), while the sink refers to a point in the system from which the fluid is to be removed (withdrawn). Therefore, based on this definition, a production probe or well represents a sink, while an injection probe or well represents a source. Furthermore, we define the flow rate imposed at a production (or sink) location as positive, i.e., $q > 0$, and the flow rate imposed at an injection location as negative, i.e., $q < 0$. However, the term sink or source has been used interchangeably in the ground water, heat conduction, and petroleum literature. For instance,

Developments in Petroleum Science, Volume 57
ISSN 0376-7361, DOI: 10.1016/S0376-7361(10)05708-0

both Carslaw and Jaeger (1959) and Gringarten and Ramey (1973) refer to heat liberation and fluid withdrawal from the system (reservoir) as a source, respectively. The term line source is often used for production wells. Therefore, we will also not be very strict on the usage of the term sink or source.

Let us consider pressure diffusion for a single-phase constant viscosity and slightly compressible fluid in a three-dimensional infinite anisotropic homogeneous porous domain bounded by $-\infty < x < \infty$, $-\infty < y < \infty$, and $-\infty < z < \infty$, in which there is no sink or source. The permeabilities in the principal directions are denoted by k_x, k_y, and k_z and the mobilities are defined as $\lambda_x = \frac{k_x}{\mu}$, $\lambda_y = \frac{k_y}{\mu}$, and $\lambda_z = \frac{k_z}{\mu}$, where μ is the fluid viscosity. Single-phase pressure diffusion for this homogeneous porous domain (Muskat, 1937a) can be written as

$$\lambda_x \frac{\partial^2 \mathcal{P}}{\partial x^2} + \lambda_y \frac{\partial^2 \mathcal{P}}{\partial y^2} + \lambda_z \frac{\partial^2 \mathcal{P}}{\partial z^2} = \varphi \frac{\partial \mathcal{P}}{\partial t}, \quad -\infty < x, y, z < \infty, t > 0, \tag{2.1}$$

subject to the initial condition

$$\mathcal{P}(x, y, z, 0) = h(x, y, z), \quad -\infty < x, y, z < \infty, t = 0, \tag{2.2}$$

and the boundary condition

$$\mathcal{P}(x, y, z, t) = 0, \quad x, y, z \to \pm\infty, t > 0, \tag{2.3}$$

where p_o is the initial pressure when the system is at rest or a reference pressure, $\mathcal{P} = p_o - p(x, y, z, t)$, $h(x, y, z) = p_o - p(x, y, z, 0)$, $p(x, y, z, 0)$ is the initial pressure distribution imposed at $t \leq 0$, and $\varphi = \phi c_t$ is the storativity. For Equation (2.1), the inner boundary condition is not given above because we will solve the initial boundary-value problem described by Equations (2.1)–(2.3) to obtain the fundamental solution or the Green function for the unit impulse, where the pressure is related to the Green function as

$$\mathcal{P}(x, y, z, t) = \int_V \mathrm{d}V \; G(x, y, z, t; x', y', z', t') h(x', y', z'), \quad V \in R^3, \tag{2.4}$$

where V the volume of the domain.

Let us consider an instantaneous point source located at $\mathbf{r}' = \{x', y', z'\}$, where $\mathbf{r} \in R^3$, $t \in R$, and $\mathbf{r}$ is the spatial vector in the infinite anisotropic homogenous medium defined above. The spatial position vector is $\mathbf{r} =$

$x\mathbf{i} + y\mathbf{j} + z\mathbf{k}$ and $r = |\mathbf{r}| = \sqrt{x^2 + y^2 + z^2}$, where $\mathbf{i}$, $\mathbf{j}$, and $\mathbf{k}$ are the unit vectors in the x, y, and z directions, respectively. Thus, the Green function for an instantaneous point source (Carslaw & Jaeger, 1959) can be written from Equation (2.1) as

$$\left(\lambda_x \frac{\partial^2}{\partial x^2} + \lambda_y \frac{\partial^2}{\partial y^2} + \lambda_z \frac{\partial^2}{\partial z^2} - \varphi \frac{\partial}{\partial t}\right) G(\mathbf{r}, t; \mathbf{r}', t')$$
$$= -\delta(\mathbf{r} - \mathbf{r}')\delta(t - t'), \quad -\infty < \mathbf{r} < \infty, t > 0 \tag{2.5}$$

with the initial condition

$$G(\mathbf{r}, t; \mathbf{r}', t') = 0, \quad -\infty < \mathbf{r} < \infty, t \leq 0 \tag{2.6}$$

and the boundary condition

$$G(\mathbf{r}, t; \mathbf{r}', t') = 0, \quad \mathbf{r} \to \pm\infty, t \geq 0, \tag{2.7}$$

where $\delta(\mathbf{r})$ is Dirac delta function of strength unity (in any consistent unit system). Notice from Equation (2.5) that the unit of G is $\frac{1}{L^3}$ and it is for an instantaneous point source of strength unity. Likewise to Carslaw and Jaeger (1959), who give a definition of the strength of a point source for the heat conduction problem, we can define the strength of a point source for the pressure diffusion considered here as the pressure change to which the amount of fluid (volume) expanded (or produced) would raise a unit pore volume of the porous medium. The unit of the strength of the point source is $pL^3 = \frac{mL}{t^2}$, and its strength can be expressed as $\frac{Q}{\varphi}$, where Q is the amount of fluid (volume) produced instantaneously.

We define the Fourier transform of $f(x)$ with respect to x as

$$F(\alpha) = \int_{-\infty}^{\infty} \mathrm{d}x \, \mathrm{e}^{\mathrm{i}\alpha x} f(x) \tag{2.8}$$

and the inverse Fourier transform as

$$f(x) = \frac{1}{2\pi} \int_{-\infty}^{\infty} \mathrm{d}x \, \mathrm{e}^{-\mathrm{i}\alpha x} F(\alpha). \tag{2.9}$$

Applying the Fourier transform given by Equation (2.8) successively to Equations (2.5)–(2.7), in the x, y, and z directions, we obtain

$$\left(|\nu|^2 + \varphi \frac{\partial}{\partial \tau}\right) \overline{\overline{\overline{G}}}(\boldsymbol{\xi}, \tau) = \delta(t), \tag{2.10}$$

where

$$\nu^2 = \lambda_x \alpha^2 + \lambda_y \beta^2 + \lambda_z \gamma^2. \tag{2.11}$$

The use of a single vector $\boldsymbol{\xi}$ and τ are justified because the Green function depends only $\mathbf{r} - \mathbf{r}'$ and $t - t'$. The Fourier transform of the initial condition becomes

$$\overline{\overline{G}}(\boldsymbol{\xi}, 0) = 0, \quad \tau \leq 0. \tag{2.12}$$

The first-order initial value problem given by Equation (2.10) has a solution in the form

$$\overline{\overline{G}}(\boldsymbol{\xi}, \tau) = \mathrm{H}(\tau)\ \exp\left(-\frac{|\nu|^2}{\varphi}\,\tau\right), \quad \tau > 0. \tag{2.13}$$

The inverse Fourier transform of Equation (2.13) gives the Green function as

$$\begin{aligned} G(\boldsymbol{\xi}, \tau) = {} & \mathrm{H}(\tau)\frac{1}{(2\pi)^3}\int_{-\infty}^{\infty} \mathrm{d}\alpha \exp\left[-\left(\mathrm{i}|\alpha x| + \frac{\lambda_x}{\varphi}|\alpha|^2\tau\right)\right] \\ & \times \int_{-\infty}^{\infty} \mathrm{d}\beta \exp\left[-\left(\mathrm{i}|\beta y| + \frac{\lambda_y}{\varphi}|\beta|^2\tau\right)\right] \\ & \times \int_{-\infty}^{\infty} \mathrm{d}\gamma \exp\left[-\left(\mathrm{i}|\gamma z| + \frac{\lambda_z}{\varphi}|\gamma|^2\tau\right)\right]. \end{aligned} \tag{2.14}$$

The Green function is obtained by integrating Equation (2.14) and then by replacing x by $x - x'$, y by $y - y'$, z by $z - z'$ and τ by $t - t'$ as

$$\begin{aligned} G(\mathbf{r}, t; \mathbf{r}', t') = {} & \frac{\varphi^{3/2}}{8\sqrt{\pi^3 \lambda_x \lambda_y \lambda_z (t - t')^3}} \exp\left\{-\frac{\varphi}{4(t - t')}\right. \\ & \left.\times \left[\frac{(x - x')^2}{\lambda_x} + \frac{(y - y')^2}{\lambda_y} + \frac{(z - z')^2}{\lambda_z}\right]\right\}. \end{aligned} \tag{2.15}$$

Note that the Green function given by Equation (2.15) is the same as one given by (Carslaw & Jaeger, 1959; Gringarten & Ramey, 1973). Note also that the Heaviside unit-step function H is omitted for convenience.

The unit impulse response (function) for a point source located at $\mathbf{r}' = \{0, 0, 0\}$ and $t' = 0$ from Equation (2.15) can be written as

$$g(\mathbf{r}, t) = \frac{\varphi^{3/2}}{8\sqrt{\pi^3 \lambda_x \lambda_y \lambda_z t^3}} \exp\left[-\frac{\varphi}{4t}\left(\frac{x^2}{\lambda_x} + \frac{y^2}{\lambda_y} + \frac{z^2}{\lambda_z}\right)\right] \tag{2.16}$$

and its Laplace transform is given as

$$\bar{g}(s, \mathbf{r}) = \frac{\varphi}{4\pi\sqrt{\lambda_x \lambda_y \lambda_z}} \frac{1}{\sqrt{\frac{x^2}{\lambda_x} + \frac{y^2}{\lambda_y} + \frac{z^2}{\lambda_z}}} \times \exp\left\{-\sqrt{\varphi s\left[\frac{x^2}{\lambda_x} + \frac{y^2}{\lambda_y} + \frac{z^2}{\lambda_z}\right]}\right\}. \tag{2.17}$$

In this book we use *unit impulse response* or *impulse response* interchangeably.

2.2. Point-Source Solutions

For wireline formation tester (WFT) probes and packers, partially penetrated wells in thick reservoirs, and reservoirs with pinchout boundaries, radial spherical and/or hemispherical flow regimes are often observed (Culham, 1974; Gerard & Horne, 1985; Goode & Thambynayagam, 1992; Kuchuk, 1994; Moran & Finklea, 1962; Raghavan & Clark, 1975). For convenience, the radial spherical flow regime is called the spherical flow regime in this book. For isotropic media ($\lambda = \lambda_x = \lambda_y = \lambda_z$), the impulse for a unit volume (for instance, Q=1 cc) response at $r = r_w$ (a spherical wellbore) for a point source of strength $\frac{1}{\varphi}$, located at $\mathbf{r}' = \{0, 0, 0\}$ and $t' = 0$ in an infinite isotropic homogeneous system, from Equation (2.16), can be written as

$$g_w(t) = \frac{\sqrt{\varphi}}{8\sqrt{\pi^3 \lambda^3 t^3}} \exp\left(-\frac{\varphi r_w^2}{4\lambda t}\right) \tag{2.18}$$

and its Laplace transform as

$$\bar{g}_w(s) = \frac{1}{4\pi r_w \lambda} \exp\left(-r_w\sqrt{\frac{\varphi s}{\lambda}}\right). \tag{2.19}$$

For a constant flow rate q at the spherical well ($r = r_w$), the wellbore pressure is obtained from the point-source solution given by Equation (2.18) by integrating with respect to time and can be written as

$$p_w(t) = p_o - \frac{q}{4\pi r_w \lambda} \operatorname{erfc}\left(\frac{r_w}{2}\sqrt{\frac{\varphi}{\lambda t}}\right), \tag{2.20}$$

where, erfc is the complementary error function. Equation (2.20) for large times can be approximated as

$$p_w(t) = p_o - \frac{q}{4\pi r_w \lambda} + \frac{q\sqrt{\varphi}}{4\pi^{3/2}\lambda^{3/2}} \frac{1}{\sqrt{t}}. \tag{2.21}$$

2.2.1. Spherical flow regime for drawdown tests

Equation (2.21) can be written for a drawdown test in oil field units with the skin factor for a spherical well in an infinite medium (Culham, 1974) as

$$p_w(t) = p_o - \frac{141.2 q\mu}{2 r_w k_s}(1 + S) + m_{sp}\frac{1}{\sqrt{t}}, \tag{2.22}$$

where the spherical slope is defined as

$$m_{sp} = \frac{2453 q\mu\sqrt{\mu\phi c_t}}{k_s^{3/2}} = \frac{2453 q\mu\sqrt{\mu\phi c_t}}{k_h\sqrt{k_v}}, \tag{2.23}$$

the spherical permeability

$$k_s = \left(k_h^2 k_v\right)^{1/3}, \tag{2.24}$$

S is the skin factor

$$S = \frac{4\pi r_w k_s \Delta p_s}{141.2 q\mu}, \tag{2.25}$$

and Δp_s is the pressure drop due to skin.

Figure 2.1 presents a linear plot of p_w as a function of $\frac{1}{\sqrt{t}}$. As can be seen from this plot, after the effects of skin and wellbore storage, a spherical

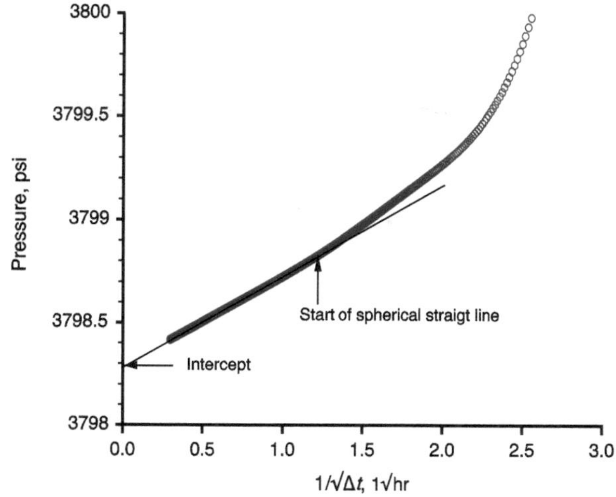

Figure 2.1 A typical spherical plot of a drawdown test for a spherical well in an infinite system.

straight line is observed. The spherical permeability and skin factor can be estimated from the slope m_{sp} and intercept $b = p_o - \frac{141.2q\mu}{2r_w k_s}(1 + S)$ of the spherical straight line if the initial reservoir pressure is known.

Figure 2.2 presents pressure derivatives with respect to $\ln(t)$ for a spherical well in an infinite system. The derivative without skin and wellbore storage effects exhibits a spherical flow regime ($m = -1/2$) starting from very early times (about 2 seconds for a reasonably good permeable formation). On the hand, the derivative with skin and wellbore storage exhibits three distinct flow periods: (1) A positive unit-slope ($m = 1$) storage-dominated flow regime at very early time, (2) A transition period, and (3) A spherical flow regime ($m = -1/2$). It should be pointed out that both derivatives merge into a single curve as expected. As in the spherical pressure plot, a spherical permeability can be obtained from the spherical flow regime. The spherical derivative from Equation (2.22) for the drawdown case can be expressed in terms of the logarithmic derivative $[\ln(t)]$ as

$$m_{sp} = \frac{\mathrm{d}p_w(t)}{\mathrm{d}\sqrt{\frac{1}{t}}} = 2t\sqrt{t}\frac{\mathrm{d}p_w(t)}{\mathrm{d}t} = 2\sqrt{t}\frac{\mathrm{d}p_w(t)}{\mathrm{d}\ln(t)} \tag{2.26}$$

when a spherical flow regime ($m = -1/2$) is observed as in Figure 2.2. To calculate the spherical permeability, the derivative value $[\frac{\mathrm{d}p_w(t)}{\mathrm{d}\ln(t)}]$ should be

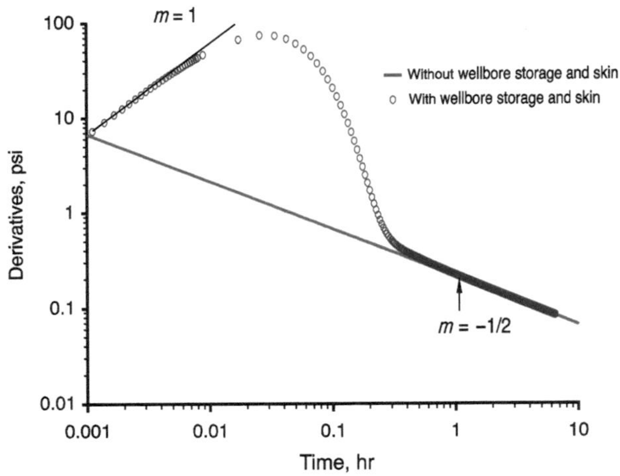

Figure 2.2 Pressure-derivatives of a drawdown test for a spherical well in an infinite system.

taken from the $m = -1/2$ straight line on a log-log plot, it then should be multiplied by $2\sqrt{t}$ to obtain the spherical slope defined by Equation (2.23).

2.2.2. Spherical flow regime for buildup tests

Let us assume that a spherical well is produced at a constant rate q until shut-in. At any time after shut-in, the buildup pressure of the system can be written from the well-known Horner (1951) superposition equation and Equation (2.22) as

$$p_{ws}(\Delta t) = p_o - m_{sp}\left(\frac{1}{\sqrt{\Delta t}} - \frac{1}{\sqrt{t_p + \Delta t}}\right), \tag{2.27}$$

where t_p is the producing time and the spherical straight line slope m_{sp} is given by Equation (2.23). As in the drawdown case, a linear plot of the wellbore pressure versus the spherical Horner time $\left[\left(\frac{1}{\sqrt{\Delta t}} - \frac{1}{\sqrt{t_p+\Delta t}}\right)\right]$ may yield a spherical straight line with a slope of m_{sp}, from which the spherical permeability, the reservoir pressure, and the total skin factor can be obtained.

Normally, the buildup derivative is obtained by taking the derivative of the shut-in pressure with respect to the Horner time $\left(t_H = \frac{t_p+\Delta t}{\Delta t}\right)$ or the superposition time for the variable rate case before the buildup test regardless of flow geometry and plotted on a log-log graph. In order to obtain m_{sp} from the log-log derivative plot, the following expression should be used when a spherical flow regime ($m = -1/2$) is observed:

$$
\begin{aligned}
m_{sp} &= \frac{\mathrm{d}p_{ws}(\Delta t)}{\mathrm{d}\left(\frac{1}{\sqrt{\Delta t}} - \frac{1}{\sqrt{t_p+\Delta t}}\right)} \\
&= \frac{2\Delta t^{3/2}(t_p+\Delta t)^{3/2}}{\left(\sqrt{t_p+\Delta t}-\sqrt{\Delta t}\right)\left(2\Delta t + t_p + \sqrt{\Delta t}\sqrt{t_p+\Delta t}\right)} \frac{\mathrm{d}p_{ws}(\Delta t)}{\mathrm{d}\Delta t} \\
&= \frac{2t_p\Delta t(t_p+\Delta t)}{\left(\sqrt{t_p+\Delta t}-\sqrt{\Delta t}\right)\left(2\Delta t + t_p + \sqrt{\Delta t}\sqrt{t_p+\Delta t}\right)} \frac{\mathrm{d}p_{ws}(\Delta t)}{\mathrm{d}\ln(t_H)},
\end{aligned}
\tag{2.28}
$$

where the value $\frac{\mathrm{d}p_{ws}(\Delta t)}{\mathrm{d}\ln(t_H)}$ should be obtained from the $m = -1/2$ straight line on a log-log derivative plot.

2.3. Line-Source Solutions

The Green function for an instantaneous line source of strength unity located at $\{x', y'\}$ is obtained by integrating the Green function for the point source given by Equation (2.15) from $-\infty$ to ∞ along the vertical z' line as

$$
G(\mathbf{r}, t; \mathbf{r}', t') = \frac{\varphi^{3/2}}{8\sqrt{\pi^3\lambda_x\lambda_y\lambda_z(t-t')^3}} \int_{-\infty}^{\infty} \mathrm{d}z' \exp\left\{-\frac{\varphi}{4(t-t')} \times \left[\frac{(x-x')^2}{\lambda_x} + \frac{(y-y')^2}{\lambda_y} + \frac{(z-z')^2}{\lambda_z}\right]\right\}. \tag{2.29}
$$

The evaluation of the integral in Equation (2.29) yields

$$
G(\mathbf{r}, t; \mathbf{r}', t') = \frac{\varphi}{4\pi\sqrt{\lambda_x\lambda_y}(t-t')} \times \exp\left\{-\frac{\varphi}{4(t-t')}\left[\frac{(x-x')^2}{\lambda_x} + \frac{(y-y')^2}{\lambda_y}\right]\right\}. \tag{2.30}
$$

The unit impulse response for an instantaneous line source located at $\mathbf{r}' = \{0, 0\}$ and $t' = 0$ from Equation (2.30) can be written as

$$g(\mathbf{r}, t) = \frac{\varphi}{4\pi\sqrt{\lambda_x \lambda_y t}} \exp\left[-\frac{\varphi}{4t}\left(\frac{x^2}{\lambda_x} + \frac{y^2}{\lambda_y}\right)\right] \tag{2.31}$$

and its Laplace transform is given as

$$\bar{g}(s, \mathbf{r}) = \frac{\varphi}{2\pi\sqrt{\lambda_x \lambda_y}} \mathrm{K}_0\left[\sqrt{\varphi s\left(\frac{x^2}{\lambda_x} + \frac{y^2}{\lambda_y}\right)}\right], \tag{2.32}$$

where K_0 is the modified Bessel function.

For isotropic media ($\lambda = \lambda_x = \lambda_y$), the unit-impulse response at $r = r_w$ (a cylindrical well) for a line-source of strength $\frac{1}{h\varphi}$, located at $\mathbf{r}' = \{0, 0\}$ and $t' = 0$ in an infinite isotropic homogeneous system bounded by two no-flow boundaries parallel to the $x - y$ planes, from Equation (2.31), can be written as

$$g_w(t) = \frac{1}{4\pi\lambda h t} \exp\left(-\frac{\varphi r_w^2}{4t\lambda}\right) \tag{2.33}$$

and its Laplace transform as

$$\bar{g}_w(s) = \frac{1}{4\pi\lambda h} \mathrm{K}_0\left(r_w\sqrt{\frac{\varphi s}{\lambda}}\right). \tag{2.34}$$

For a constant flow rate q, the pressure at a wellbore ($r = r_w$) with the skin factor can be obtained by integrating Equation (2.33) with respect to time as

$$p_w(t) = p_o - \frac{q\mu}{4\pi kh}\left[\mathrm{E}_1\left(\frac{\phi\mu c_t r_w^2}{4kt}\right) + 2S\right], \tag{2.35}$$

where p_o is the initial formation pressure, E_1 is the Exponential integral, h is the formation thickness, and S is the skin (damage) factor given as

$$S = \frac{2\pi kh \Delta p_s}{q\mu}. \tag{2.36}$$

Equation (2.35) is the well-known Exponential integral solution (Horner, 1951; Theis, 1937) at the wellbore for a drawdown test in an infinite system.

For $4kt/(\phi\mu c_t r_w^2) > 100$ (Matthews & Russell, 1967), the exponential integral solution is given as

$$p_w(t) = p_o + \frac{q_r\mu}{4\pi kh}\left[\ln\left(\frac{e^{\gamma}\phi\mu c_t r_w^2}{4kt}\right) - 2S\right], \tag{2.37}$$

where $\gamma = 0.577215665\ldots$ is Euler's constant.

2.3.1. Radial flow regime for drawdown tests

The wellbore pressure for a well producing at a constant rate q in an infinite radial system can be written from Equation (2.37) (in oil-field units) as

$$p_w(t) = p_o - m_r\left[\log(t) + b_{rs}\right], \tag{2.38}$$

where the radial slope is defined as

$$m_r = \frac{162.6q\mu}{kh} \tag{2.39}$$

and

$$b_{rs} = \log\left(\frac{k}{\phi\mu c_t r_w^2}\right) - 3.2275 + 0.8686S. \tag{2.40}$$

It should be pointed out that Earlougher (1977) defines the slope given by Equation (2.39) as negative. Therefore, there will be differences in appearance between the formulas given by Earlougher (1977) and in this book.

Figure 2.3 presents a semilog plot of p_w as a function of t for the test given in Figure 1.23. As can be seen from this plot, after the effects of skin and wellbore storage, a radial semilog-straight line is observed, from which kh and, if h is known, the permeability k can be obtained from the slope m_r. The wellbore pressure p_w at $t = 1$ hr is given as

$$p_{1hr} = p_o - m_r b_{rs} \tag{2.41}$$

from which the skin factor is obtained as

$$S = 1.5113\left[\frac{p_o - p_{1hr}}{m_r} - \log\left(\frac{k}{\phi\mu c_t r_w^2}\right) + 3.2275\right]. \tag{2.42}$$

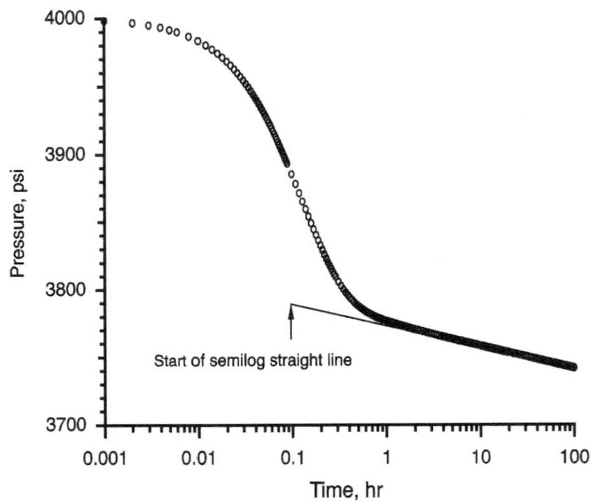

Figure 2.3 Semilog (MDH) plot of the wellbore pressure for a drawdown test.

The above analysis procedure is called the *semilog method* or MDH for drawdown tests. The semilog straight line period is called the *infinite acting radial flow regime.*

Taking the derivative of Equation (2.38) with respect to $\ln(t)$ gives

$$\frac{\mathrm{d}\Delta p_w(t)}{\mathrm{d}\ln(t)} = \frac{m_r}{\ln(10)}. \tag{2.43}$$

Figure 2.4 presents a log-log plot of the pressure derivative for the drawdown test. As shown in Figure 2.4, the logarithmic derivative exhibits an infinite acting radial flow regime for the same period as in the semilog plot after 10 hours. As can be seen in Figure 2.4, the pressure derivative before 10 hr is distorted by the wellbore storage effect. The log-log derivative given in Figure 2.4 is usually called the Bourdet derivative (Bourdet, Ayoub, & Pirard, 1989; Bourdet, Whittle, Douglas, & Pirard, 1983) and is the basis for system identification and flow regime analysis. The derivative of pressure or pressure change data from multirate, interference, or buildup tests in any type of formation or reservoir with respect to an appropriate time function such as $\ln(t)$, logarithm of Horner time, or superposition time is plotted on a log-log plot.

2.3.2. Radial flow regime for buildup tests

Let us assume that a vertical well is producing at a constant rate q until shut-in. At any time after shut-in, the buildup pressure of the system can

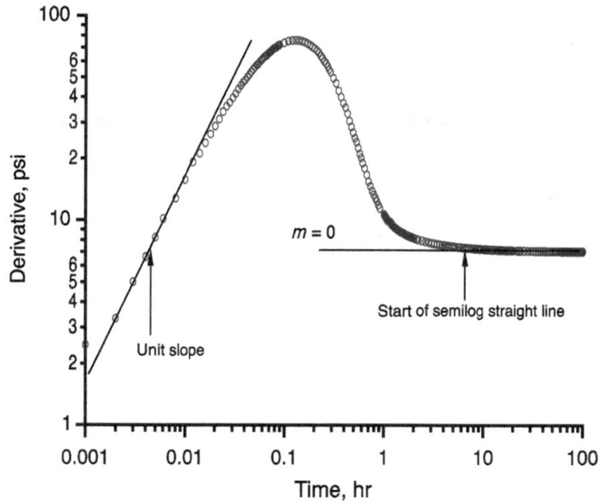

Figure 2.4 Logarithmic derivative plot of the wellbore pressure for a drawdown test.

be written from the well-known Horner (1951) superposition equation and Equation (2.41) as

$$p_{ws}(\Delta t) = p_o - m_r \log\left(\frac{t_p + \Delta t}{\Delta t}\right), \tag{2.44}$$

where t_p is the producing time and the semilog straight line slope m_r is given as

$$m_r = \frac{162.6 q\mu}{kh}, \tag{2.45}$$

which is same as that is given by Equation (2.39) for the drawdown case, where the term $\frac{t_p+\Delta t}{\Delta t}$ is called the Horner time (t_H). As in the drawdown case, a linear plot of the wellbore pressure versus the Horner time (t_H) may yield a semilog straight line with a slope of m_r, from which kh, or the permeability with a known h, and the initial or extrapolated reservoir pressure at $t_H = 1$ are obtained. Using permeability, the total skin factor is obtained from

$$S = 1.5113\left[\frac{p_{1hr} - p_w(\Delta t = 0)}{m_r} - \log\left(\frac{k}{\phi\mu c_t r_w^2}\right) + 3.2275\right]. \tag{2.46}$$

The pressure (p_{1hr}) at one hour must be obtained from the semilog straight line.

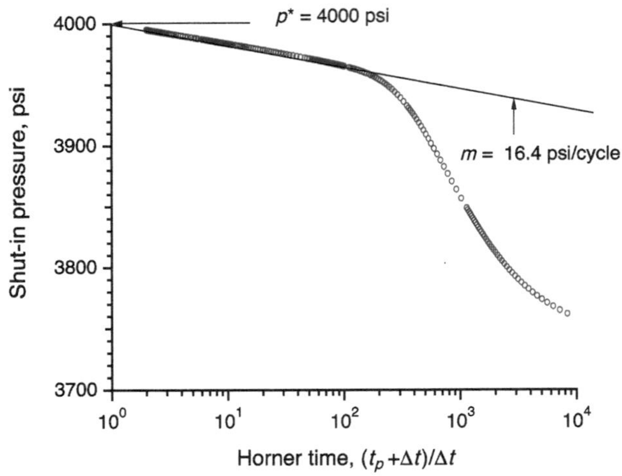

Figure 2.5 Horner plot of shut-in pressure for a buildup test.

Figure 2.3 presents the Horner plot as a function of Horner time (t_H) of the buildup test given by Figure 1.23. As expected for this case, the Horner plot yields a semilog straight line with a slope of 16.4 psi/cycle and an extrapolated pressure $p^\star$ of 4000 psi. Note that the shut-in pressure is dominated by the wellbore storage effect until $t_H = 80$, after which the semilog straight line becomes apparent.

This analysis procedure is called the *Horner method*. The semilog straight line period in the Horner plot is called the *infinite acting radial flow regime*. As discussed above, the wellbore pressure at early times is usually affected by wellbore storage and other wellbore conditions and geometry, such as partial penetration. Thus, the wellbore pressure at early times usually deviates from the semilog straight line, as shown in Figure 2.5.

The pressure derivative for the buildup case is obtained by taking the derivative of Equation (2.44) with respect to the logarithm of the Horner time (t_H). Thus,

$$\frac{\mathrm{d}p_{ws}(\Delta t)}{\mathrm{d}\ln(t_H)} = \frac{(t_p + \Delta t)}{t_p}\Delta t\frac{\mathrm{d}p_{ws}(\Delta t)}{\mathrm{d}\Delta t} = \frac{(t_p + \Delta t)}{t_p}\frac{\mathrm{d}p_{ws}(\Delta t)}{\mathrm{d}\ln(\Delta t)} = \frac{m_r}{\ln(10)}. \tag{2.47}$$

As for the drawdown test, a log-log plot of the pressure derivative of a buildup test for a fully completed vertical well in an infinite isotropic medium also becomes flat (constant, indicating an infinite acting flow regime after the effect of wellbore distortions, as shown in Figure 2.6.

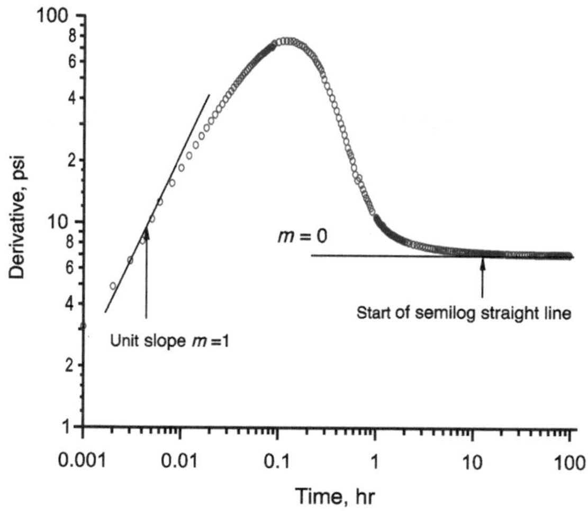

Figure 2.6 Log-log plot of the logarithmic derivative of the pressure for the buildup test.

2.4. Skin Factor

Given a thin damage formation zone with reduced or enhanced permeability around the wellbore, as shown in Figure 2.7, that causes a steady-state pressure drop between the formation and wellbore. Based on this steady-state pressure drop, the skin or skin factor is defined by van Everdingen (1953) and Hurst (1953) as

$$S = \frac{2\pi kh}{q\mu}\Delta p_s, \tag{2.48}$$

where q is the constant flow rate across the skin region. As shown by Hawkins (1956), the skin (S) can be treated as a zone of a finite radius r_s with a skin zone permeability k_s as

$$S = \left(\frac{k}{k_s} - 1\right)\ln\left(\frac{r_s}{r_w}\right), \tag{2.49}$$

where k is the formation permeability and r_w is the wellbore radius.

In addition to the damage skin, there are many sources of steady-state pressure drops in the vicinity of the wellbore as pressure transients diffuse away from the wellbore. These steady-state pressure drops can also be expressed as skin factors. For instance: (1) Geometric skin due to flow convergence near the wellbore such as partial penetration, (2) Non-Darcy skin due to visco-inertial effects, etc. The total skin, for instance,

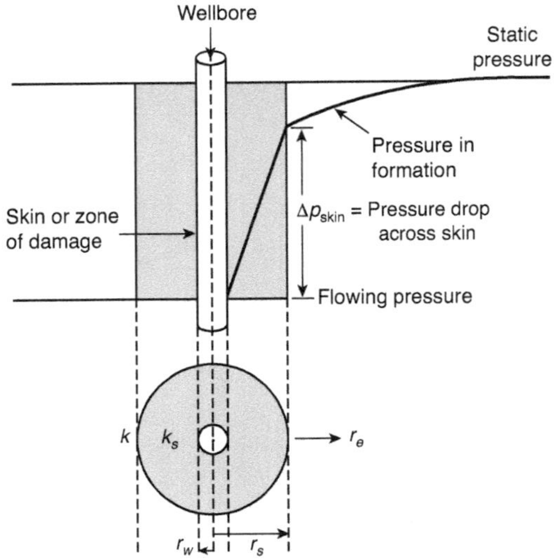

Figure 2.7 Schematic of wellbore skin effect.

normally obtained from the infinite acting radial flow regime may have many components (Kabir, 2009) such as:

$$S_t = S_d + S_p + S_{\text{perf}} + S_\theta + S_g + S_{nD}, \tag{2.50}$$

where S_d = damage skin, S_p = due to partial penetration (limited entry), S_{perf} = due to perforation efficiency, S_θ = due to well deviation, S_g = due to wellbore and near-wellbore formation geometry, and S_{nD} = due to turbulence and inertial effects (non-Darcy). It should be pointed out that the length factors are included in these dimensionless skins. For example, conventionally the total skin due to partial penetration and damage zone is expressed as $S_t = S_d/b + S_p$, where the penetration ratio $b = h_w/h$, h is the thickness of the producing zone, and h_w is the thickness of the perforated interval.

2.5. Wellbore Storage

The wellbore storage affects the wellbore and observation well pressure behavior significantly in oil and gas reservoirs, as shown in Figure 2.8. As can be seen in this figure, the early-time wellbore pressure is distorted by wellbore storage and skin effects when they are compared with the pressures without wellbore storage and skin effects. After wellbore storage effects, the late-time wellbore pressure behavior is normally affected by boundaries, faults, fractures, and interference from other wells. Figure 2.8 also shows the

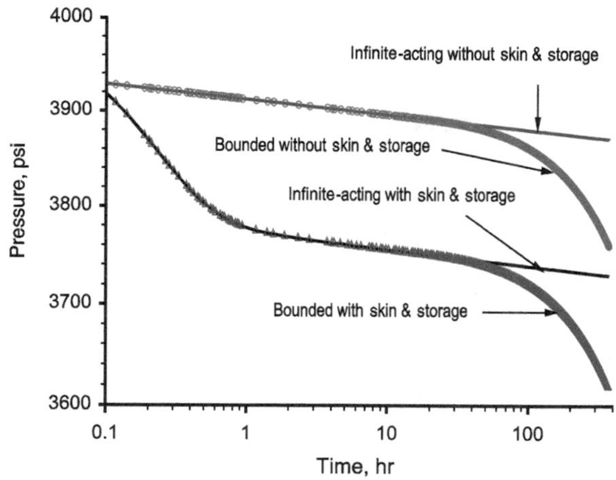

Figure 2.8 Wellbore storage, skin, and no-flow boundary effects on the wellbore pressure.

effect of a no-flow circular boundary. It should be stated that early-time and late-time are very relative. For instance, a nearby no-flow boundary due to a sealing fault could affect the wellbore pressure within minutes. The wellbore storage effect normally masks the early time pressure characteristics due to boundaries and geological features if they are in the vicinity of the wellbore, about a few hundred feet if the downhole shut-in is not applied.

When a constant or variable flow rate is applied at the surface or at any point in the wellbore, except the sandface, the rate change cannot propagate to the sandface instantaneously due to the compressibility of the fluid in the wellbore, tool, or chamber. This causes the sandface flow rate to increase slowly (decrease for buildup tests) to become equal to a constant rate or zero. The period from the start of the rate pulse to the time at which the sandface flow rate becomes approximately equal to the surface rate, is called the *wellbore storage dominated flow period.* This leads to the sandface rate $q_{sf}(t)$ being different from $q_m(t)$, the imposed (applied or measured) rate everywhere in the wellbore except the sandface. To estimate the difference between the two, it is useful to define C, the storage coefficient, as the product of the wellbore volume (V_w) and the effective wellbore fluid compressibility (c_w) as

$$C = V_w c_w. \tag{2.51}$$

If the wellbore storage is constant (no leak from the wellbore in which the compressibility fluid is constant—independent of time and pressure), then the sandface flow rate (q_{sf}) for a producing well in terms of the measured flow rate (q_m) at the tool location or at the surface and due to the wellbore

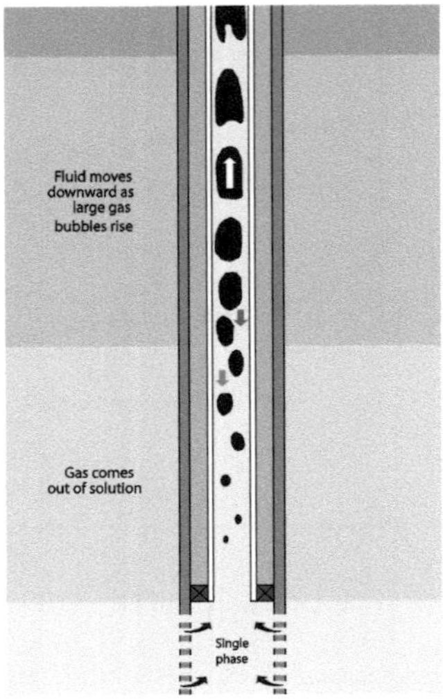

Figure 2.9 A schematic of multiphase fluid flow in the production string creating a variable wellbore storage condition in the wellbore (after Ayestaran (1987)).

storage (van Everdingen & Hurst, 1949) can be expressed as

$$q_{sf}(t) = q_m(t) + C\frac{\mathrm{d}p_w(t)}{\mathrm{d}t}, \tag{2.52}$$

where q_m is the measured flow rate at any location in the wellbore or tool string from the sandface to the wellhead and p_w is the wellbore pressure. The wellbore storage equation given by Equation (2.52) is the fundamental wellbore storage formulation and normally covers most well testing tool configuration and well types. However, it does not apply if the phase segregation is taking place in the wellbore and/or the liquid level is changing in the well. There are also other exceptions; further details of wellbore effects including nonisothermal flow are given by Kabir (2009).

For naturally flowing oil wells, the presence of gas in the wellbore makes compressibility vary, which in turn produces a variable wellbore storage condition in the wellbore. The pressure gauge could therefore experience a multiphase fluid flow while the fluid flow in the formation is single phase, as shown in Figure 2.9. Particularly, if a well is shut in for a buildup test, the multiphase fluid in the production string will segregate according to their densities and create a variable wellbore storage as a function of time. The phase segregation phenomenon can create considerable humps in the wellbore pressure and discontinuities in derivatives.

Figure 2.10 presents a horizontal buildup test example (Kuchuk, 1997). As can be seen from this figure, the test is almost totally dominated by the changing wellbore storage, although both pressure change and derivative

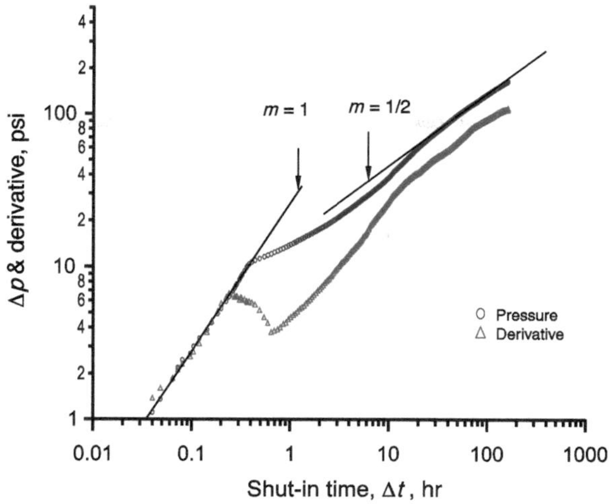

Figure 2.10 The pressure change and derivative for a buildup test given by Kuchuk (1997).

exhibit a perfect unit-slope ($m = 1$) period until 0.3 hr. After the unit-slope period, the phase segregation in the production string distorts the unit-slope period significantly until about 30 hr. Finally, both pressure change and derivative show a linear flow regime ($m = 1/2$) towards the end of the test. Practically, this 160-hr buildup test does not provide information about the well productivity and formation, or the reservoir pressure.

The wellbore storage equation given by Equation (2.52) is a simple material balance for a slightly compressible fluid in the wellbore. As discussed above, the evolution of solution gas in the wellbore fluid makes the wellbore fluid compressibility vary, which in turn produces a variable wellbore storage condition in the wellbore. When the well is shut in, gas, oil, and water will segregate, and gas will occupy the top section of the production string. As the wellbore shut-in pressure increase, some of the gas may dissolve in wellbore oil, therefore there may be a slight but continuous change in fluid compressibility until the system reaches an equilibrium state.

Figure 2.11 presents the wellbore storage coefficient as a function of time given by (Meunier, Wittmann, & Stewart, 1985) for a buildup test shown by Figure 2.12. The wellbore storage coefficient is computed from Equation (2.52), in which $q_m(t) = 0$ because the well was shut in, using the measured afterflow rate $q_{sf}(t)$ and pressure given by Figure 2.12. As can be seen from Figure 2.11, the wellbore storage coefficient, C, decreases with time as the shut-in pressure increases. After 0.7 hr, the wellbore storage coefficient is not computable because the afterflow rate cannot be measured due to the spinner type flowmeter threshold.

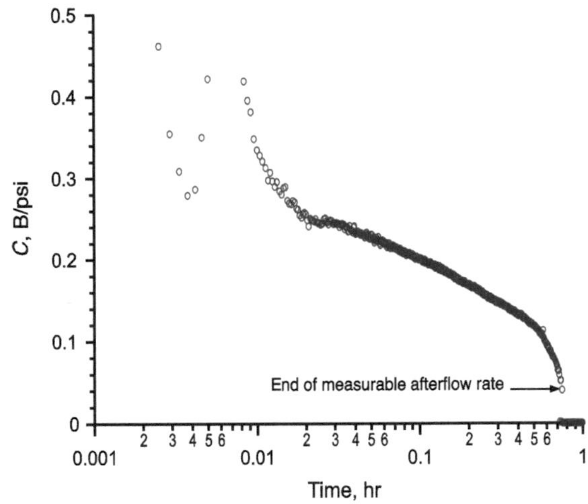

Figure 2.11 Change in wellbore storage coefficient during a buildup test (after Meunier et al. (1985)).

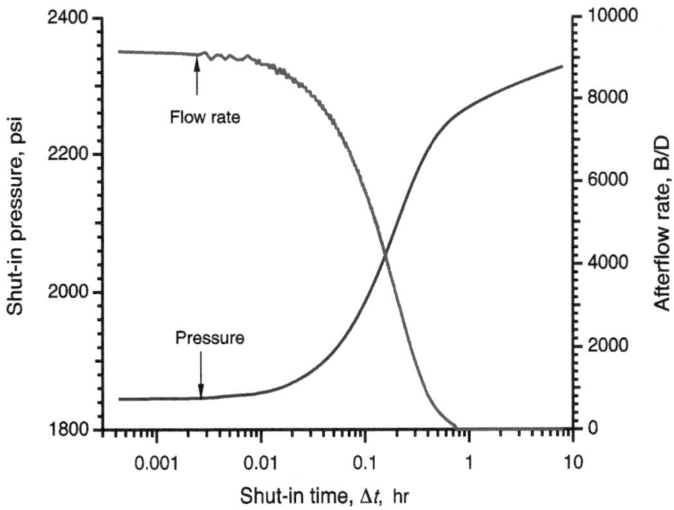

Figure 2.12 Pressure and afterflow rate measurements for the buildup tested given by Meunier et al. (1985).

2.6. Flow Regime Identification

Flow regime identification that leads to model identification is one of the crucial steps in pressure transient test interpretation. Above, we

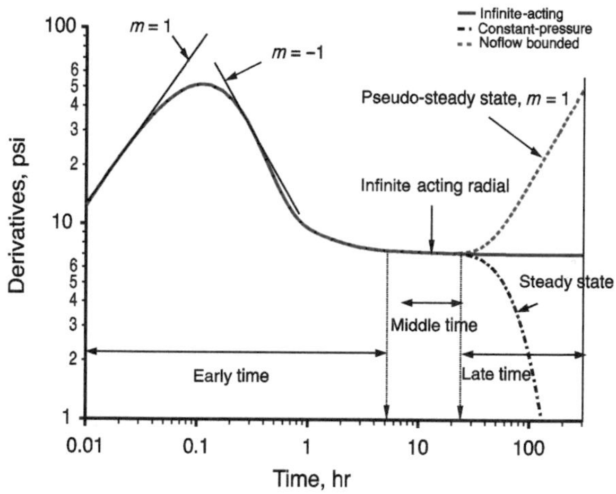

Figure 2.13 Flow regimes for an infinite acting, and constant-pressure and no-flow bounded systems with wellbore storage and skin.

have introduced a few of the flow regimes. In general, skin and wellbore storage affect most of the early time behavior of a test and therefore mask flow regimes. Downhole flow rate measurements for drawdown tests and downhole shut-in for buildup tests minimize skin and wellbore storage effects. Flow regimes are identified by searching for a similar "signature" of the observed response from a library of available drawdown type curves with geological information. Conventionally, the flow regimes are defined as follows:

1. Early time period,
2. Middle time period, and
3. Late time period.

As shown in Figure 2.13, the derivatives look like an unsymmetrical bell-shape curve with four distinct periods: (1) A unit-slope ($m = 1$) period, (2) A maximum arch interval, (3) A negative slope to ($m = -1$) period, and (4) A transition period before the infinite acting radial flow regime during the early time period. The first three periods are due to skin and wellbore-storage effects, while the sandface flow rate is increasing slowly.

The storage dominated unit-slope flow period is given by Agarwal, Al-Hussainy, and Ramey (1970) as

$$\begin{aligned} p_{wD}(t_D) &= \frac{t_D}{C_D} \\ \frac{\mathrm{d}p_{wD}(t_D)}{\mathrm{d}\ln(t_D)} &= \frac{t_D}{C_D}, \end{aligned} \tag{2.53}$$

where p_{wD} is the dimensionless wellbore pressure, which is defined as

$$p_{wD} = \frac{kh}{141.2q\mu}[p_o - p_w(t)], \tag{2.54}$$

t_D is the dimensionless time, which is defined as

$$t_D = \frac{0.0002637kt}{\phi\mu c_t r_w^2}, \tag{2.55}$$

and C_D is the dimensionless wellbore storage, which is defined as

$$C_D = \frac{5.6146C}{2\pi\phi c_t h r_w^2}. \tag{2.56}$$

Equation (2.53) is only an approximation of p_{wD}, for which the pressure drop due to the formation is assumed to be small and neglected. In other words, at the moment the rate pulse is applied at the surface, there will be flow from the formation into the wellbore after a few seconds, but the pressure drop due to the formation will be small compared to the pressure change due to the wellbore fluid expansion or compression during the early time.

After the skin and wellbore-storage dominated the early time period, the derivatives exhibit an infinite-acting radial flow regime during the middle time period if there is no inhomogeneity or heterogeneity around the wellbore, say with in a few hundred feet. The middle time period also implies a middle area or volume between the area from the wellbore to a distance, where its effect is masked by the wellbore and skin, and the nearest boundary. Then the derivatives for the constant-pressure and no-flow outer boundary cases deviate from the infinite-acting radial behavior during the late time period, where the pseudosteady-state or steady-state flow regimes are observed.

In general, there can be more than a few flow regimes in each period. For instance, a system may exhibit linear, spherical, and radial flow in the early time period. Therefore, the above three flow periods shown in Figure 2.13 are only for a well in a circular reservoir with either a no-flow or constant pressure boundary. For a horizontal well, many more flow regimes can be observed. In general, there are many flow regimes that can be observed, but not during a single test. The following are the most common flow regimes.

Early time period:

1. Positive unit-slope wellbore-storage dominated period,
2. Negative unit-slope wellbore-storage dominated period (see Figure 2.14),

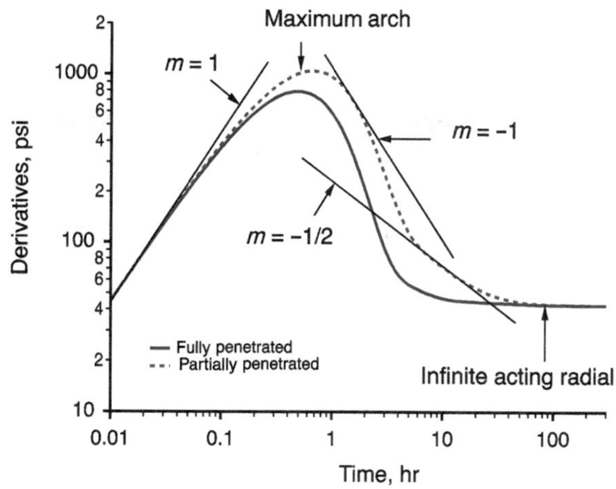

Figure 2.14 Flow regimes for an infinite acting and a partially-penetrated vertical well with wellbore storage and skin.

3. Radial,
4. Linear,
5. Trilinear,
6. Bilinear,
7. Spherical,
8. Hemispherical,
9. Ellipsoidal,
10. Elliptical.

Middle time period:

1. Unit-slope storage due to inner composite regions and fractures
2. Channel Linear,
3. Radial Composite,
4. Semi-radial,
5. Dual porosity,
   (a) Pseudosteady-state matrix flow, and
   (b) Transient matrix flow.
6. Dual permeability.

Late time period:

1. Channel Linear,
2. Steady state,
3. Pseudo-steady state.

Normally, depending on well geometry and geology, any combination of these flow regimes can appear at any time. For instance, for a partially penetrated well in a bounded thick reservoir, one can observe, as shown in

Figure 2.14, these flow regimes:

1. Positive unit-slope ($m = 1$) wellbore-storage dominated period,
2. Negative unit-slope ($m = -1$) wellbore-storage dominated period,
3. Spherical ($m = -1/2$),
4. Second radial (pseudoradial).

As shown in Figure 2.14, the first radial flow is totally masked (covered) by the wellbore storage effect. If k_x/k_y is large, then instead of a spherical flow regime, we may observe an ellipsoidal flow regime.

CHAPTER 3

Convolution

Contents

3.1. Introduction

In this chapter we present a fundamental treatment of the convolution integral and its applications in pressure transient formation and well testing. It mainly deals with simultaneously measured pressure and flow rate at the wellbore or surface, and pressure measurements at different spatial locations in the formation.

The convolution integral is used to solve linear parabolic partial differential equations applicable to heat and mass transfer, and pressure diffusion in porous media with time dependent boundary conditions. The convolution integral is based on Duhamel's principle and was first introduced by Duhamel (1833) through his work on heat conduction in a solid with a variable boundary temperature. In other words, the convolution integral or Duhamel's superposition theorem is used to handle time-dependent boundary conditions in pressure diffusion (transients). The convolution integral is of the Volterra type integral equations of the first kind.

Developments in Petroleum Science, Volume 57
ISSN 0376-7361, DOI: 10.1016/S0376-7361(10)05709-2

In pressure transient testing, the convolution integral is a mathematical expression of the wellbore pressure in terms of the measured flow rate and the constant-rate pressure behavior of the reservoir as a forward problem. As an inverse problem (interpretation of which may include deconvolution given in Chapter 4), if the wellbore pressure with the corresponding flow rate is measured, it is then possible to identify the system and to estimate its parameters. During the last 80 years, much effort has been devoted to measuring the wellbore (downhole) pressure response of the system to a surface production or injection rate schedule (function). As we discussed in Chapter 1, pressure transient testing techniques, tools, and gauges have improved significantly during the last four decades. The pressure gauge resolution is now better than 0.01 psi with a few psi absolute accuracy.

Although the flow rate measurements are also moving downhole, in the pressure case that took place mostly in the 1950s, most well testing rate measurements are still made at the surface. It should be pointed out that the accuracy of surface rate measurements has improved significantly during the last ten years compared to choke-regulated or production tank measurements. However, surface flow-rate measurements have three main drawbacks: First, the fluid rate seen by the pressure sensor is quite different from that measured at the wellhead or in the tank; second, there is a considerable wellbore volume between the pressure sensor and the wellhead where the flow rate is measured; and third, the pressure and rate measurements do not correspond to the same time span. These difficulties with surface flow-rate measurements limit the use of accurate measurements of downhole pressure as a system response. In any case, if downhole flow-rate measurements are not possible, then surface rate should be measured accurately to perform any meaningful interpretation of downhole transient pressure measurements. As discussed in Chapter 2, for a slightly compressible fluid in the wellbore, the sandface flow rate can be obtained from the van Everdingen and Hurst (1949) wellbore material-balance formula (see Equation (2.52)) if the surface flow rate is constant or measured. However, a constant-rate schedule is seldom attained for flow tests (drawdown or injection). The assumption of slightly compressible fluid in the flow string is valid only for injection wells. For naturally flowing oil wells, the presence of gas as the fluid rises in the tubing causes wellbore-fluid compressibility to increase considerably, which in turn produces a variable-wellbore-storage condition in the production string. For buildup tests, the sandface flow rate goes to zero asymptotically after some time if there is no leak into the annulus or from the wellhead valve.

The second main topic in this chapter is the pressure-pressure convolution for pressure measurements among wells, among probes, and among packer and probes of multiprobe wireline formation testers (WFT). As is well known, multiprobe WFTs provide the capability to conduct controlled local production and vertical interference tests using wireline systems. The tools can be set repeatedly at different locations in a single trip, and the formation can be probed in detail through pressure transients. Therefore, heterogeneity may be identified through sequential interference tests at different locations of the wellbore.

The convolution integral (Duhamel's or the superposition theorem) is the main topic of this chapter because it has been widely used in reservoir engineering to derive solutions for partial differential equations with time-dependent boundary conditions. For example, multiple-rate testing is a special application of the superposition theorem. To understand the interpretation and testing aspects of simultaneously measured pressure and flow-rate measurements, it is essential to understand the mathematical nature of convolution with respect to fluid flow in porous media. A brief introduction to the convolution integral will be presented in the next section.

3.2. Convolution Integral

The first kind linear Volterra integral equation as a convolution integral can be written as

$$\begin{aligned}\psi(t) = (f * \kappa)(t) &= \int_0^t \mathrm{d}\tau f(\tau)\kappa(t-\tau) \\ = (\kappa * f)(t) &= \int_0^t \mathrm{d}\tau \kappa(\tau) f(t-\tau), \quad 0 \le t < \infty, \end{aligned} \tag{3.1}$$

where $*$ is the convolution product. The function (functional) $\psi(t)$ is a convolution of the functions $f(t)$ and $\kappa(t)$. These three real-valued functions are scalar functions. The convolution kernel $\kappa(t)$ is called the *unit impulse response* or *influence function* and is continuous in the domain $0 \le t$. The convolution kernel $\kappa(t)$ in pressure transient testing is a solution of the pressure diffusion equation at the spatial position of interest including wellbore storage and skin. The forcing function $f(t)$ may be continuous or piecewise continuous and is usually a time-dependent boundary condition.

For instance, flow-rate measurements at the surface are normally piecewise continuous. If the impulse response or the convolution kernel $\kappa(t)$ of the system is known, then the response of the system given by Equation (3.1) is output (result) to an input, the forcing function $f(t)$, as shown in Figure 3.1. In the well testing terminology, normally the wellbore pressure is output, the production rate is input, and the reservoir model is the convolution kernel. Figure 3.1 shows a schematic representation of the convolution operation as an input and output system.

Figure 3.2 presents the dimensionless wellbore pressure ($\psi = p_D$) (output) as a convolution of the dimensionless impulse response, where $\kappa(t)$

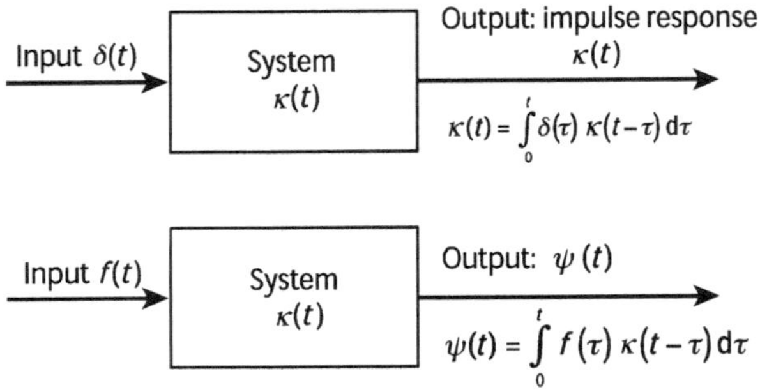

Figure 3.1 A schematic representation of the convolution operation.

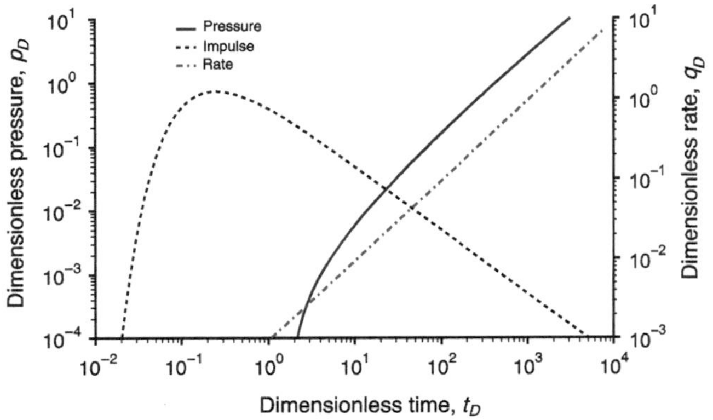

Figure 3.2 The dimensionless pressure, impulse response, and flow rate.

is substituted with $g_D(t_D)$ given as

$$g_D(t_D) = \frac{1}{2t_D} \exp\left(-\frac{1}{4t_D}\right), \tag{3.2}$$

which is the impulse response for 1D radial flow, the forcing function $f(t)$ substituted with $q_D(t_D) = \alpha t_D$, the dimensionless flow rate, where t_D is dimensionless time and α is a positive constant. The same figure also presents $\kappa = g_D$ and $f = q_D$ as a function of dimensionless time. As can be seen from this figure, g_D approaches zero as $t_D \to \infty$, the dimensionless flow rate q_D [forcing function $f(t)$] subjects the dimensionless wellbore pressure to an increase as q_D increases. Moreover, the effect of the rapid increase in q_D until t_D of 0.4 (about a few seconds for most formations) has little effect on the behavior of the dimensionless wellbore pressure p_D.

A further illustration is shown in Figure 3.3, where the wellbore pressure change Δp as a convolution of the 1D radial line-source impulse response in

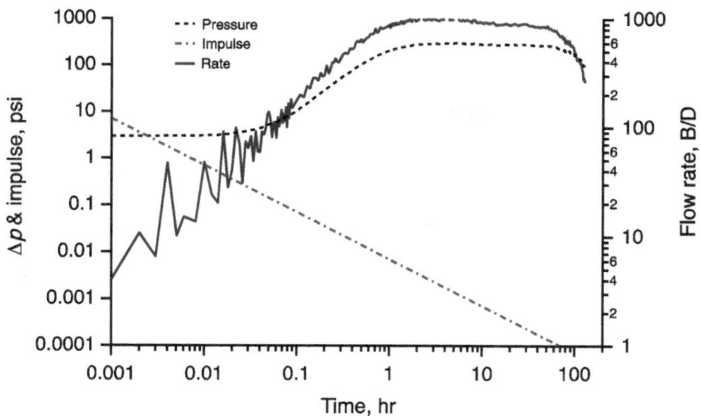

Figure 3.3 The wellbore pressure, impulse response, and flow rate.

field units is given as

$$\kappa(t) = g_w(t) = \frac{141.2\mu}{kh}\frac{1}{2t}\exp\left(-\frac{1}{4}\frac{r_w^2}{0.0002637\eta t}\right), \tag{3.3}$$

where $\eta = \frac{k}{\phi\mu c_t}$ and the rate data $f(t)$ is shown in the figure. As can be seen from this figure, as in the previous case, $\kappa(t) = g_w(t)$ approaches zero as $t \to \infty$. On the other hand, the function $f(t)$ forces the wellbore pressure to increase as $f(t)$ increases and then both decrease. As shown in Figure 3.3, the rapid oscillations in rate $f(t)$ at early times are smoothed significantly by the convolution operator, so that Δp is very smooth. In other words, the convolution integral is a strong smoothing operator.

The convolution integral given by Equation (3.1) has many useful properties with respect to integration and differentiation. In the following paragraphs these properties will be presented (see Bracewell (1973) for additional details). In this book we assume that κ is fundamentally a *heat* or *pressure diffusion kernel* and is continuously differentiable. As we said above, it is a solution to the pressure diffusion equation. The forcing function (flow rate) f is a continuous or piecewise-continuous function for the convolution integral of the pressure diffusion. Furthermore, the pressure diffusion kernel κ should satisfy (Cannon, 1984)

$$\kappa(t) > 0, \quad \lim_{t\to 0}\kappa(t) = 0, \quad |\kappa'(t)| < M, \quad M \in R \text{ for } t > 0. \tag{3.4}$$

Further we assume that the forcing function f satisfies

$$\lim_{t\to 0} f(t) = 0. \tag{3.5}$$

although it can be nonzero at $t = 0$ (for instance see Eq. 2a in Kuchuk and Ayestaran (1985)). Throughout this book if not stated otherwise, it will be assumed that the convolution kernel $\kappa(t)$ and $f(t)$ satisfy the equalities or inequalities given in Equations (3.4) and (3.5). Under such assumptions, the following hold for the convolution integral (Bracewell, 1973).

Commutative:

$$\begin{aligned}\psi(t) = f * \kappa = \kappa * f = \int_0^t \mathrm{d}\tau f(\tau)\kappa(t-\tau) \\ = \int_0^t \mathrm{d}\tau f(t-\tau)\kappa(\tau).\end{aligned} \tag{3.6}$$

Associative:

$$f * (\kappa * h) = (\kappa * f) * h. \tag{3.7}$$

Distributive:

$$f * (\kappa + h) = f * \kappa + f * h. \tag{3.8}$$

Shift Property:

$$\psi(t - t_0) = f(t - t_0) * \kappa(t) = f(t) * \kappa(t - t_0). \tag{3.9}$$

Differentiation:

$$\begin{aligned}\psi'(t) = \int_0^t \mathrm{d}\tau f'(\tau)\kappa(t-\tau) = \int_0^t \mathrm{d}\tau f(t-\tau)\kappa'(\tau) \\ \frac{\mathrm{d}[f(t) * \kappa(t)]}{\mathrm{d}t} = \frac{\mathrm{d}f(t)}{\mathrm{d}t} * \kappa(t) = f(t) * \frac{\mathrm{d}\kappa(t)}{\mathrm{d}t}.\end{aligned} \tag{3.10}$$

$$\psi''(t) = \int_0^t \mathrm{d}\tau f'(\tau)\kappa'(t-\tau) = \int_0^t \mathrm{d}\tau f'(t-\tau)\kappa'(\tau). \tag{3.11}$$

Integration:

$$\begin{aligned}\psi(t) = \int_0^t \mathrm{d}\tau f(\tau)\kappa'(t-\tau) = \int_{-\infty}^{\infty} \mathrm{d}\tau f(\tau)\kappa'(t-\tau) \\ = \int_{-\infty}^{t} \mathrm{d}\tau f(\tau)\kappa'(t-\tau);\end{aligned} \tag{3.12}$$

$$\int_0^t \mathrm{d}\xi\,\psi(\xi) = \int_0^t \mathrm{d}\xi \left[\int_0^\xi \mathrm{d}\tau f(\tau)\kappa(\xi-\tau)\right]$$
$$= \int_0^t \mathrm{d}\xi\; F(\xi)\kappa(t-\xi) = \int_0^t \mathrm{d}\xi\; f(\xi)\mathcal{P}(t-\xi), \quad (3.13)$$

where

$$F(t) = \int_0^t \mathrm{d}\tau f(\tau) \quad (3.14)$$

and

$$\mathcal{P}(t) = \int_0^t \mathrm{d}\tau\kappa(\tau); \quad (3.15)$$

$$\int_0^t \mathrm{d}\tau\delta(\tau)\kappa(t-\tau) = \int_0^t \mathrm{d}\tau\delta(t-\tau)\kappa(\tau) \equiv \kappa(t); \quad (3.16)$$

and

$$\int_0^t \mathrm{d}\tau H(\tau)\kappa'(t-\tau) = \int_0^t \mathrm{d}\tau H(t-\tau)\kappa'(\tau) \equiv \kappa(t), \quad (3.17)$$

where $\delta(t)$ and $H(t)$ are the Dirac delta and Heaviside unit-step functions.

Laplace Transform:

$$\mathcal{L}\left[\int_0^t \mathrm{d}\tau f(\tau)\kappa(t-\tau)\right] = \bar{f}(s)\bar{\kappa}(s). \quad (3.18)$$

Fourier Transform:

$$\mathcal{F}\left[\int_0^t \mathrm{d}\tau f(\tau)\kappa(t-\tau)\right] = \bar{f}(w)\bar{\kappa}(w). \quad (3.19)$$

Scaling:

$$|a|\psi\left(\frac{t}{a}\right) = \int_0^t \mathrm{d}\tau f\left(\frac{\tau}{a}\right)\kappa\left(\frac{t-\tau}{a}\right), \quad (3.20)$$

where a is a constant. The forcing function f has to be continuously differentiable for the some of properties of the convolution integral given above.

3.3. Discrete Convolution

Let us consider a partition of the time interval $\{0, t\}$ into discrete time intervals as $0 = t_0 < t_1 < t_2 < \cdots < t_n < t_{n+1} = t$, and let $\kappa(t)$ be a constant in the interval $\{(t_{n+1} - t_{i+1}), (t_{n+1} - t_i)\}$ and equal to $\kappa(t_{n+1} - t_i)$, then the convolution integral given by Equation (3.1) can be written as

$$\psi(t_{n+1}) = \sum_{i=0}^{n} \kappa(t_{n+1} - t_i) \int_{t_i}^{t_{i+1}} f(\tau)\mathrm{d}\tau, \quad n = 0, 1, \ldots, N-1, \quad (3.21)$$

where $f(t_{n=0}) = 0$ and N is the number of total data points of ψ as well as f and κ. In general, these three functions can contain different total numbers of data points. Let $F(t)$ be the definite integral of $f(t)$, then the integral given in Equation (3.21) can be written as

$$\psi(t_{n+1}) = \sum_{i=0}^{n} \kappa(t_{n+1} - t_i)[F(t_{i+1}) - F(t_i)], \quad n = 0, 1, \ldots, N-1. \quad (3.22)$$

If all intervals of $\{t_i, t_{i+1}\}$ are of the same length, i.e. $\{t_i, t_{i+1}\}$ is the same for all i, where $i = 0, 1, \ldots, n$, then Equation (3.22) can be written as

$$\begin{aligned}\psi_{n+1} &= \sum_{i=0}^{n} \kappa_{n+1-i}[F_{i+1} - F_i] \\ &= \sum_{i=0}^{n} \mathcal{P}_{n+1-i}[f_{i+1} - f_i], \quad n = 0, 1, \ldots, N-1, \quad (3.23)\end{aligned}$$

where $\mathcal{P}$ is the definite integral of κ and is normally the constant-rate pressure behavior of the system. To obtain the second equality in Equation (3.23), the commutative property of the convolution integral given by Equation (3.6) is used. Let us write a few terms of the second equality in Equation (3.23) explicitly for N data points as

$$\begin{aligned}\psi_1 &= \mathcal{P}_1 f_1, \quad n = 0, \\ \psi_2 &= \mathcal{P}_1(f_2 - f_1) + \mathcal{P}_2 f_1, \quad n = 1, \\ \psi_3 &= \mathcal{P}_1(f_3 - f_2) + \mathcal{P}_2(f_2 - f_1) + \mathcal{P}_3 f_1, \quad n = 2, \quad (3.24) \\ &\vdots \\ \psi_N &= \mathcal{P}_1(f_N - f_{N-1}) \;\ldots\;\ldots\;\ldots\;\ldots\;\ldots + \mathcal{P}_N f_1, \quad n = N-1,\end{aligned}$$

where $f(t_{i=n=0}) = 0$. The above system can be written more concisely as a system of linear algebraic equations as

$$\mathbf{A}\mathbf{x} = \mathbf{b}, \tag{3.25}$$

where the coefficient matrix **A** is defined by

$$\mathbf{A} = \begin{bmatrix} f_1 & 0 & 0 & 0 & \dots & 0 \\ (f_2 - f_1) & f_1 & 0 & 0 & \dots & 0 \\ (f_3 - f_2) & (f_2 - f_1) & f_1 & 0 & \dots & 0 \\ \vdots & & & & \dots & \\ (f_N - f_{N-1}) & (f_{N-1} - f_{N-2}) & (f_{N-3} - f_{N-4}) & \dots & \dots & f_1 \end{bmatrix}, \tag{3.26}$$

which is a lower triangular N-by-N matrix, the vector **x** is given as

$$\mathbf{x} = [\mathcal{P}_1,\ \mathcal{P}_2,\ \mathcal{P}_3,\ \dots, \mathcal{P}_N]^T, \tag{3.27}$$

the vector **b** is given as

$$\mathbf{b} = [\psi_1,\ \psi_2,\ \psi_3,\ \dots, \psi_N]^T, \tag{3.28}$$

and T denotes the transpose of a vector.

The convolution integral given by Equation (3.1) can be written in many different discrete forms as in Equation (3.23) using different integration rules, for instance see Bostic, Agarwal, and Carter (1980), Kuchuk and Ayestaran (1985), Thompson and Reynolds (1986), Kuchuk, Carter, and Ayestaran (1990a), and Raghavan (1993).

3.4. Duhamel's (Superposition) Theorem and Pressure-Rate Convolution

The origin of the integral equations goes back to works of J. Fourier (1768-1830) and Abel (1823) (see (Corduneanu, 1991; Linz, 1985)). Both of these works are earlier than the publication of Duhamel's (superposition) theorem (Duhamel, 1833). Duhamel's theorem is already given by Equation (3.23) from Equation (3.1). Because of its historical importance in heat conduction and pressure diffusion, next we will present the derivation of the convolution integral Equation (3.1) from Duhamel's theorem. This derivation was first presented by van Everdingen and Hurst (1949) in the petroleum literature, although the convolution integral given by Equation (3.1) was presented earlier by Muskat (1937a).

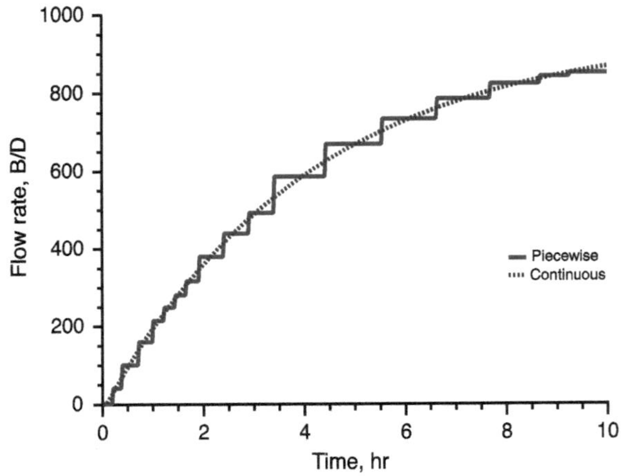

Figure 3.4 Step-wise production history.

Figure 3.4 presents a continuous flow-rate function. As shown in this figure, let us assume that the continuous flow rate can be approximated by a series of constant-rate sequences (piecewise continuous) at a value equal to q_i for each time between t_i and t_{i+1} with $q_0 = 0$ at $t_0 = 0$ as

$$q(t) = \sum_{i=0}^{n} q_i \theta_i(t), \tag{3.29}$$

where n is the number of rate steps and θ is a binary operator such that

$$\theta_i(t) = \begin{cases} 1 & \text{if } t_i \leq t < t_{i+1} \\ 0 & \text{otherwise.} \end{cases} \tag{3.30}$$

The piecewise (step-wise) continuous flow-rate function (history) shown in Figure 3.4 can also be re-written as a sum of rate changes over time as

$$q(t) = \sum_{i=0}^{n} (q_{i+1} - q_i) H(t - t_i), \tag{3.31}$$

where $H(t)$ is the Heaviside function.

Next, we derive the convolution integral from Duhamel's theorem.

If we assume single-phase flow of a slightly compressible fluid, the diffusivity equation leads to a linear relationship between flow rate and pressure changes. As a result of this linearity, the cumulative pressure change observed under variable-rate conditions is the sum of the pressure changes caused by each rate change since the time of each rate change. Let us assume that the system is at rest with an initial reservoir pressure p_o when $t \leq 0$, and then the production (flow) starts instantaneously at $t_0 = 0^+$ at a rate q_1. This is the first rate at an infinitesimally small time. For this system, the pressure as a function of time for a piecewise continuous production history given in Figure 3.4 can be written as

$$\begin{aligned} p(t) &= p_o - q_1 p_u(t) - (q_2 - q_1)\, p_u(t - t_1) \\ &\quad - (q_3 - q_2)\, p_u(t - t_2) - \cdots . \\ p(t_{n+1}) &= p_o - \sum_{i=0}^{n} (q_{i+1} - q_i)\, p_u(t_{n+1} - t_i), \end{aligned} \tag{3.32}$$

where $p_u(t)$ is the unit-rate ($q = 1$) pressure response (change) of the system. The second equality in Equation (3.32) is the same as the third equality Equation (3.23). We develop an integral form of Equation (3.32) by defining $\tau = t_i$, $\Delta\tau = t_{i+1} - t_i$ and $\Delta q = q_{i+1} - q_i$ and then rewriting Equation (3.32) as

$$p(t_{n+1}) = p_o - \sum_{i=0}^{n} \frac{\Delta q}{\Delta \tau} p_u(t_{n+1} - \tau)\, \Delta\tau. \tag{3.33}$$

If we assume that $q(t)$ is differentiable for $t > 0$, then as $n \to \infty$ and $t_{i+1} - t_i \to 0$ (infinitesimal time increments), then Equation (3.33) becomes

$$\begin{aligned} p(t) &= p_o - \int_0^t \frac{\mathrm{d}q}{\mathrm{d}\tau}(\tau) p_u(t - \tau)\mathrm{d}\tau \\ &= p_o - \int_0^t q(t - \tau)\frac{\mathrm{d}p_u}{\mathrm{d}\tau}(\tau)\mathrm{d}\tau, \end{aligned} \tag{3.34}$$

where the second equality of Equation (3.34) follows from the commutative property of the convolution integral given by Equation (3.6).

If we apply integration by parts to the integral given in the first equality of Equation (3.34), we can obtain another form of the convolution integral, which is given as

$$p(t) = p_o - \left[q(t) p_u(0) - q(0) p_u(t) - \int_0^t q(\tau) \frac{\mathrm{d}p_u}{\mathrm{d}\tau}(t - \tau)\mathrm{d}\tau \right]. \tag{3.35}$$

or can be reduced to

$$p(t) = p_o + \int_0^t q(\tau)\frac{\mathrm{d}p_u}{\mathrm{d}\tau}(t - \tau)\mathrm{d}\tau, \tag{3.36}$$

because $p_u(0) = 0$ and we have assumed $q(0) = 0$ in deriving Equation (3.32). Furthermore, if we note that the following relationship holds between the derivatives of $p_u(t-\tau)$ with respect to t and τ

$$\frac{\mathrm{d}p_u(t-\tau)}{\mathrm{d}t} = -\frac{\mathrm{d}p_u(t-\tau)}{\mathrm{d}\tau}, \tag{3.37}$$

and then using Equation (3.37) in Equation (3.36), we obtain

$$p(t) = p_o - \int_0^t q(\tau)\frac{\mathrm{d}p_u(t-\tau)}{\mathrm{d}t}\mathrm{d}\tau. \tag{3.38}$$

Equation (3.38) can also be written in terms of pressure change as

$$\begin{aligned} p(t) &= p_o - \int_0^t q_N(\tau)\frac{\mathrm{d}\Delta p_c(t-\tau)}{\mathrm{d}t}\mathrm{d}\tau \\ &= p_o - \int_0^t \frac{\mathrm{d}q_N(\tau)}{\mathrm{d}t}\Delta p_c(t-\tau)\mathrm{d}\tau, \end{aligned} \tag{3.39}$$

where $q_N = q/q_r$ is the normalized flow rate, q_r is a reference constant flow rate, and Δp_c is the constant flow-rate pressure change. It should be stated that both $p_u(t)$ and Δp_c can be a function of wellbore storage and skin.

Equations (3.34), (3.38) and (3.39) represent the pressure-rate Duhamel's principle (convolution integral) and express the transient pressure as a convolution of the derivative of the flow-rate and unit-rate pressure response Equation (3.34), flow rate and the derivative of the unit-rate pressure response Equation (3.38), the normalized flow rate and the derivative of the constant-rate pressure response Equation (3.39), or the derivative of the normalized flow rate and the constant-rate pressure response Equation (3.39). The convolution over time incorporates all continuous and discrete flow-rate variations. If we use $\frac{\mathrm{d}p_u(t)}{\mathrm{d}t}$ (the time-derivative of the unit-rate pressure response) $\equiv g(t)$ (the unit-rate impulse response), then the convolution integral given by Equation (3.34) as the output (pressure) of a system to the input (flow rate) in a more general form can be written as

$$p(\mathbf{r},t) = p_o - \int_0^t q_{sf}(\tau)\; g(\mathbf{r}, t-\tau)\; \mathrm{d}\tau \tag{3.40}$$

and in the Laplace domain as

$$\bar{p}(\mathbf{r},s) = \frac{p_o}{s} - \bar{q}_{sf}(s)\bar{g}(\mathbf{r},s), \tag{3.41}$$

where $\mathbf{r}$ is the spatial position vector, t is time, the impulse response g is a time-independent solution of the system due to an impulse source (sink) in the system, and q_{sf} is the sandface flow rate. In reality, this rate could be any measured rate at any position in the wellbore, including the surface. Equation (3.40) gives the relationship between flow rate (at an internal boundary surface) and pressure (at any point $\mathbf{r}$ in the system) and it will be called the *pressure-rate* $(p\text{–}r)$ *convolution* or formulation throughout this book.

The unit of $g(t)$ (the impulse response) is $(p/(L^3/t))/t$, where p is pressure, t is time, and L is length. In other words, it is pressure per unit flow rate per unit time $\Rightarrow p/L^3$. The unit of $\bar{g}(s)$ is the unit of $tg(t) \Rightarrow pt/L^3$ and the unit of the Laplace transform variable s is $1/t$.

For most pressure transient well tests and wireline formation testers (WFT), pressure gauges and flow-rate metering devices are not located at the sandface (except for observation probes of WFTs), but they are located somewhere in the wellbore or in the tool string. Therefore there is a finite (small or large) volume of a compressible fluid between the sandface and the location of the pressure gauges and the flow-rate metering devices. In fact, pressure gauges and flow-rate metering devices may also not be closely located at the sandface. For most well tests these days, pressure gauges are located above the producing zones in the wellbore, and the flow rate is usually measured at the surface. It should be stated that permanent downhole pressure gauges and flow-rate metering devices that are normally located above the producing zones in the wellbore have been increasingly used. For wireline formation testers, the pressure is measured downhole close to the sandface, and the flow rate is usually measured downhole by various metering devices that are not far from the pressure sensors. Next, we will express the pressure-rate convolution in terms of simultaneously measured flow rate and pressure in well test and wireline formation pressure transient testing applications.

The sandface flow rate (q_{sf}) in terms of the measured flow rate (q_m) at the tool location or at the surface and due to the wellbore storage (van Everdingen & Hurst, 1949) can be expressed as

$$q_{sf}(t) = q_m(t) + C\frac{\mathrm{d}p_w(t)}{\mathrm{d}t}, \tag{3.42}$$

where q_m is the measured flow rate at any location in the wellbore or tool string, including the sandface, and C is the wellbore storage coefficient given as $C = c_w V_w$, where c_w is the compressibility of the wellbore fluid and V_w is the total volume in which the pressure sensor resides and directly communicates with the sandface hydraulically. In other words, this is the volume of compressible fluid in the wellbore or in the tool flowline that is in contact with the formation (sandface) and the pressure sensor. Note that, in general, neither the sandface flow rate q_{sf} nor the measured flow rate q_m need be constant in Equation (3.42). It should be pointed out that the wellbore storage coefficient C defined above is only valid for the wellbore or tool volume containing single-phase compressible fluids. For other types of wellbore storage phenomena see, for instance, (Earlougher, 1977; Fair, 1981; 1996; Hasan & Kabir, 1995).

The initial condition for Equation (3.42) is given as

$$p_w = p_o \quad \text{at } t = 0, \tag{3.43}$$

assuming that the formation is in hydraulic communication with the wellbore before the onset of any production (or injection) from the formation. If the wellbore pressure is different from the initial formation pressure at the start of production, the above formulation has to be modified, for instance see Kuchuk and Wilkinson (1991).

In the Laplace domain, Equation (3.42) becomes (Kuchuk & Wilkinson, 1991)

$$\bar{q}_{sf}(s) = \bar{q}_m(s) + C\left[s\bar{p}_w(s) - p_o\right], \tag{3.44}$$

and also in terms of the wellbore pressure change as

$$\bar{q}_{sf}(s) = \bar{q}_m(s) - sC\overline{\Delta p}_w(s), \tag{3.45}$$

where $\Delta p_w = p_o - p_w$ and p_w is the wellbore pressure (in the wellbore volume V_w), which can be written from Equation (3.41) as

$$\bar{p}_w(r_w, s) = \frac{p_o}{s} - \frac{\bar{q}_m(s)\bar{g}_w(r_w, s)}{1 + Cs\bar{g}_w(r_w, s)}, \tag{3.46}$$

and the wellbore impulse response $g_w = g_f + \delta(t)\Delta p_{u,\text{skin}}$, g_f is the impulse response due to the formation, $\delta(t)$ is the Dirac delta function, and $\Delta p_{u,\text{skin}}$ is the pressure drop for the unit flow rate due to skin factor S. Substitution of Equation (3.46) in Equation (3.44) yields

$$\bar{q}_{sf}(s) = \frac{\bar{q}_m(s)}{1 + Cs\bar{g}_w(r_w, s)}. \tag{3.47}$$

Substitution of Equation (3.47) in Equation (3.41) gives the pressure change at any time and spatial location in the system as

$$\Delta\bar{p}(\mathbf{r}, s) = \left[\frac{\bar{q}_m(s)}{1 + Cs\bar{g}_w(r_w, s)}\right]\bar{g}(\mathbf{r}, s). \tag{3.48}$$

If q_m the measured flow rate is constant, i.e. $q = q_m$ for all t, the above equation becomes

$$\Delta\bar{p}(\mathbf{r}, s) = \frac{q}{[1 + Cs\bar{g}_w(r_w, s)]}\frac{\bar{g}(\mathbf{r}, s)}{s}. \tag{3.49}$$

On the other hand, the wellbore pressure at r_w in terms of the measured flow rate and the wellbore storage for a given formation response can be directly obtained by substitution of Equation (3.42) in Equation (3.40) as

$$\Delta p_w(r_w, t) = \int_0^t \left[q_m(\tau) + C\frac{\mathrm{d}p_w(\tau)}{\mathrm{d}\tau}\right] g_w(r_w, t-\tau)\,\mathrm{d}\tau, \tag{3.50}$$

where g_w includes the skin effect as stated above. Equation (3.50) is an integral-differential equation and not very useful for computational purposes (Kuchuk, 1990a). However, it can be written in a much more compact form directly from Equation (3.48) as

$$p_w(r_w, t) = p_o - \int_0^t q_m(\tau)\, g_{wf}(r_w, t-\tau)\,\mathrm{d}\tau, \tag{3.51}$$

where g_{wf} is the impulse response in the wellbore, including the pressure change due to wellbore mechanical (damage) skin and wellbore storage effects, and can be expressed as

$$g_{wf}(t) = \mathcal{L}^{-1}\left[\frac{\bar{g}_w(r_w, s)}{1 + Cs\bar{g}_w(r_w, s)}\right], \tag{3.52}$$

where $\mathcal{L}^{-1}$ denotes the inverse Laplace transform operator.

The convolution equation in dimensionless form can be written as

$$p_{wD}(\mathbf{r}_D, t_D) = \int_0^{t_D} q_D(\tau)\, g_{wfD}(\mathbf{r}_D, t_D - \tau)\,\mathrm{d}\tau, \tag{3.53}$$

where g_{wfD} is the dimensionless impulse response in the wellbore, including the skin value and wellbore storage effects, and can be expressed as

$$g_{wfD}(\mathbf{r}_D, t_D) = \mathcal{L}^{-1}\left[\frac{\bar{g}_{wD}(\mathbf{r}_D, s)}{1 + C_D s\bar{g}_{wD}(1, s)}\right] \tag{3.54}$$

and the Laplace transform of the dimensionless wellbore pressure is given as

$$\bar{p}_{wD}(\mathbf{r}_D, s) = \frac{\bar{q}_D(s)\bar{g}_{wD}(\mathbf{r}_D, s)}{1 + C_D s \bar{g}_{wD}(1, s)}, \tag{3.55}$$

where the dimensionless pressure and time (in field units) are defined as

$$p_{wD} = \frac{kh}{141.2 q_r \mu}[p_o - p_w(t)] \tag{3.56}$$

and

$$t_D = \frac{0.0002637kt}{\phi \mu c_t r_w^2}, \tag{3.57}$$

the dimensionless wellbore storage is

$$C_D = \frac{5.6146C}{2\pi \phi c_t h r_w^2}, \tag{3.58}$$

the dimensionless wellbore radius, $r_D = r/r_w$, q_r is a reference rate, $r_D = 1$ at the wellbore, and s should be taken as the dimensionless Laplace domain variable. The convolution integrals given by Equations (3.51) and (3.53) with Equations (3.52) and (3.54) and their Laplace transforms given by Equations (3.48) and (3.55) provide a general framework for interpreting simultaneously measured pressure data with an arbitrarily varying flow rate as a function of time for both pressure transient well and wireline formation testing.

3.5. Wellbore Pressure for Certain Variable Sandface Flow-Rate Schedules

The wellbore pressure in Equation (3.51) or its dimensionless form given by Equation (3.53) can directly be obtained for certain wellbore and reservoir geometries if the flow rate (the inner time-dependent boundary condition) can be approximated by simple functions (schedules). In this section, the analytical convolution integration will be performed for certain flow-rate functions (schedules) with the line source (radial) and point source (spherical) impulse responses to obtain the wellbore pressure. The dimensionless pressure drop due to mechanical skin S is given as $Sq_D(t_D)$ and should be added to the $p_{wD}(t_D)$ values below.

3.5.1. Polynomial rate functions

Let us assume that the flow rate can be approximated by an mth-degree polynomial (generalized power series of t_D) as

$$q_D(t_D) = \sum_{k=0}^{m} \beta_k t_D^k, \tag{3.59}$$

where β is a constant.

Line-Source Radial Flow: The line source impulse response (without skin and wellbore storage) is well known as

$$g_{wfD}(t_D) = \frac{1}{2t_D} \exp\left(-\frac{1}{4t_D}\right) \tag{3.60}$$

which corresponds to the $\beta_0 = 1$ and $m = 0$ case in Equation (3.59). Substituting Equations (3.59) and (3.60) in Equation (3.53) yields the dimensionless wellbore pressure as

$$p_{wD}(t_D) = \int_0^{t_D} \sum_{k=0}^{m} \beta_k (t_D - \tau)^k \exp\left(-\frac{1}{4\tau}\right) \frac{d\tau}{2\tau}. \tag{3.61}$$

Using the binomial expansion for the series given in Equation (3.61), we obtain

$$p_{wD}(t_D) = \int_0^{t_D} \sum_{k=0}^{m} \beta_k \sum_{i=0}^{k} (-1)^i \binom{k}{i} t_D^{k-i} \tau^i \exp\left(-\frac{1}{4\tau}\right) \frac{d\tau}{2\tau}. \tag{3.62}$$

Evaluating the integral (Mathematica, 2009) in Equation (3.62) gives

$$p_{wD}(t_D) = \sum_{k=0}^{m} \beta_k \sum_{i=0}^{k} (-1)^i \binom{k}{i} t_D^{k-i} 2^{-(2i+1)} \Gamma\left(-i, \frac{1}{4t_D}\right), \tag{3.63}$$

where Γ is the incomplete gamma function. The convolution of the polynomial flow-rate variations was first treated by Streltsova (1988). Although the appearance of Equation (3.63) is different from that given by Streltsova (1988), fundamentally they are same.

Linear Flow-Rate Function: For $m = 1$ in Equation (3.59) (the flow rate changes linearly with time) the flow rate can simply be written as

$$q_D(t_D) = \beta_1 t_D. \tag{3.64}$$

The dimensionless wellbore pressure can be written directly from Equation (3.63)(Kuchuk, 1990b; Streltsova, 1988) as

$$p_{wD}(t_D) = \beta_1 \left[\frac{1}{2} \mathrm{E}_1 \left(\frac{1}{4t_D} \right) \left(\frac{1}{4} + t_D \right) - \frac{t_D}{2} \exp\left(-\frac{1}{4t_D} \right) \right]. \quad (3.65)$$

Point-Source Spherical Flow: The point source impulse response (without skin and wellbore storage) can be written as

$$g_{wfD}(t_D) = \frac{1}{2\sqrt{\pi}\, t_D^{3/2}} \exp\left(-\frac{1}{4t_D} \right). \quad (3.66)$$

Substituting Equations (3.59) and (3.66) in Equation (3.53) and using, the binomial expansion yields the dimensionless wellbore pressure as

$$p_{wD}(t_D) = \frac{1}{2\sqrt{\pi}} \int_0^{t_D} \sum_{k=0}^{m} \beta_k \sum_{i=0}^{k} (-1)^i \binom{k}{i} t_D^{k-i} \tau^i \times \exp\left(-\frac{1}{4\tau} \right) \frac{d\tau}{\tau^{3/2}}. \quad (3.67)$$

Evaluating the integral (Mathematica, 2009) in Equation (3.67) gives

$$p_{wD}(t_D) = \frac{1}{2\sqrt{\pi}} \sum_{k=0}^{m} \beta_k \sum_{i=0}^{k} (-1)^i \binom{k}{i} t_D^{k-i} 2^{1-2i} \Gamma\left(\frac{1}{2} - i, \frac{1}{4t_D} \right). \quad (3.68)$$

Linear Flow-Rate Function: When the flow rate changes linearly as a function of time, the dimensionless wellbore pressure can be written directly from Equation (3.68) (Kuchuk, 1990b) as

$$p_{wD}(t_D) = \beta_1 \left[\left(\frac{1}{2} + t_D \right) \operatorname{erfc}\left(\frac{1}{2\sqrt{t_D}} \right) - \sqrt{\frac{t_D}{\pi}} \exp\left(-\frac{1}{4t_D} \right) \right]. \quad (3.69)$$

3.5.2. Exponential flow rate

Hurst (1953) and van Everdingen (1953) expressed the sandface flow rate as

$$q_D(t_D) = 1 - \exp(-\beta t_D), \quad (3.70)$$

where β is a constant.

Line-Source Radial Flow: Using the line source impulse response given by Equation (3.60) and the flow rate Equation (3.70) in Equation (3.53) yields the dimensionless wellbore pressure as

$$p_{wD}(t_D) = \int_0^{t_D} \{1 - \exp[-\beta(t_D - \tau)]\} \exp\left(-\frac{1}{4\tau}\right) \frac{\mathrm{d}\tau}{2\tau}. \tag{3.71}$$

With the series expansion for $\exp(-\beta\tau)$, Equation (3.71) can be written as

$$p_{wD}(t_D) = [1 - \exp(-\beta t_D)]p_D(t_D) - \frac{\exp(-\beta t_D)}{2} \times \int_0^{t_D} \sum_{k=1}^{\infty} \frac{(\beta\tau)^k}{k!} \exp\left(-\frac{1}{4\tau}\right) \frac{\mathrm{d}\tau}{\tau}, \tag{3.72}$$

where

$$p_D(t_D) = \frac{1}{2}\mathrm{E}_1\left(\frac{1}{4t_D}\right). \tag{3.73}$$

Evaluating (Mathematica, 2009) the integral gives

$$p_{wD}(t_D) = [1 - \exp(-\beta t_D)]p_D(t_D) - \frac{\exp(-\beta t_D)}{2} \sum_{k=1}^{\infty} \frac{(\beta t_D)^k}{k!} \mathrm{E}_{k+1}\left(\frac{1}{4t_D}\right), \tag{3.74}$$

where E_{k+1} is the exponential integral function. With less than 0.1% error when $t_D > 100$, Equation (3.74) can be further simplified (Kuchuk, 1990b) as

$$p_{wD}(t_D) = [1 - \exp(-\beta t_D)]p_D(t_D) - \exp(-\beta t_D)[\mathrm{J}_0(\sqrt{\beta}] - 1]p_D(t_D) - \frac{\exp(-\beta t_D)}{2} \exp\left(-\frac{1}{4t_D}\right) [\mathrm{E}_\mathrm{i}(\beta t_D) - \ln(\beta t_D) - \gamma], \tag{3.75}$$

where E_i is the exponential integral function, J_0 is the Bessel function of the first kind, and γ is the Euler constant $0.57722\ldots$.

Point-Source Spherical Flow: Using the point source impulse response given by Equation (3.66) and the flow rate Equation (3.70) in Equation (3.53) yields the dimensionless wellbore pressure as

$$p_{wD}(t_D) = \frac{1}{2\sqrt{\pi}} \int_0^{t_D} \{1 - \exp[-\beta(t_D - \tau)]\} \exp\left(-\frac{1}{4\tau}\right) \frac{\mathrm{d}\tau}{\tau^{3/2}}. \tag{3.76}$$

Using the series expansion for $\exp(-\beta\tau)$, Equation (3.76) can be written as

$$p_{wD}(t_D) = [1 - \exp(-\beta t_D)]p_D(t_D) - \frac{1}{2\sqrt{\pi}}\exp(-\beta t_D) \times \int_0^{t_D} \sum_{k=1}^{\infty} \frac{(\beta\tau)^k}{k!}\exp\left(-\frac{1}{4\tau}\right)\frac{d\tau}{\tau^{3/2}}, \tag{3.77}$$

where

$$p_D(t_D) = \text{erfc}\left[\frac{1}{2\sqrt{t_D}}\right]. \tag{3.78}$$

Evaluating (Mathematica, 2009) the integral gives

$$p_{wD}(t_D) = [1 - \exp(-\beta t_D)]p_D(t_D) - \frac{1}{\sqrt{\pi}}\exp(-\beta t_D)\sum_{k=1}^{\infty}\frac{\beta^k}{2^{2k}k!}\Gamma\left(\frac{1}{2} - k, \frac{1}{4t_D}\right). \tag{3.79}$$

The summation given in Equation (3.79) is strongly convergent and also p_{wD} can be written (Kuchuk, 1990b) as

$$p_{wD}(t_D) = 1 - \frac{\beta}{\beta+1}\left[e^{t_D}\text{erfc}(\sqrt{t_D}) + \frac{2\mathbf{F}(\sqrt{\beta t_D})}{\sqrt{\pi\beta}}\right] - \frac{1}{\beta+1}e^{-\beta t_D}. \tag{3.80}$$

where $\mathbf{F}$ is Dawson's integral (Abramowitz & Stegun, 1972), defined as $\mathbf{F}(x) = e^{-x^2}\int_0^x e^{\tau^2} d\tau$.

3.6. Logarithmic Convolution (Superposition or Multirate) Analysis

The convolution integral or Duhamel's (superposition) theorem was known to Hurst (1934), Muskat (1934), and others in the petroleum literature in the 1930s. However, multirate or variable-rate well testing was first introduced by Odeh and Selig (1963), where they modified the Horner time $\left(\frac{t_p+\Delta t}{\Delta t}\right)$ to account for the rate variations before buildup tests. Russell (1963) introduced two-rate flow tests by using the logarithmic approximation of the unit pressure response in an infinite 1D-radial reservoir.

Odeh and Jones (1965) generalized two earlier methods (Odeh & Selig, 1963; Russell, 1963) again by using the logarithmic approximation for the exponential integral solution for each constant-rate flow period (drawdown test), referred to as *Pressure Drawdown Analysis, Variable-Rate case*. In the first Society of Petroleum Engineering Monograph, Matthews and Russell (1967) referred to the method *Multiple-Rate Flow Test Analysis*. Multirate analysis is also called superposition of variable rate or multirate.

In this book multirate analysis will be called *Logarithmic Convolution Analysis* because the convolution integral kernel $p_u(t)$ in Equation (3.34) or Δp_c in Equation (3.39) is assumed to be a logarithmic type. This assumption excludes the interpretation of pressure data with multirate or variable rate if $p_u(t)$ or Δp_c includes the wellbore storage effect. As will be discussed in Chapter 5, pressure data with multirate or variable rate can be interpreted without difficulty by using a nonlinear parameter estimation method.

To derive the logarithmic convolution method, let us rewrite Equation (3.39) in a discrete form for the wellbore pressure change as

$$\Delta p_w(t) = \sum_{i=0}^{n} \left[q_N\left(t_{i+1}\right) - q_N\left(t_i\right) \right] \Delta p_c\left(t - t_i\right) \qquad n = 0, 1, \ldots, N-1, \tag{3.81}$$

where N is the number of data points in Δp_w and $q_N = q_m/q_r$ (the normalized flow rate). q_m and q_r are the measured and reference constant flow rate, respectively. Δp_c, the constant flow-rate pressure change with skin effect but without wellbore storage effect, at the wellbore in an 1D radial infinite system is the well-known exponential integral solution (Horner, 1951; Theis, 1937) given by

$$\Delta p_c(t) = \frac{q_r \mu}{4\pi kh} \left[\mathrm{E}_1 \left(\frac{\phi \mu c_t r_w^2}{4kt} \right) + 2S \right]. \tag{3.82}$$

For $4kt/(\phi \mu c_t r_w^2) > 100$, the exponential integral solution becomes

$$\Delta p_c(t) = \frac{q_r \mu}{4\pi kh} \left[\ln \left(\frac{\mathrm{e}^{\gamma} \phi \mu c_t r_w^2}{4kt} \right) + 2S \right], \tag{3.83}$$

where $\gamma = 0.5772\ldots$ is Euler's constant. Substituting Equation (3.83) in Equation (3.81) gives

$$\Delta p_w(t) = \frac{q_r \mu}{4\pi kh} \sum_{i=0}^{n} \left[q_N\left(t_{i+1}\right) - q_N\left(t_i\right) \right] \times \left\{ \ln \left[\frac{\mathrm{e}^{\gamma} \phi \mu c_t r_w^2}{4k(t - t_i)} \right] + 2S \right\}. \tag{3.84}$$

Finally, the logarithmic convolution equation from Equation (3.84) can be written in oil field units (Earlougher, 1977; Kuchuk, 1990a; Matthews & Russell, 1967) as

$$J_w(t) = \frac{\Delta p_w(t)}{q_N(t)} = m_r t_{lct}\left[t, q_N(t)\right] + b, \tag{3.85}$$

where J_w is called *rate-normalized pressure*, the logarithmic convolution time t_{lct} is defined as

$$t_{lct} = \frac{1}{q_N(t_{n+1})} \sum_{i=0}^{n} \left[q_N(t_{i+1}) - q_N(t_i)\right] \log(t_{n+1} - t_i), \tag{3.86}$$

the slope is given as $m_r = 162.6 q_r \mu / kh$ and the intercept as $b = m_r[\log(k/\phi\mu c_t r_w^2) + 3.2275 + 0.87S]$, $q_N = q_m/q_r$ is the normalized measured flow rate, q_m is the measured flow rate any place in the wellbore from the sandface to surface. All rates have to be treated as downhole rates; i.e. qB. It should be pointed out that Equation (3.85) is slightly different from those given by Matthews and Russell (1967) and Earlougher (1977).

Equations (3.85) and (3.86) can be also expressed in many different forms depending on the integration scheme chosen, see (Earlougher, 1977; Ehlig-Economides, Joseph, Erba, & Vik, 1986; Fetkovich & Vienot, 1984; Jargon & van Poollen, 1965; Kabir & Kuchuk, 1985; Kuchuk, 1990a; 1990b; Kuchuk & Ayestaran, 1985; McEdwards, 1981; Meunier, Wittmann, & Stewart, 1985; Odeh & Jones, 1965; Raghavan, 1993; Simmons, 1990; Thompson & Reynolds, 1986). These authors called J_w *rate-normalized pressure*, while Gladfelter, Tracy, and Wilsey, 1955; Ramey, 1976a; Winestock and Colpitts, 1965 called it *reciprocal productivity index*. The logarithmic convolution time t_{lct} is also called *superposition time*.

Without wellbore storage effects, a linear plot of J_w vs. t_{lct} should yield a straight line with slope m_r and intercept b, from which the permeability and skin can be estimated. For the illustration of the logarithmic rate convolution technique, the wellbore pressure and flow rate given by Figure 3.5 for a drawdown test from a fully penetrating vertical well in an infinite reservoir bounded by no-flow boundaries at the top and bottom will be used. It is assumed that both pressure and flow rate are measured downhole.

Figure 3.6 presents a linear plot of the rate-normalized pressure, J_w, calculated from Equation (3.85), versus the logarithmic convolution time, t_{lct}, calculated from Equation (3.86) for the wellbore pressure and flow-rate data given in Figure 3.5. As can be seen in Figure 3.6, the rate-normalized pressure as a function of the logarithmic convolution time exhibits a straight line for the almost entire test duration because all rate variations are accounted for. A few points at early times fall outside the straight line because the logarithmic approximation of the exponential integral is not valid during

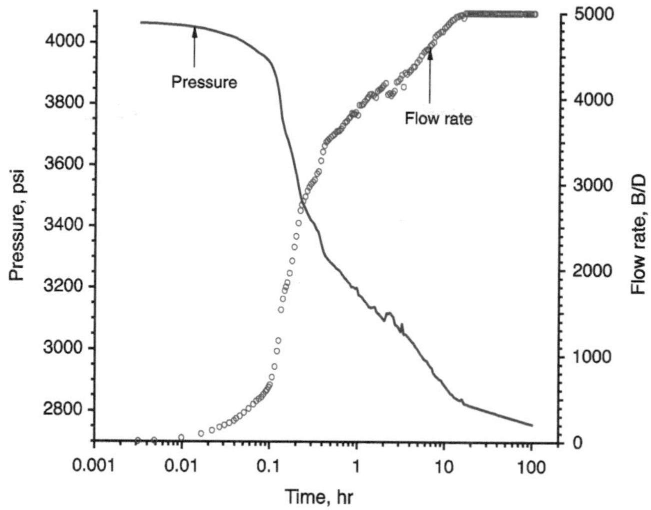

Figure 3.5 The wellbore pressure and flow-rate measurements for a hypothetical drawdown test.

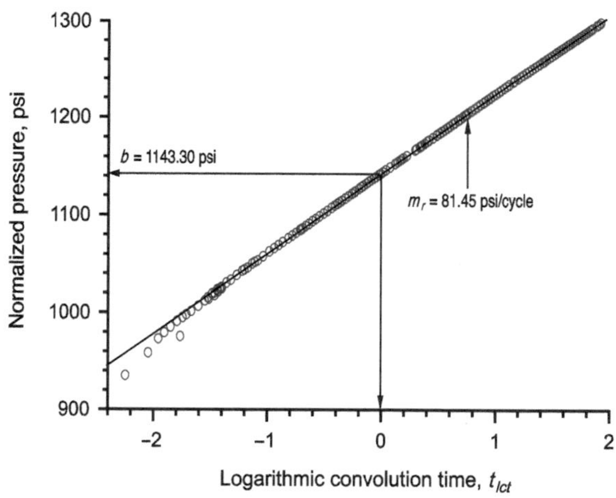

Figure 3.6 Logarithmic convolution plot for the drawdown test.

such early times. From the slope $m_r = 81.45$ psi/cycle and intercept b at $t_{lct} = 0 = 1143.30$ psi given in Figure 3.6, the permeability and skin can be calculated.

The logarithmic convolution method is simple and easy to use, and in many respects it is similar to semilog methods. Computing the normalized pressure, J_w, and the logarithmic convolution time, t_{lct}, is simple, given the widespread use of computers. The logarithmic convolution performs reasonably well for a fully penetrated well in a homogeneous reservoir with high enough permeability (greater than 1 md) and with negligible wellbore

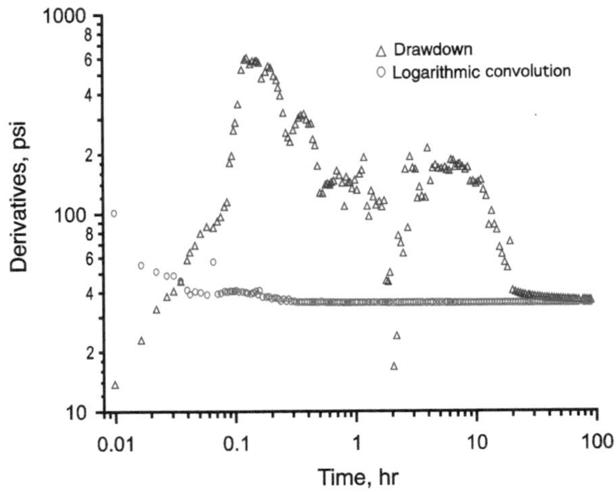

Figure 3.7 Comparison of derivatives of the drawdown pressure and logarithmic convolution.

storage between the tool and the sandface. Other convolution techniques also can be developed as diagnostic tools for different flow geometries such as linear, bilinear, spherical, etc. However, as shown by Ehlig-Economides et al. (1986), Kuchuk (1990a; 1990b), each of these distinct flow regimes can be diagnosed from logarithmic convolution derivatives as in the normal pressure derivative case. Thus, there is no need to obtain a different convolution for each well and reservoir geometry and set of boundary conditions. The logarithmic convolution derivative is obtained by taking the derivative of Equation (3.85) with respect to the logarithmic convolution time $t_{lct}(t, q_{mD})$ as

$$\frac{\mathrm{d}J_w(t)}{\mathrm{d}t_{lct}\left[t, q_N(t)\right]} = \frac{m_r}{\ln(10)}. \tag{3.87}$$

Figure 3.7 compares derivatives of the drawdown and the logarithmic convolution. The the logarithmic convolution derivative computed from Equation (3.87) (the derivative of the rate-normalized pressure (J_w) with respect to t_{lct}, the logarithmic convolution time) exhibits an infinite acting period almost from the beginning of the test, as shown in Figure 3.7, while the drawdown derivative with respect to $\ln(t)$ exhibits an infinite-acting period towards the end of the test.

As stated above, if the logarithmic convolution technique given by Equation (3.85) is used for analysis of multirate tests, for which flow rates are measured at the surface, wellbore storage will affect the behavior of the normalized pressure, J_w, negatively. i.e. the logarithmic convolution technique is not a very useful system identification and flow regime analysis. Let us consider four flow (multirate) tests as shown in Figure 3.8, where the

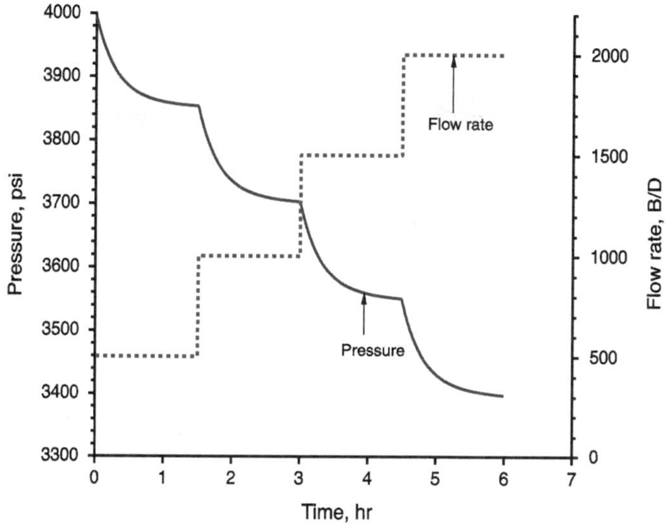

Figure 3.8 The wellbore pressure and surface flow-rate measurements for a hypothetical multirate test.

flow rates shown in the figure are measured at the surface while the flow pressure is measured downhole. It was assumed that the well is a vertical and fully completed well in an 1D infinite radial reservoir with a value of the skin equal to 13.5 and a value wellbore storage coefficient equal to 0.043 B/psi. Each flow test is run for about 1.5 hr with a very short stabilized sandface flow-rate period (Figure 3.8).

Figure 3.9 presents a linear plot of the rate-normalized pressure, J_w, versus the logarithmic convolution time, t_{lct}, for the multirate test data given by Figure 3.8. As can be seen in Figure 3.9, the rate-normalized pressure as a function of the logarithmic convolution time does not exhibit an identifiable straight line. The J_w is not a single-valued function of t_{lct}. It should be noticed that as the wellbore storage effect becomes negligible towards the end of each flow period (the start of a stabilized sandface flow rate), all the curves merge into a single line. As stated above, the logarithmic convolution technique can be used with surface multirate measurements provided the effect of the wellbore storage is negligible on the flowing wellbore pressure.

In summary, surface flow-rate measurements for a multirate test have three main shortcomings: (1) The fluid seen by the pressure sensor is quite different from what is measured at the wellhead or in the tank, (2) There is normally a considerable wellbore volume between the pressure sensor and the wellhead where the flow rate is measured, and (3) Both pressure and rate measurements do not belong to the same time span. In other words, a multirate test basically consists of sequential constant rate drawdowns during which only transient downhole pressure is continuously measured, and flow rates are usually measured intermediately at the end of each flow period. For

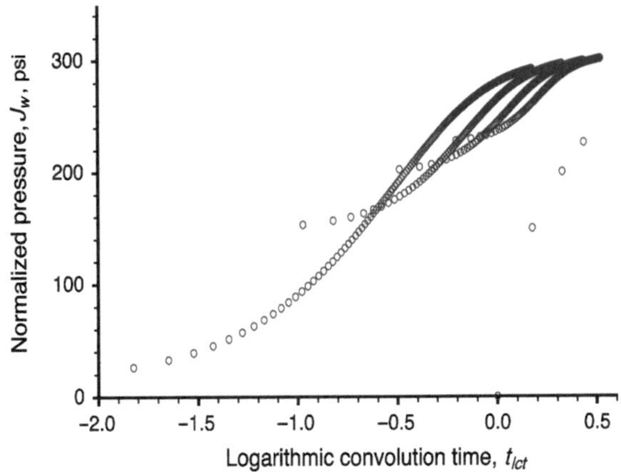

Figure 3.9 Logarithmic convolution plot for the drawdown test.

reasonable high permeability formations with little formation damage, the flow rate becomes constant after a few hours (when the wellbore storage effect becomes negligible) during each drawdown. However, pressure measurements are strongly affected by wellbore storage, and the flow rate varies considerably until it stabilizes during each drawdown. Furthermore, if the flow rates rapidly fluctuate (the flow rate given in Figure 3.8 is very smooth), the test cannot be analyzed using the logarithmic convolution procedure because the logarithmic approximation of the constant-pressure response will not be valid to be used in the convolution integral Equation (3.51). In the situation as stated above, one has to use a nonlinear least-squares parameter estimation technique.

3.7. Rate-Pressure Convolution

For most pressure transient well test applications, the pressure-rate convolution given by Equation (3.51) or Equation (3.53) is sufficient. However, some applications (Kuchuk, Hollaender, Gok, & Onur, 2005) may require to perform a rate-pressure convolution that can be written as

$$\begin{aligned} q_w(r_w, t) &= \int_0^t \frac{\mathrm{d}\Delta p_m(\tau)}{\mathrm{d}\tau} q_{wf}(r_w, t-\tau)\mathrm{d}\tau \\ &= \int_0^t \Delta p_m(t-\tau)\frac{\mathrm{d}q_{wf}(r_w, \tau)}{\mathrm{d}\tau}\mathrm{d}\tau, \end{aligned} \tag{3.88}$$

where q_w is the wellbore flow rate, q_{wf} is the rate impulse response due to the constant-pressure wellbore boundary condition in the wellbore, including the pressure drop due to wellbore mechanical (damage) skin and wellbore storage effects, and $\Delta p_m = p_o - p_m$, where p_m is the measured pressure.

The Laplace transform of Equation (3.51) can be written as

$$\bar{p}_w(s) = p_o/s - \bar{q}_m(s)\bar{g}_{wf}(s) \tag{3.89}$$

and solving Equation (3.89) for $\bar{q}_m$ gives

$$\bar{q}_m(s) = \frac{p_o/s - \bar{p}_w(s)}{\bar{g}_{wf}(s)}, \tag{3.90}$$

where g_{wf} given by Equation (3.52) is the impulse pressure response in the wellbore, including the skin and wellbore storage effects. In the time domain, Equation (3.90) in the convolution form can be written as

$$q_m(t) = \int_0^t [p_o - p_w(\tau)]\kappa(t - \tau)\mathrm{d}\tau, \tag{3.91}$$

where

$$\kappa(t) \equiv \mathcal{L}^{-1}\left[\frac{1}{g_{wf}(t)}\right], \tag{3.92}$$

The terms q_m and p_w are flow-rate and pressure measurements in the wellbore, respectively. As can be seen from Equations (3.88) and (3.91), the measured flow rate and pressure in the wellbore can be expressed in many different ways in terms of the pressure impulse response due to the constant-rate wellbore boundary condition or the rate impulse response due to the constant-pressure boundary condition.

3.8. Pressure-Pressure Convolution

Pressure-pressure convolution mainly deals with pressure transient measurements acquired at different spatial locations in the formation or in the reservoir. Spatial locations can be distributed both in the vertical and lateral directions. For instance, spatial locations are distributed laterally for interference and pulse tests for a multiwell system. On the other hand, they are distributed vertically for vertical interference tests and tests conducted with multiprobe wireline formation testers. These multiprobe tests are called *Interval Pressure Transient Tests* or IPTTs. For horizontal wells, they are distributed laterally for all interference and pulse tests, and IPTTs.

For some interference or pulse tests, or IPTTs, there can be large uncertainties in both measured and estimated flow rates. For instance, in interference testing the measured flow rates could be inaccurate due to unfavorable surface separator conditions. Given the large uncertainties in flow-rate measurements, Goode, Pop, and Murphy (1991) presented a formulation that combines the pressure data recorded at two different locations (for instance, at horizontal and vertical probe locations or active well and observation well locations) to eliminate a need for flow-rate data. This formulation is called *pressure-pressure* (p–p) *convolution*. Because pressure measurements at an observation location (probe, packer, or well) are usually acquired accurately using high-resolution quartz pressure gauges, then p–p convolution becomes quite attractive for model (flow regime) identification and parameter estimation without flow-rate measurements.

3.8.1. Pressure-pressure convolution for multiwell pressure transient testing

Multiwell (multiple-well) pressure transient testing is used to characterize a reservoir section among the wells. The most common multiwell testing is interference and pulse testing between two wells: one is the active (producing) well and the other one is the observation well. These tests are used to obtain the formation mobility k/μ and storativity ϕc_t. In an interference test, the production rate is changed from one (q_1) value to another value (q_2) in the active well, while the observation well is shut in. But in principle both wells can be production wells or are shut in at the same time. In most interference tests, $q_2 = 0$. In other words, after a production period, the active well is also shut in. In a pulse test, the production rate is changed from one value (q_1) to another value (q_2) for a short period of time repeatedly (more than a few times) in the active well, while the observation well is shut in. The duration of each pulse t_L is kept the same, and $q_2 = 0$ in practice. However, the duration of each pulse t_L and/or its associated flow rate can be different. In most applications of interference and pulse testing, flow rates are measured at the surface, and wellbore pressures are measured downhole. Therefore as well documented by Jargon (1976), Kamal and Brigham (1976), Ogbe and Brigham (1984), Prats and Scott (1975), and Tongpenyai and Raghavan (1981), wellbore storage significantly affects both interference and pulse test measurements. Therefore in interference and pulse tests interpretation, we face two problems simultaneously: (1) Uncertain flow-rate measurements and (2) wellbore storage effects both in active and observation wells. Therefore, pressure-pressure convolution and deconvolution (to be covered in Chapter 4) can be used to minimize these two effects on interference and pulse test interpretation.

Like wireline formation testing, single-well vertical interference tests are performed (Burns, 1969; Hirasaki, 1974; Prats, 1970), as shown in

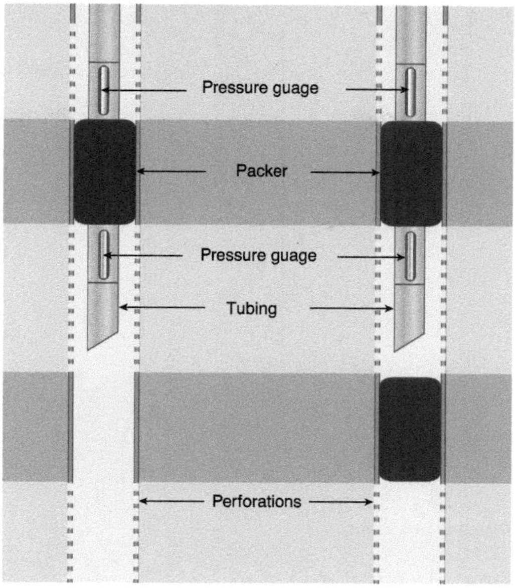

Figure 3.10 A schematic of a single-well vertical interference test setup with single or two packers for zonal isolation.

Figure 3.10, to obtain vertical permeability and determine vertical flow barriers in thick and/or multilayer reservoirs. As for interference and pulse tests, pressure-pressure convolution and deconvolution can also be used for vertical interference test interpretation to minimize uncertain flow-rate measurements and wellbore storage effects both in the tested zone and observation zone in the wellbore.

As shown previously, the pressure distribution at any spatial coordinate $\mathbf{r}$ and time t in the Laplace domain is given by Equation (3.48) as

$$\Delta\bar{p}(\mathbf{r}, s) = \left[\frac{\bar{q}_m(s)}{1 + Cs\bar{g}_w(r_w, s)}\right]\bar{g}(\mathbf{r}, s), \tag{3.93}$$

where q_m is the measured flow rate, g_w is the unit-impulse response including the skin S effect, and all the four quantities (q_m, g_w, S, and C) are at the sink/source location.

For the formulation of p-p convolution, let us suppose that we have a 2D reservoir with an active well at the origin of the coordinate system ($\{x = 0, y = 0\}$) and two observation wells at two distinct spatial locations $\mathbf{r}_1$ and $\mathbf{r}_2$, as shown in Figure 3.11, where $r_w \leq |\mathbf{r}_1| = r_1 < |\mathbf{r}_2| = r_2$; r_1 and r_2 are the distances from the observation well to the center of the coordinate system, and r_w is the active wellbore radius (Figure 3.11). We have chosen a 2D reservoir for simplicity, and further assumed that the point

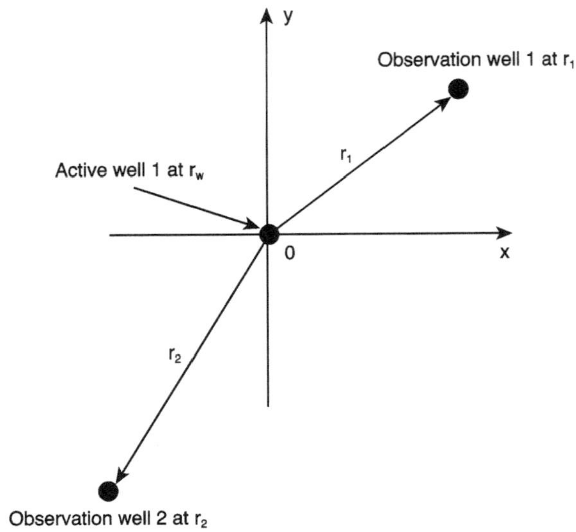

Figure 3.11 A schematic representation of a multiwell system with one active and two observation wells.

or line source wells at spatial locations $\mathbf{r}_1$ and $\mathbf{r}_2$ are free of wellbore storage and skin effects. In reality, p–p convolution can be applied between any two or more wells (vertical-horizontal, vertical-slanted, slanted-horizontal, etc., with skin and wellbore storage).

As shown by Goode et al. (1991), Equation (3.93) can be written at the two distinct spatial locations $\mathbf{r}_1$ and $\mathbf{r}_2$ (Figure 3.11) as

$$\Delta\bar{p}(\mathbf{r}_1, s) = \left[\frac{\bar{q}_m(s)}{1 + Cs\bar{g}_w(r_w, s)}\right]\bar{g}(\mathbf{r}_1, s) \tag{3.94}$$

and

$$\Delta\bar{p}(\mathbf{r}_2, s) = \left[\frac{\bar{q}_m(s)}{1 + Cs\bar{g}_w(r_w, s)}\right]\bar{g}(\mathbf{r}_2, s), \tag{3.95}$$

where $\bar{g}(\mathbf{r}_1, s)$ and $\bar{g}(\mathbf{r}_2, s)$ are the Laplace transforms of impulse responses at the spatial locations $\mathbf{r}_1$ and $\mathbf{r}_2$. In the above two equations, Δp should be obtained at each location either by shifting all pressure measurements to the datum reference pressure or by using the initial pressures p_{o1} and p_{o2} at $\mathbf{r}_1$ and $\mathbf{r}_2$. Using these two equations, the quantity $\frac{\bar{q}_m(s)}{1+Cs\bar{g}_w(r_w,s)}$ can be eliminated and yields the pressure change at $\mathbf{r}_2$ as

$$\Delta\bar{p}(\mathbf{r}_2, s) = \Delta\bar{p}(\mathbf{r}_1, s)\bar{\mathcal{G}}(\mathbf{r}_1, \mathbf{r}_2, s) \tag{3.96}$$

and in the time domain as

$$\Delta p(\mathbf{r}_2, t) = \int_0^t \mathrm{d}\tau \, \Delta p(\mathbf{r}_1, t-\tau) \mathcal{G}(\mathbf{r}_1, \mathbf{r}_2, \tau) \tag{3.97}$$

or

$$p(\mathbf{r}_2, t) = p_{o2} - \int_0^t \mathrm{d}\tau \, \Delta p(\mathbf{r}_1, t-\tau) \mathcal{G}(\mathbf{r}_1, \mathbf{r}_2, \tau), \tag{3.98}$$

where the $\mathcal{G}$-function is defined by

$$\mathcal{G}(\mathbf{r}_1, \mathbf{r}_2, t) = \mathcal{L}^{-1}\left[\frac{\bar{g}(\mathbf{r}_2, s)}{\bar{g}(\mathbf{r}_1, s)}\right], \tag{3.99}$$

$\mathcal{L}^{-1}$ denotes the inverse Laplace transform operator and g is the impulse response. We can also write

$$\Delta \bar{p}(\mathbf{r}_1, s) = \Delta \bar{p}(\mathbf{r}_2, s) \bar{\mathcal{G}}(\mathbf{r}_1, \mathbf{r}_2, s) \tag{3.100}$$

and in the time domain as

$$\Delta p(\mathbf{r}_1, t) = \int_0^t \mathrm{d}\tau \, \Delta p(\mathbf{r}_2, t-\tau) \mathcal{G}(\mathbf{r}_1, \mathbf{r}_2, \tau) \tag{3.101}$$

or

$$p(\mathbf{r}_1, t) = p_{o1} - \int_0^t \mathrm{d}\tau \, \Delta p(\mathbf{r}_2, t-\tau) \mathcal{G}(\mathbf{r}_1, \mathbf{r}_2, \tau), \tag{3.102}$$

where the $\mathcal{G}$-function is defined by

$$\mathcal{G}(\mathbf{r}_1, \mathbf{r}_2, t) = \mathcal{L}^{-1}\left[\frac{\bar{g}(\mathbf{r}_1, s)}{\bar{g}(\mathbf{r}_2, s)}\right]. \tag{3.103}$$

For a two-well system: one active well at r_w and one observation well at $\mathbf{r}_1$, the p–p convolution can be written as

$$\Delta \bar{p}(\mathbf{r}_1, s) = \Delta \bar{p}(r_w, s) \bar{\mathcal{G}}(r_w, \mathbf{r}_1, s) \tag{3.104}$$

and in the time domain as

$$\Delta p(\mathbf{r}_1, t) = \int_0^t d\tau \Delta p(r_w, t-\tau)\mathcal{G}(r_w, \mathbf{r}_1, \tau) \tag{3.105}$$

or

$$p(\mathbf{r}_1, t) = p_{o1} - \int_0^t d\tau \Delta p(r_w, t-\tau)\mathcal{G}(r_w, \mathbf{r}_1, \tau), \tag{3.106}$$

where the $\mathcal{G}$-function is defined by

$$\mathcal{G}(r_w, \mathbf{r}_1, t) = \mathcal{L}^{-1}\left[\frac{\bar{g}(\mathbf{r}_1, s)}{\bar{g}(r_w, s)}\right]. \tag{3.107}$$

The above convolution integrals given by Equation (3.97) through Equation (3.106) are called the pressure-pressure (p-p) convolution (formulation) and are quite general and can be applied to interval interference tests conducted along the wellbore with multiprobe or packer-multiprobe formation testers as well as multiwell interference tests. Although we have presented three different p-p convolution formulations between Wells 2 and 1 $\{\mathbf{r}_2, \mathbf{r}_1\}$, Wells 1 and 2 $\{\mathbf{r}_1, \mathbf{r}_2\}$, and Well 1 and the active well $\{\mathbf{r}_1, r_w\}$, we could have also written Well 2 and the active well $\{\mathbf{r}_2, r_w\}$, the active well and Well 1 $\{r_w, \mathbf{r}_1\}$, and the active well and Well 2 $\{r_w, \mathbf{r}_2\}$. Normally for given two spatial coordinates $\mathbf{r}_1$ and $\mathbf{r}_2$, if the length $r_1 \equiv |\mathbf{r}_1|$ to the source/sink (assuming it is the origin of the coordinate system) is shorter than the length $r_2 \equiv |\mathbf{r}_2|$, the formulation given by Equation (3.97) or Equation (3.98) should be used because a measurable pressure change will be observed first at $\mathbf{r}_1$ and at time t_1, and it will be observed at $\mathbf{r}_2$ and at time t_2, where $t_1 < t_2$. This is only true for isotropic homogenous formations. For anisotropic homogenous and heterogeneous formations, the spatial pressure change is not directly proportional to the distance $|\mathbf{r}|$. In other words, if $|\mathbf{r}_1| = r_1 < |\mathbf{r}_2| = r_2$, it is always true that $\Delta p(\mathbf{r}_1, t) > \Delta p(\mathbf{r}_2, t)$ in homogenous formations. But it is not necessarily true that $\Delta p(\mathbf{r}_1, t) > \Delta p(\mathbf{r}_2, t)$ in anisotropic homogenous and heterogeneous formations. For instance if the permeability in the x direction k_x is much greater than the permeability in the y direction k_y, equi-pressure contour lines will be long-elongated ellipses, where in $a \gg b$ (major and minor axes of ellipses). In other words, $\Delta p(x = a \& y = 0, t) = \Delta p(x = 0 \& y = b, t)$ even though $a \gg b$ (see Figure 3.11). Therefore, for anisotropic homogenous and heterogeneous formations, where we first observe a measurable pressure change that should be denoted $\mathbf{r}_1$, thus $\Delta p(\mathbf{r}_1, t)$ in Equations (3.97) and (3.98). If the source/sink location is at $r_1 = r_w$, then Equations (3.97) and (3.98) should be used, as in Equations (3.105) and (3.106).

For a vertical interference test with a dual-packer and observation probe(s) within the wellbore, $\Delta p(\mathbf{r}_1, t)$ can be the pressure at the packed-interval, while $\Delta p(\mathbf{r}_2, t)$ can represent the pressure at the vertical observation probe(s). For the multiprobe wireline formation tester, $\Delta p(\mathbf{r}_1, t)$ can be the pressure at the sink or horizontal probe, while $\Delta p(\mathbf{r}_2, t)$ can be the pressure at the vertical observation probe(s).

However, it should be stated that the pressure diffusion equation implies that the pressure in the fluid-filled porous medium due to a pressure or rate pulse in a well will change (drop or rise depending on whether it is a source/sink) instantaneously everywhere in the medium although not at the same rate. Furthermore, it implies that the speed of pressure diffusion is infinite. Therefore, there should be no restriction on whether $\mathbf{r}_1$ or $\mathbf{r}_2$ is used in the left and right side of the above p-p convolution equations. In reality, a local excitation (source/sink) in a porous medium is not instantaneously felt throughout the medium. It takes time for a pressure disturbance to propagate from its source to other positions in the medium due to the finite speed of sound and signal propagations. Therefore, between any two $\Delta p(\mathbf{r}, t)$ values, the one that has the largest measurable pressure change at any location at the smallest time should be used as an input (in the right side of p-p convolution equations) and the other one should be used as an output of the system (in the left side of p-p convolution equations).

The p-p convolution is referred to as computing $\Delta p(\mathbf{r}_2, t)$ at the spatial location at $\mathbf{r}_2$ with measured or known $\Delta p(\mathbf{r}_1, t)$ at the spatial location $\mathbf{r}_1$ for a given $\mathcal{G}$-function of the system under consideration; i.e., convolving $\Delta p(\mathbf{r}_1, t)$ and $\mathcal{G}$ to obtain $\Delta p(\mathbf{r}_2, t)$. On the other hand, computing $\mathcal{G}$ from Equation (3.97) with measured $\Delta p(\mathbf{r}_1, t)$ and $\Delta p(\mathbf{r}_2, t)$ is called *p-p deconvolution*.

It is worth noting that the unit of the $\mathcal{G}$-function in the p-p convolution equations (Equation (3.97) to Equation (3.102)) is inverse time; in oilfield units, its unit is 1/hr. On the other hand, the unit of the impulse function g in the p-r convolution equation (Equation (3.93)) is pressure per flow rate per time; in oilfield units, its unit is (psi/B/D)/hr. For model identification purposes, it is the standard method to use log-log diagnostic plots of $tg(t)$, which is the Bourdet, Whittle, Douglas, and Pirard (1983) derivative of the unit-rate drawdown response vs. time. Similarly, for the case of p-p convolution problem, we can use log-log diagnostic plots of $t\mathcal{G}(t)$ vs. time for model identification purposes. The unit of $tg(t)$ response is psi/(B/D), while the unit of $t\mathcal{G}(t)$ response is unitless (dimensionless) because it is a ratio of two unit-impulse responses with the same unit. Recent p-r and p-p deconvolution algorithms encode the p-r or p-p convolution equation in terms of the logarithm of $tg(t)$ or the $t\mathcal{G}(t)$ response as a function of $\ln(t)$, ensuring that the positivity of these responses are explicitly constrained at the price of working with a nonlinear p-r or p-p convolution equation (see for instance, Baygun, Kuchuk, and Arikan (1997), Onur,

Ayan, and Kuchuk (2009a), and von Schroeter, Hollaender, and Gringarten (2004)).

Similar to the functionality of the flow-rate response q in the p-r convolution equation (Equation (3.93)), the pressure response $\Delta p(\mathbf{r}_1, t)$ in the p-p convolution equation (Equation (3.97) or Equation (3.98)) is typically chosen as the pressure response measured at the source/sink location (i.e., $\mathbf{r}_1$ represents the spatial location of the source/sink) convolved with the $\mathcal{G}$-function to compute the pressure response $\Delta p(\mathbf{r}_2, t)$ at an observation location. However, the derivation of Equation (3.97) is very general, so one can chose the pressure response $\Delta p(\mathbf{r}_1, t)$ as the pressure response measured at an observation location and convolve it with the $\mathcal{G}$-function to compute the pressure response $\Delta p(\mathbf{r}_2, t)$ at another observation location.

It is also important to note that the $\mathcal{G}$-function in Equation (3.97) is independent of tool or wellbore storage at a source/sink location, while the derivation of Equation (3.97) assumes that skin and storage at the observation location are negligible. If the wellbore storage volume (well volume at the observation well) is considerable, then the flow from the observation well into the formation towards the active well cannot be neglected. In the next section, we will present solutions that include wellbore storage effects at the observation well when the wellbore storage is significant.

One of the important applications of the p-p convolution equation given by Equation (3.97) or Equation (3.98) is to use it for parameter estimation by history matching measured $\Delta p(\mathbf{r}_2, t)$ data with measured $\Delta p(\mathbf{r}_1, t)$ data convolved with the $\mathcal{G}$-function response for a reservoir model. As to be discussed later, one can use one of the recent deconvolution algorithms (Levitan, 2005; Pimonov, Ayan, Onur, & Kuchuk, 2009a; Pimonov, Onur, & Kuchuk, 2009b; von Schroeter et al., 2004) to compute the $\mathcal{G}$-function [or $t\mathcal{G}(t)$] directly from measured responses $\Delta p(\mathbf{r}_1, t)$ and $\Delta p(\mathbf{r}_2, t)$. One can then identify the interpretation model(s) to be used in the parameter estimation problem from the signature of the computed $\mathcal{G}$ or $t\mathcal{G}$-function and perform parameter estimation directly from the deconvolved $\mathcal{G}$ or $t\mathcal{G}$-function. Therefore, it is useful to derive the forms of $\mathcal{G}$ or $t\mathcal{G}$ functions from applications of various p-p formulations and understand their signatures because they provide crucial information regarding uniqueness and non-uniqueness issues (i.e., identification of the parameters that can be determined uniquely) in parameter estimation based on the p-p convolution.

First, we start with a well-known two-well interference test problem including wellbore storage and skin at both the sink (production) and observation wells and derive the $\mathcal{G}$-function for the case where $\Delta p(\mathbf{r}_1, t)$ represents the production well pressure response and $\Delta p(\mathbf{r}_2, t)$ represents the observation well response in Equation (3.97). Then, we provide equations for the convolution responses that apply during a radial flow period for a vertical well in a single layer system for various p-p convolution options that could be used in parameter estimation when analyzing multiwell interference or pulse tests. These equations indicate which parameters can be uniquely determined from p-p convolution. It is shown that p-p convolution analysis may sometimes not provide unique estimates for some of the parameters, which can be uniquely determined from the p-r convolution analysis.

3.8.2. Pressure-pressure convolution for two-well interference test

Let us consider a fully penetrating vertical active (producing) well located at the origin, as shown in Figure 3.11, producing at $q_m(t)$ (flow rate) with wellbore storage and skin effects in an infinite reservoir with a uniform thickness h, porosity ϕ, and permeability k; and an observation well with wellbore storage and skin located at a spatial position $|\mathbf{r}_1| = r_1 = r_o$, the radial distance from the producing well (Well 1 in Figure 3.11) (Ogbe & Brigham, 1984; Tongpenyai & Raghavan, 1981). Although the full solutions for the pressure response at both active and observation wells with finite radii and wellbore storage and skin in a two-well system were given by Tongpenyai and Raghavan (1981), without loss of generality, we will present the final solutions for line-source active and observation wells (infinitesimally small wellbore radii for both well).

First, the pressure change at the active well for this system can be written as

$$\Delta p_w(t) = \int_0^t q_w(\tau) g_{ww}(t-\tau)\,\mathrm{d}\tau + \int_0^t q_o(\tau) g_{wo}(t-\tau)\,\mathrm{d}\tau \qquad (3.108)$$

and in the Laplace domain

$$\overline{\Delta p}_w(s) = \bar{q}_w(s)\bar{g}_{ww}(s) + \bar{q}_o(s)\bar{g}_{wo}(s), \qquad (3.109)$$

where the subscripts w and o denote the producing well and observation well, respectively. g_{ww} is the impulse response at the active well due to its production, and g_{wo} is the impulse response at the active well due to the injection rate q_o from the sandface of the observation well, where the well contains a volume of compressible fluids (wellbore storage), into the formation; i.e. q_o is an unintended flow that occurs because of compressible fluids stored in the observation well volume. q_m is the production from the active well that is deliberately applied to conduct an interference test. It should be noted that we have not assumed anything about the signs (injection or production) of the flow rates q_w and q_o at both wells in Equations (3.108) and (3.109). As stated in Chapter 2, according to the sign convention in this book, the flow rate is positive for production (flow from the formation into the wellbore) and negative for injection (flow from the wellbore into the formation).

Second the pressure response at the observation well can be written as

$$\Delta p_o(t) = \int_0^t q_w(\tau) g_{ow}(t-\tau)\,\mathrm{d}\tau + \int_0^t q_o(\tau) g_{oo}(t-\tau)\,\mathrm{d}\tau \tag{3.110}$$

and in the Laplace domain

$$\overline{\Delta p}_o(s) = \bar{q}_w(s)\bar{g}_{ow}(s) + \bar{q}_o(s)\bar{g}_{oo}(s), \tag{3.111}$$

g_{ow} is the impulse response at the observation well due to production at the active well, and g_{oo} is the impulse response at the observation well due its injection.

Because both wells have wellbore storage, the flow rate at producing well can be written from Equation (3.42) as

$$q_w(t) = q_m(t) + C_w \frac{\mathrm{d}p_w(t)}{\mathrm{d}t} \tag{3.112}$$

and in the Laplace domain from Equation (3.45) as

$$\bar{q}_w(s) = \bar{q}_m(s) - sC_w \overline{\Delta p}_w(s), \tag{3.113}$$

where q_m is the measured flow rate (surface or downhole) from the active well and C_w is the wellbore storage of the active well. For the observation well, the flow rate can be written as

$$q_o(t) = C_o \frac{\mathrm{d}p_o(t)}{\mathrm{d}t}, \tag{3.114}$$

as stated above, q_o is an injection rate and is negative because it is the flow from the observation well into the formation, and in the Laplace domain we obtain

$$\bar{q}_o(s) = -sC_o \overline{\Delta p}_o(s), \tag{3.115}$$

where C_o is the wellbore storage coefficient at the observation well.

Substituting Equations (3.113) and (3.115) in Equation (3.111) gives the Laplace transform of the pressure change at the observation well

$$\overline{\Delta p}_o(s) = \bar{g}_{ow}(s)\left[\bar{q}_m(s) - sC_w \overline{\Delta p}_w(s)\right] - sC_o \overline{\Delta p}_o(s)\bar{g}_{oo}(s) \tag{3.116}$$

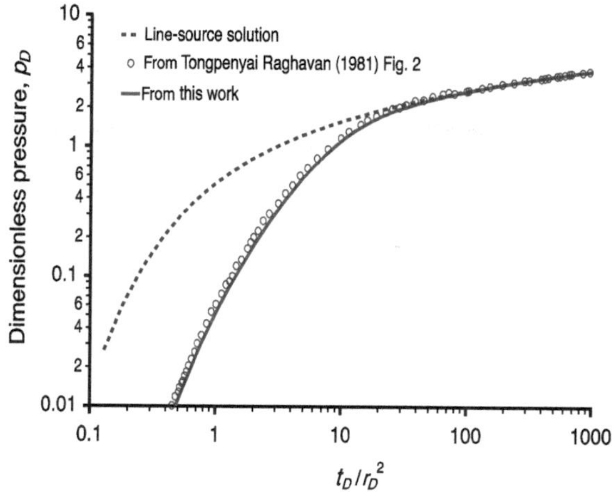

Figure 3.12 Comparison of dimensionless pressures as a function of t_D/r_D^2 from Fig. 2 of Tongpenyai and Raghavan (1981) and from Equation (3.118) (labeled as from this work) for an observation well in a two-well system.

and similarly at the producing well

$$\overline{\Delta p}_w(s) = \bar{g}_{ww}(s)\left[\bar{q}_m(s) - sC_w\overline{\Delta p}_w(s)\right] - sC_o\overline{\Delta p}_o(s)\bar{g}_{wo}(s). \quad (3.117)$$

Solving the above two equations yields

$$\overline{\Delta p}_o = \frac{\bar{q}_m\bar{g}_{ow}}{1 + s(C_w\bar{g}_{ww} + C_o\bar{g}_{oo}) + s^2C_wC_o(\bar{g}_{ww}\bar{g}_{oo} - \bar{g}_{wo}\bar{g}_{ow})} \quad (3.118)$$

for the observation well and

$$\overline{\Delta p}_w(s) = \frac{\bar{q}_m[\bar{g}_{ww} + sC_o(\bar{g}_{ww}\bar{g}_{oo} - \bar{g}_{wo}\bar{g}_{ow})]}{1 + s(C_w\bar{g}_{ww} + C_o\bar{g}_{oo}) + s^2C_wC_o(\bar{g}_{ww}\bar{g}_{oo} - \bar{g}_{wo}\bar{g}_{ow})} \quad (3.119)$$

for the producing well. Normally, the reciprocity principle holds between two fully penetrated line-source or cylindrical-source vertical wells (Carter, Kemp, Pierce, & Williams, 1974; McKinley, Vela, & Carlton, 1968; von Schroeter & Gringarten, 2009), thus $\bar{g}_{wo} = \bar{g}_{ow}$ in Equations (3.119) and (3.118).

As can be seen from Figure 3.12, the dimensionless pressure obtained from Equation (3.118) compares very well with that given in Fig. 2 of Tongpenyai and Raghavan (1981), given the fact that we have digitized the dimensionless pressure from a 2-by-2.5-Inch graph. As in the Tongpenyai and Raghavan (1981) paper, we have used $r_D = 500$, the dimensionless distance from the active well to the observation well; $S_w = S_o = 5$ (skin) for both wells, $C_D = 10^3$ (dimensionless wellbore storage) for the active well, and $C_D = 10^5$ for the observation well.

Using $\bar{g}_{wo} = \bar{g}_{ow}$ and dividing Equation (3.118) by Equation (3.119) yields the observation well pressure as

$$\overline{\Delta p_o} = \frac{\overline{\Delta p_w}(s)\bar{g}_{ow}(s)}{\bar{g}_{ww}(s) + sC_o\left[\bar{g}_{ww}(s)\bar{g}_{oo}(s) - \bar{g}_{wo}^2(s)\right]} \tag{3.120}$$

and in the time domain

$$\Delta p_o(t) = \int_0^t \mathrm{d}\tau\, \Delta p_w(t-\tau)\mathcal{G}(\tau) \tag{3.121}$$

where the $\mathcal{G}$-function is defined by

$$\mathcal{G}(t) = \mathcal{L}^{-1}\left\{\frac{\bar{g}_{ow}(s)}{\bar{g}_{ww} + sC_o\left[\bar{g}_{ww}(s)\bar{g}_{oo}(s) - \bar{g}_{wo}^2(s)\right]}\right\}. \tag{3.122}$$

It should be noted that $\mathcal{G}$-function given by Equation (3.122) is independent of the producing well wellbore storage C_w. If $C_o = 0$ then $\bar{g}_o = \bar{g}_{ow}$, $\bar{g}_w = \bar{g}_{ww}$, and Equation (3.122) reduces to Equation (3.103), which can be written in full notation as

$$\mathcal{G}(r_w, r_o, s) = \mathcal{L}^{-1}\left[\frac{\bar{g}_o(r_o, s)}{\bar{g}_w(r_w, s)}\right]. \tag{3.123}$$

The Laplace transform of the unit-impulse response $g(r_w, s)$ in Equation (3.123) for a fully penetrated vertical cylindrical-source well can be written (in the oil-field units) as

$$\bar{g}_w(r_w, s) = \frac{141.2\mu}{kh}\left[\frac{\mathrm{K}_0\left(r_w\sqrt{\frac{s}{\eta}}\right)}{r_w\sqrt{\frac{s}{\eta}}\mathrm{K}_1\left(r_w\sqrt{\frac{s}{\eta}}\right)} + S\right] \tag{3.124}$$

and $g(r_o, s)$ for the observation well

$$\bar{g}_o(r_o, s) = \frac{141.2\mu}{kh}\frac{\mathrm{K}_0\left(r_o\sqrt{\frac{s}{\eta}}\right)}{r_w\sqrt{\frac{s}{\eta}}\mathrm{K}_1\left(r_w\sqrt{\frac{s}{\eta}}\right)}, \tag{3.125}$$

where k in md denotes the formation permeability; μ is the viscosity in cp;

r_w is the wellbore radius in ft; r_o is the distance between the two wells in ft; h in ft denotes the thickness of the reservoir; η is the diffusivity constant given by

$$\eta = \frac{0.0002637k}{\phi c_t \mu}, \tag{3.126}$$

where ϕ is the porosity of the reservoir; c_t is the total compressibility of the rock and fluid in psi^{-1}. K_0 and K_1 denote the modified Bessel function of the second kind, and s is the Laplace transform with respect to the time t in hr; $S = S_w$ denotes the skin factor at the active (producing) well and $S_o = 0$, skin at the observation well. Substitution of Equations (3.124) and (3.125) in Equation (3.123) and simplifying the resulting equation gives the $\mathcal{G}$-function as

$$\mathcal{G}(r_w, r_o, t) = \mathcal{L}^{-1}\left[\frac{\mathrm{K}_0\left(r_o\sqrt{\frac{s}{\eta}}\right)}{\mathrm{K}_0\left(r_w\sqrt{\frac{s}{\eta}}\right) + S\, r_w\sqrt{\frac{s}{\eta}}\mathrm{K}_1\left(r_w\sqrt{\frac{s}{\eta}}\right)}\right] \tag{3.127}$$

and for line-source wells as

$$\mathcal{G}(r_w, r_o, t) = \mathcal{L}^{-1}\left[\frac{\mathrm{K}_0\left(r_o\sqrt{\frac{s}{\eta}}\right)}{\mathrm{K}_0\left(r_w\sqrt{\frac{s}{\eta}}\right) + S}\right]. \tag{3.128}$$

As stated above, the $\mathcal{G}$-function in Equation (3.128) is independent of the wellbore storage C_w at the producing well, but it depends on its skin factor. In addition, Equation (3.128) indicates that the $\mathcal{G}$-function is a unique function of the diffusivity constant and the skin factor. This means that one can only uniquely estimate the value of diffusivity constant η and skin by history matching the observation well pressure data with the *p*-*p* convolution integral given by Equation (3.121) (the convolution of the producing well pressure data $\Delta p_w(r_w, t)$ and $\mathcal{G}$-function given by Equation (3.128)). As discussed by Onur, Gok, and Yamanlar (2001), the individual values of permeability and porosity cannot not be estimated directly from Equation (3.121) with a known total compressibility and viscosity. On the other hand, as shown by Jargon (1976) and Onur et al. (2001), it is possible to estimate the individual values of permeability and porosity of the formation, and the values of wellbore storage and skin at the production well by history matching only observation well pressures, by using the *p*-*r* convolution of flow-rate data at the producing well. However, it should be stated that while the *p*-*p* convolution is useful without a reliable rate data, it is better to use the *p*-*r* convolution when reliable rate data are available.

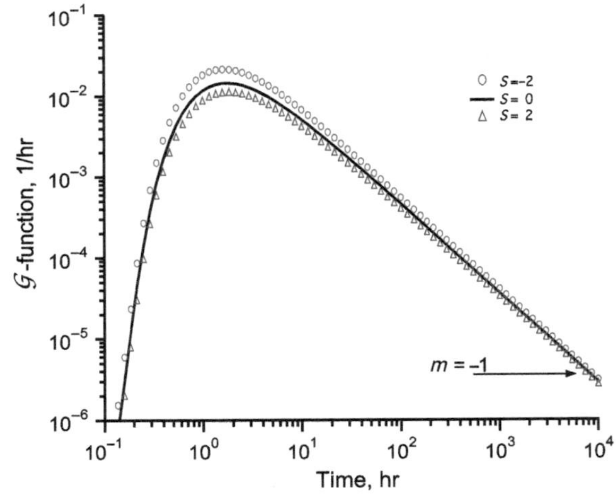

Figure 3.13 Behavior of $\mathcal{G}$-functions for a two-well interference test: $\eta = 1.3 \times 10^4$ md psi/cp, $r = 328$ ft, and $r_w = 0.354$ ft.

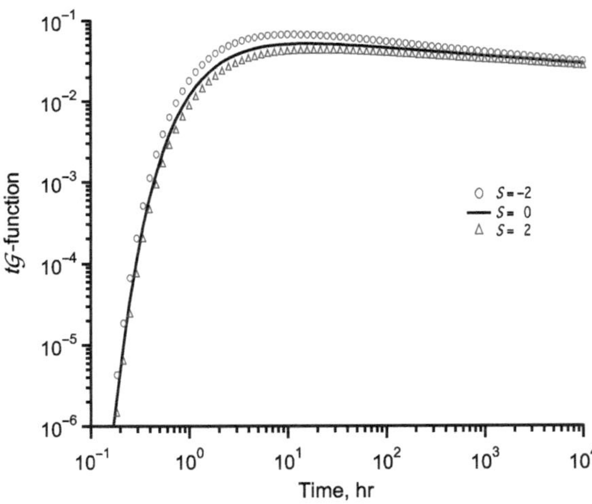

Figure 3.14 Behavior of $t\mathcal{G}$ (logarithmic derivative) for a two-well interference test: $\eta = 1.3 \times 10^4$ md psi/cp, $r = 328$ ft, and $r_w = 0.354$ ft.

Figures 3.13 and 3.14 present the behaviors of the $\mathcal{G}$ and $t\mathcal{G}$ functions computed from Equation (3.128) with $\eta = 1.3 \times 10^4$ md psi/cp and three different values of skin factor (S) at the producing well for the two-well interference problem as discussed above. The $\mathcal{G}$-function asymptotically approaches a -1 slope line, and hence the $t\mathcal{G}$ function asymptotically approaches a constant (or zero slope) value, indicating a radial flow regime at late times. The beginning of the radial flow occurs roughly when

$t \geq 25r^2/\eta$ ($\approx$ 270 hr) for the example shown in Figures 3.13 and 3.14, but weakly depends on the skin factor at the producing well.

The large time approximation for the $\mathcal{G}$-function is often useful for determining a few formation parameters. Thus, using the large time (or s small) approximation for the modified Bessel functions ($K_0(x) \approx -\ln(e^{\gamma} x/2)$, where the Euler constant $\gamma = 0.57722\cdots$) in Equation (3.128), an approximate $\mathcal{G}$-function for the radial flow period (Goode et al., 1991) can be written as

$$\mathcal{G}(t) = \mathcal{L}^{-1}\left[\frac{D_o - \ln(s)}{D_w - \ln(s)}\right], \tag{3.129}$$

where D_o and D_w are constants given by

$$D_o = \ln\left(\frac{4\eta}{e^{2\gamma} r^2}\right), \tag{3.130}$$

and

$$D_w = \ln\left(\frac{4\eta e^{2S}}{e^{2\gamma} r_w^2}\right). \tag{3.131}$$

Taking the inverse Laplace transform of Equation (3.131)(Goode et al., 1991) and keeping only the first two leading terms gives

$$\mathcal{G}(t) \approx \frac{(D_w - D_o)}{tY^2(t)}\left[1 - \frac{\pi^2}{2Y^2(t)}\cdots\right], \tag{3.132}$$

where

$$Y(t) = D_w + \ln t + \gamma. \tag{3.133}$$

By taking the derivative of the natural logarithm of the $\mathcal{G}$-function, Equation (3.132), with respect to the natural logarithm of time, we obtain

$$\frac{d\ln\mathcal{G}(t)}{d\ln t} \approx -1 - \frac{2}{Y(t)}, \tag{3.134}$$

which confirms that the radial flow for the $\mathcal{G}$-function can be identified by a slope of -1 on a log-log plot, as t approaches infinity. Similarly, we can show that the radial flow for the $t\mathcal{G}$ function is identified nearly by a constant

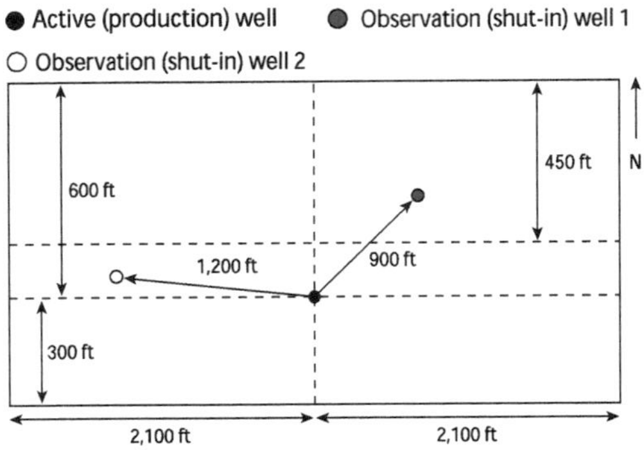

Figure 3.15 A schematic representation for the well-reservoir configuration considered for a two-well interference test in a closed rectangular reservoir (Onur et al., 2009a).

(a slope of zero) on a log-log plot that can be expressed as

$$\frac{\mathrm{d}\ln[t\mathcal{G}(t)]}{\mathrm{d}\ln t} \approx -\frac{2}{Y(t)}\cdots, \tag{3.135}$$

which indicates that the radial flow for the $t\mathcal{G}$ function can be identified by a straight line with a slope of zero, as t approaches infinity. However, we should note from Equation (3.135) that the $t\mathcal{G}$ function during the radial flow will be a weak function of time and decrease monotonically with time by the following relation

$$t\mathcal{G}(t) \approx \frac{(D_w - D_o)}{Y^2(t)} = \frac{\ln\left(\frac{r^2}{r_w^2 \mathrm{e}^{-2S}}\right)}{\left[\ln\left(\frac{4\eta t}{\mathrm{e}^{\gamma} r_w^2 \mathrm{e}^{-2S}}\right)\right]^2}, \tag{3.136}$$

where the second equality follows from Equations (3.130), (3.131) and (3.133).

Recently, Onur et al. (2009a) have presented diagnostic log-log plots for model (flow regime) identification from $\mathcal{G}$- and $t\mathcal{G}$-functions in comparison with the g and tg unit-impulse responses for a two-well interference test in closed rectangular (homogeneous-isotropic) reservoirs, as shown in Figure 3.15. For the well-reservoir system shown in Figure 3.15, log-log diagnostic plots of g_{ww}, tg_{ww}, g_{oo}, and tg_{oo} versus t (Figure 3.16) in comparison with log-log diagnostic plots of $\mathcal{G}$ and $t\mathcal{G}$ versus t (Figure 3.17) are shown in Figures 3.16 and 3.17, respectively. Both wells are fully penetrating vertical wells with identical wellbore radii, and the observation well response is free of wellbore storage and skin effects; i.e. $C_o = 0.0$ B/psi and $S_o = 0.0$. Other input well and reservoir parameters are: $\phi = 0.2$,

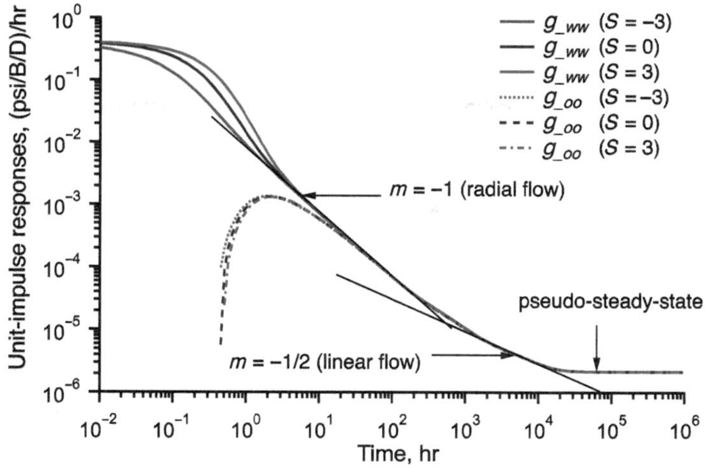

Figure 3.16 Unit-impulse responses g_{ww} and g_{oo} versus time for a two-well interference test in a closed rectangular reservoir.

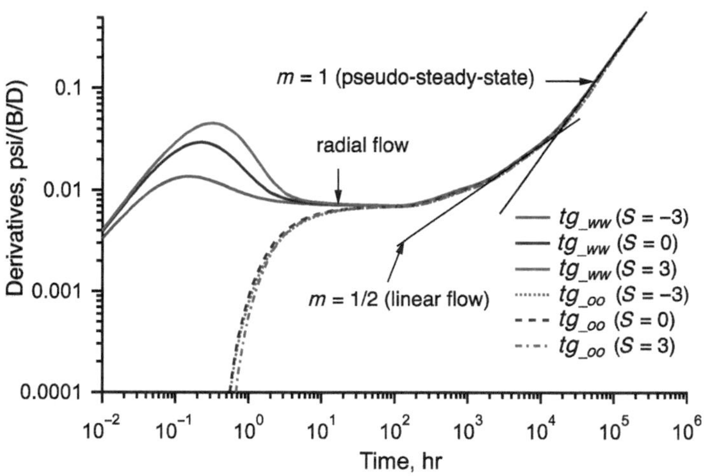

Figure 3.17 Derivatives [tg_{ww} and tg_{oo} (time × unit-impulse responses)] versus time for a two-well interference test in a closed rectangular reservoir.

$k = 125$ md, $h = 40$ ft, $c_t = 10^{-5}$ psi^{-1}, $\mu = 0.5$ cp, $r_w = 0.354$ ft, $C_w = 0.1$ B/psi. The skin factor at the producing well is assumed: $S_w = S = -3, 0,$ and 3.

As can be seen from Figure 3.16, from the behaviors of the unit-impulse responses (g_{ww} and g_{oo}), a radial flow regime is identified by a -1 slope line, a linear flow regime due to the no-flow north and south boundaries of the

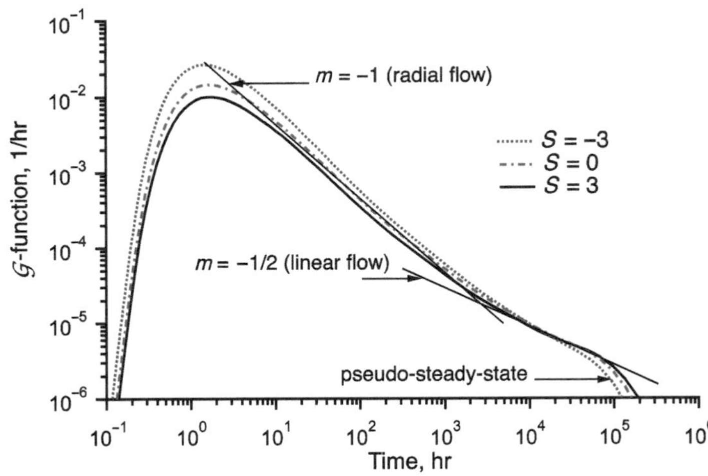

Figure 3.18 $\mathcal{G}$-functions versus time for a two-well interference test in a closed rectangular reservoir.

reservoir is identified by a $-1/2$ slope line, and a pseudo-steady-state flow regime is identified by a zero slope line. On the other hand, the derivatives, tg_{ww} and tg_{oo}, shown in Figure 3.17, exhibit the expected flow regimes: a radial flow (0 slope line), a linear flow (1/2 slope line), and a pseudo-steady-state flow (a unit-slope). Notice that the skin factor at the producing well has an effect on g_{ww}, g_{oo}, tg_{ww}, and tg_{oo} responses only at very early times ($t \leq 5$ hr) for this example. On the other hand, the skin factor at the producing well has a profound effect on the $\mathcal{G}$- and $t\mathcal{G}$-functions for all times, as shown in Figures 3.18 and 3.19. As can be seen from these figures, the graphs of both $\mathcal{G}$- and $t\mathcal{G}$-functions display the radial and linear flow regimes with the same slope lines as in both g and tg unit-impulse responses (Figures 3.16 and 3.17). However, the $\mathcal{G}$- and $t\mathcal{G}$-function display these flow regimes at a slightly later time than those from g and tg unit-impulse responses. This is because the $\mathcal{G}$- and $t\mathcal{G}$-functions are based on the ratio of the impulse responses at both the observation well and the producing well Equation (3.128) and, hence, the same flow regimes are not observed at the same time intervals as those from g and tg unit-impulse responses, as can be seen clearly in Figures 3.16–3.19. One interesting observation is that the behaviors of the $\mathcal{G}$- and $t\mathcal{G}$-functions are different from those of g and tg unit-impulse responses during the pseudo-steady state flow regime. As can be seen from Figures 3.18 and 3.19, unlike the g unit-impulse responses, the $\mathcal{G}$-function does not approach a constant value (or zero slope line) during the pseudo-steady state flow regime, rather it asymptotically approaches zero for all values of the skin factor. Unfortunately, we do not have asymptotic expressions of the $\mathcal{G}$-functions that apply to linear and pseudo-steady state flow regimes.

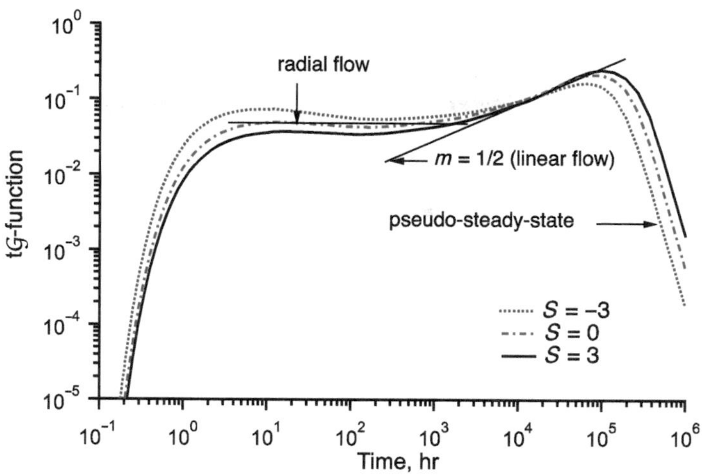

Figure 3.19 $t\mathcal{G}$-functions versus time for a two-well interference test in a closed rectangular reservoir.

3.8.3. Pressure-pressure convolution for wireline formation testers

In this section, we consider applications of pressure-pressure convolution for Interval Pressure Transient Tests (IPTTs) conducted by both dual-probe (Figure 3.20) and dual-packer (Figure 3.21) modules with observation probes.

Some of the wireline formation testers may not have a direct flow metering device or flow-rate measurements may have large uncertainties. For instance, fluid production at the sink probe or at the packer module can be obtained from a sample chamber module, flow-control module, or pumpout module. In cases where the flow-control and/or pumpout modules are used, reliable flow-rate data can often be determined from direct measurements of the piston displacement and tool characteristics (Kuchuk, Ramakrishnan, & Dave, 1994). In the case of single probe testing, Zimmerman, MacInnis, Hoppe, Pop, and Long (1990) noted that flow-rate data could also be computed from sink (production) probe pressure data using a modified form of a technique described by Samson, Fligelman, and Braester (1985). Goode and Thambynayagam (1992) presented a similar flow-rate estimation method based on the solution given by Moran and Finklea (1962). The flow-rate estimation methods given by Zimmerman et al. (1990) and Goode and Thambynayagam (1992) assume that the flow pattern around the sink probe is reasonably well approximated by a spherical flow pattern and require an iterative procedure that adjusts a lumped parameter involving horizontal mobility until the predicted volume of fluid recovered matches the recovered sample chamber volume. On the

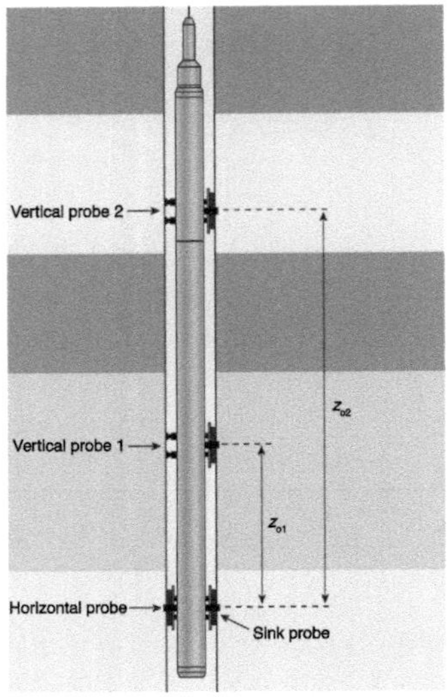

Figure 3.20 Schematic representation of a multiprobe wireline formation tester with observation probes.

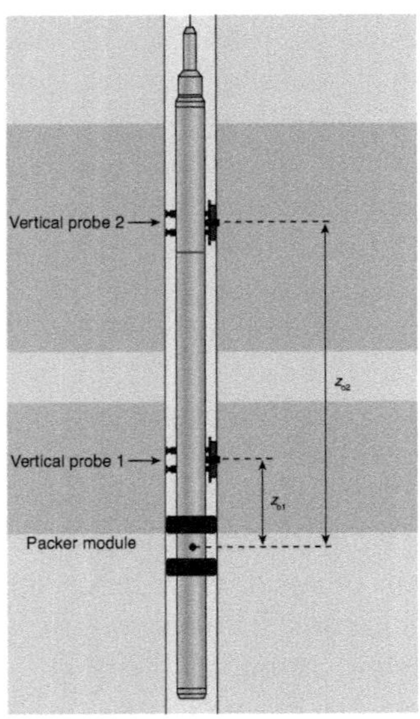

Figure 3.21 Schematic representation of wireline formation tester dual-packer module and observation probes.

other hand, the flow-rate estimation method given by Kuchuk et al. (1994) is based on the use of the tool mechanical characteristics of the packer and probe configuration, and they recommend that the estimated flow rates be used as initial guesses in an optimization procedure based on the p-r convolution that corrects the initial measured or estimated flow rates and simultaneously estimates the formation parameters such as static formation pressure, horizontal and vertical permeability.

In any case, as stated above, when uncertainties in flow-rate measurements are large for WFTs, the pressure-pressure (p-p) convolution becomes quite attractive for model (flow regime) identification and parameter estimation without flow-rate measurements because pressure measurements at probes and packer are usually acquired accurately with high-resolution quartz gauges.

3.8.3.1. Pressure-pressure convolution for dual-probe module and observation probes

The wireline formation tester (WFT) dual-probe module tool configuration consists of a number of modules, as shown in Figure 3.20, and provides a capability to conduct controlled local production and interference tests in openhole wells (Pop, Badry, Morris, Tottrup, & Jonas, 1993; Zimmerman et al., 1990). The formation pressures along the wellbore are simultaneously measured at four different locations by a sink (production), and one horizontal and two observation probes, as shown in Figure 3.20, where the horizontal probe is diametrically opposite to the sink probe. The second observation probe is called *vertical probe 1* and is vertically displaced at a distance of z_{o1} from the center of the sink probe (Figure 3.20). The third observation probe is called *vertical probe 2* and is vertically displaced at a distance of z_{o2}. In some applications, the third observation probe may not be included. The tool configuration shown in Figure 3.20 allows a number of formation pressure measurements along the wellbore, formation testing, and multiple formation fluid sampling. The tool provides a capability to conduct interval tests (IPTT) that are similar to conventional drawdown and interference tests using the wireline conveyance systems.

When the all modules are set in the wellbore, a short test, called a pretest, is conducted at all probe locations to establish communication with the formation as well as to measure reservoir pressure. Although it can be variable, 5 to 20 cc of fluid from the formation through each probe is produced during a typical pretest. After pretests, the formation fluid is produced through the sink probe and after some time, the production is stopped and the pressure builds up. During both production and buildup periods, the flow rate and the flowing and the buildup pressure are measured at the sink probe, while the formation pressure at all observation probe locations is also measured. At the sink probe, a drawdown test and a subsequent buildup test are conducted (conducting multiple drawdown and buildup tests is possible). Between the sink probe and each observation probe, basically an interference test is conducted during production and buildup periods. For instance, if the tool has a sink probe with one horizontal and one vertical observation probe, then two interference tests are conducted between the sink probe and horizontal, and between the sink probe and the vertical observation probe. The drawdown test and a subsequent buildup test at the sink probe and interference tests between the probes are called IPTTs.

Let us suppose that we conduct an IPTT with the dual-probe module tool configuration shown in Figure 3.20 but with only one horizontal and one vertical observation probe at a distance $z_{o1} = z_o$ from the center of the sink probe. The pressure response at the vertical observation probe can be written as

$$\Delta p_v(t) = \int_0^t q_s(\tau) g_{vs}(t-\tau)\,\mathrm{d}\tau + \int_0^t q_v(\tau) g_{vv}(t-\tau)\,\mathrm{d}\tau + \int_0^t q_h(\tau) g_{vh}(t-\tau)\,\mathrm{d}\tau \tag{3.137}$$

and in the Laplace domain

$$\overline{\Delta p}_v(s) = \bar{q}_s(s)\bar{g}_{vs}(s) + \bar{q}_v(s)\bar{g}_{vv}(s) + \bar{q}_h(s)\bar{g}_{vh}(s), \tag{3.138}$$

where the subscripts s, v, and h denote the sink, vertical and horizontal probes, respectively. The term g is the impulse response function with the first subscript denoting the observation point and the second the sink. For instance, g_{vs} is the impulse response at the vertical probe due to the production at the sink probe. The actual flow rates are denoted by q: q_s is the production from the sink probe that is applied deliberately to conduct the interval (inter-probe) interference test. q_h and q_v are the flow rates at the horizontal and vertical probes that take place because of compressible fluids in the tool volume that is in contact with the formation. During a production test through the sink probe, the observation probes are passive, and in each one no flow is deliberately imposed. The tool storage of each observation probe, however, causes an unintended flow to occur from the tool into the formation; it is usually insignificant for most WFTs. In any case, we will present solutions if the tool storage volumes at the observation probes are significant so that there is a considerable flow from the probes in the formation. The flow rate at each probe can be written from Equation (3.42) as

$$q_s(t) = q_m(t) + C_s \frac{\mathrm{d}p_s(t)}{\mathrm{d}t} \tag{3.139}$$

and from Equation (3.45) in the Laplace domain as

$$\bar{q}_s(s) = \bar{q}_m(s) - sC_s\overline{\Delta p}_s(s), \tag{3.140}$$

for the sink probe,

$$q_h(t) = C_h \frac{\mathrm{d}p_h(t)}{\mathrm{d}t} \tag{3.141}$$

and in the Laplace domain as

$$\bar{q}_h(s) = -sC_h\overline{\Delta p}_h(s), \tag{3.142}$$

for the horizontal probe, and

$$q_v(t) = C_v \frac{\mathrm{d}p_v(t)}{\mathrm{d}t} \tag{3.143}$$

and in the Laplace domain as

$$\bar{q}_v(s) = -sC_v\overline{\Delta p}_v(s), \tag{3.144}$$

for the vertical probe, where q_m is the measured flow rate at the sink probe, and q_h and q_v are the wellbore storage induced flow rates at the horizontal and vertical observation probes, and C_s, C_h, and C_v are the wellbore storage constants (the tool volume $\times c_w$, which is the isothermal compressibility of the fluid in the volume) for the sink, horizontal, and vertical probes. The rates q_h and q_v from the tool into the formation are due to fluid expansion in the tool volume.

Substituting Equations (3.140), (3.142) and (3.144) in Equation (3.138), we obtain the vertical probe pressure change as

$$\begin{aligned}\overline{\Delta p}_v(s) = {} & \left[\bar{q}_m(s) - sC_s\overline{\Delta p}_s(s)\right]\bar{g}_{vs}(s) - sC_v\overline{\Delta p}_v(s)\bar{g}_{vv}(s) \\ & - sC_h\overline{\Delta p}_h(s)\bar{g}_{vh}(s).\end{aligned} \tag{3.145}$$

Similar expressions can also be written for $\Delta p_h(t)$ (horizontal probe) and $\Delta p_s(t)$ (sink, probe)

$$\begin{aligned}\overline{\Delta p}_s(s) = {} & \left[\bar{q}_m(s) - sC_s\overline{\Delta p}_s(s)\right]\bar{g}_{ss}(s) - sC_v\overline{\Delta p}_v(s)\bar{g}_{sv}(s) \\ & - sC_h\overline{\Delta p}_h(s)\bar{g}_{sh}(s)\end{aligned} \tag{3.146}$$

for the sink probe, and

$$\begin{aligned}\overline{\Delta p}_h(s) = {} & \left[\bar{q}_m(s) - sC_s\overline{\Delta p}_s(s)\right]\bar{g}_{hs}(s) - sC_v\overline{\Delta p}_v(s)\bar{g}_{hv}(s) \\ & - sC_h\overline{\Delta p}_h(s)\bar{g}_{hh}(s)\end{aligned} \tag{3.147}$$

for the horizontal probe.

The above system given by Equations (3.145)–(3.147) constitutes linear algebraic equations and the solutions can be written as

$$\overline{\Delta p}_s(s) = \frac{\bar{q}_m\{\bar{g}_{ss} + [C_h(\bar{g}_{hh}\bar{g}_{ss} - \bar{g}_{hs}\bar{g}_{sh}) + C_v(\bar{g}_{ss}\bar{g}_{vv} - \bar{g}_{sv}\bar{g}_{vs})]\, s + as^2\}}{D} \tag{3.148}$$

for the sink probe, where the denominator of Equation (3.148) is given as

$$\begin{aligned} D &= 1 + (C_s\bar{g}_{ss} + C_h\bar{g}_{hh} + C_v\bar{g}_{vv})s + a_1 s^2 + a_2 s^3 \\ a &= C_h C_v[\bar{g}_{vh}(\bar{g}_{hs}\bar{g}_{sv} - \bar{g}_{hv}\bar{g}_{ss}) + \bar{g}_{vs}(\bar{g}_{hv}\bar{g}_{sh} - \bar{g}_{hh}\bar{g}_{sv}) \\ &\quad + \bar{g}_{vv}(\bar{g}_{hh}\bar{g}_{ss} - \bar{g}_{hs}\bar{g}_{sh})], \\ a_1 &= C_h C_s(\bar{g}_{hh}\bar{g}_{ss} - \bar{g}_{hs}\bar{g}_{sh}) + C_h C_v(\bar{g}_{hh}\bar{g}_{vv} - \bar{g}_{hv}\bar{g}_{vh}) \\ &\quad + C_s C_v(\bar{g}_{ss}\bar{g}_{vv} - \bar{g}_{sv}\bar{g}_{vs}), \\ a_2 &= C_h C_s C_v[\bar{g}_{vh}(\bar{g}_{hs}\bar{g}_{sv} - \bar{g}_{hv}\bar{g}_{ss}) + \bar{g}_{vs}(\bar{g}_{hv}\bar{g}_{sh} - \bar{g}_{hh}\bar{g}_{sv}) \\ &\quad + \bar{g}_{vv}(\bar{g}_{hh}\bar{g}_{ss} - \bar{g}_{hs}\bar{g}_{sh})], \end{aligned}$$

$$\overline{\Delta p}_h = \frac{\bar{q}_m[\bar{g}_{hs} + sC_v(\bar{g}_{hs}\bar{g}_{vv} - \bar{g}_{hv}\bar{g}_{vs})]}{D} \tag{3.149}$$

for the horizontal probe, and

$$\overline{\Delta p}_v = \frac{\bar{q}_m[\bar{g}_{vs} + sC_h(\bar{g}_{hh}\bar{g}_{vs} - \bar{g}_{hs}\bar{g}_{vh})]}{D} \tag{3.150}$$

for the vertical probe. Notice that the argument s (the Laplace transform variable) of all impulse responses is omitted for simplicity.

Dividing Equation (3.150) by Equation (3.149) and expressing the result in terms of the vertical probe pressure gives

$$\overline{\Delta p}_v = \frac{\overline{\Delta p}_h\,[\bar{g}_{vs} + sC_h(\bar{g}_{hh}\bar{g}_{vs} - \bar{g}_{hs}\bar{g}_{vh})]}{\bar{g}_{hs} + sC_v(\bar{g}_{hs}\bar{g}_{vv} - \bar{g}_{hv}\bar{g}_{vs})} \tag{3.151}$$

and as pressure-pressure (p-p) convolution in the time domain

$$\Delta p_v(t) = \int_0^t \mathrm{d}\tau\, \Delta p_h(t-\tau)\mathcal{G}(\tau) \tag{3.152}$$

where the $\mathcal{G}$-function is defined by

$$\mathcal{G}(t) = \mathcal{L}^{-1}\left[\frac{\bar{g}_{vs} + sC_h(\bar{g}_{hh}\bar{g}_{vs} - \bar{g}_{hs}\bar{g}_{vh})}{\bar{g}_{hs} + sC_v(\bar{g}_{hs}\bar{g}_{vv} - \bar{g}_{hv}\bar{g}_{vs})}\right]. \tag{3.153}$$

If $C_h = 0$ and $C_v = 0$, then Equation (3.153) reduces to Equation (3.103). It should be noted that the $\mathcal{G}$-function given by Equation (3.153) is independent of the sink probe wellbore storage C_s. Similarly, dividing Equation (3.148) by Equation (3.149) yields the vertical probe pressure as

$$\overline{\Delta p_v} = \frac{\overline{\Delta p_s}(s)\,[\bar{g}_{vs} + sC_h(\bar{g}_{hh}\bar{g}_{vs} - \bar{g}_{hs}\bar{g}_{vh})]}{\bar{g}_{ss} + [C_h(\bar{g}_{hh}\bar{g}_{ss} - \bar{g}_{hs}\bar{g}_{sh}) + C_v(\bar{g}_{ss}\bar{g}_{vv} - \bar{g}_{sv}\bar{g}_{vs})]\,s + as^2} \tag{3.154}$$

and in the time domain

$$\Delta p_v(t) = \int_0^t \mathrm{d}\tau\, \Delta p_s(t-\tau)\mathcal{G}(\tau) \tag{3.155}$$

where the $\mathcal{G}$-function is defined by

$$\mathcal{G}(t) = \mathcal{L}^{-1}\left\{\frac{\bar{g}_{vs} + sC_h(\bar{g}_{hh}\bar{g}_{vs} - \bar{g}_{hs}\bar{g}_{vh})}{\bar{g}_{ss} + [C_h(\bar{g}_{hh}\bar{g}_{ss} - \bar{g}_{hs}\bar{g}_{sh}) + C_v(\bar{g}_{ss}\bar{g}_{vv} - \bar{g}_{sv}\bar{g}_{vs})]\,s + as^2}\right\}. \tag{3.156}$$

If we assume that each probe is a point source (sink), the above equation can be further simplified by using $g_{sh} = g_{hs}$, $g_{sv} = g_{vs}$, and $g_{hv} = g_{vh}$ due to the principle of reciprocity. If $C_h = 0$ and $C_v = 0$ in Equation (3.156) also reduces to Equation (3.103). Notice also that $\mathcal{G}$-function given by Equation (3.156) is also independent of the sink probe wellbore storage C_s.

3.8.3.2. Pressure-pressure convolution for dual-packer and observation probe

The wireline formation tester dual-packer module and observation probe tool configuration shown in Figure 3.21 is similar to the multiprobe tool configuration shown in Figure 3.20. The packer module uses two inflatable packers that are set against the borehole wall to isolate a wellbore interval (section) of the formation (see Figure 3.21). The pumpout module, using borehole fluid, inflates the packers above the hydrostatic mud pressure. This module offers access to the formation over a wellbore area thousands of times larger than the probe mouth. It also allows fluids to be withdrawn at a higher rate without falling below the bubble point pressure and provides permeability estimation with a radius of investigation in the range of 10 to 50 ft, depending on the formation characteristics and duration of the production and buildup periods. A dual-packer-probe tool configuration shown in Figure 3.21 can have one or two observation probes. The function of these probes is the same as the probes of the multiprobe tool. It is basically measuring the pressure at the sandface that is hydraulically isolated from the wellbore by the mudcake as the formation fluid is produced through the packer interval and a subsequent buildup test.

Let us suppose that we conduct a vertical interference test with a dual-packer module and a single vertical observation probe at a distance $z_{o1} = z_o$ from the center of the packer-open interval. The pressure response at the vertical probe can be written as

$$\Delta p_v(t) = \int_0^t q_p(\tau) g_{vp}(t-\tau)\,\mathrm{d}\tau + \int_0^t q_v(\tau) g_{vv}(t-\tau)\,\mathrm{d}\tau \quad (3.157)$$

and in the Laplace domain

$$\overline{\Delta p}_v(s) = \bar{q}_s(s)\bar{g}_{vp}(s) + \bar{q}_v(s)\bar{g}_{vv}(s), \quad (3.158)$$

where the subscripts p and v denote the packer module and vertical probe, respectively. g are the impulse response functions with the first subscript denoting the observation point and the second the sink (packer). For instance, g_{vp} is the impulse response at the vertical probe due to the production at the packer. q_m is the production from the packer that is applied deliberately to conduct the inter-probe interference test. q_v is the flow rate at the vertical probe due to the tool storage and is from the probe into the formation. Therefore, the flow rate can be written from Equation (3.42) as

$$q_p(t) = q_m(t) + C_p \frac{\mathrm{d}p_p(t)}{\mathrm{d}t} \quad (3.159)$$

and in the Laplace domain as

$$\bar{q}_p(s) = \bar{q}_m(s) - sC_p\overline{\Delta p}_p(s) \quad (3.160)$$

for the packer and it is given by Equation (3.143) for the vertical probe, where q_m is the measured flow rate for the packer and C_p the packer tool wellbore storage. Substituting Equations (3.144) and (3.160) in Equation (3.158) yields

$$\overline{\Delta p}_v(s) = \bar{q}_m(s)\bar{g}_{vp}(s) - sC_p\overline{\Delta p}_p(s)\bar{g}_{vp}(s) - sC_v\overline{\Delta p}_v(s)\bar{g}_{vv}(s) \quad (3.161)$$

for the vertical probe and

$$\overline{\Delta p}_p(s) = \bar{q}_m(s)\bar{g}_{pp}(s) - sC_p\overline{\Delta p}_p(s)\bar{g}_{pp}(s) - sC_v\overline{\Delta p}_v(s)\bar{g}_{pv}(s) \quad (3.162)$$

for the packer. Solving the above two equations yields

$$\overline{\Delta p}_p(s) = \frac{\bar{q}_m[\bar{g}_{pp} + sC_v(\bar{g}_{pp}\bar{g}_{vv} - \bar{g}_{pv}\bar{g}_{vp})]}{1 + s(C_p\bar{g}_{pp} + C_v\bar{g}_{vv}) + s^2C_pC_v(\bar{g}_{pp}\bar{g}_{vv} - \bar{g}_{pv}\bar{g}_{vp})} \tag{3.163}$$

for the packer and

$$\overline{\Delta p}_v = \frac{\bar{q}_m\bar{g}_{vp}}{1 + s(C_p\bar{g}_{pp} - C_v\bar{g}_{vv}) + C_pC_v(\bar{g}_{pv}\bar{g}_{vp} - \bar{g}_{pp}\bar{g}_{vv})s^2} \tag{3.164}$$

for the vertical probe. Dividing Equation (3.164) by Equation (3.163) yields the vertical probe pressure as

$$\overline{\Delta p}_v = \frac{\overline{\Delta p}_p(s)\bar{g}_{vp}}{\bar{g}_{pp} + sC_v(\bar{g}_{pp}\bar{g}_{vv} - \bar{g}_{pv}\bar{g}_{vp})} \tag{3.165}$$

and in the time domain

$$\Delta p_v(t) = \int_0^t \mathrm{d}\tau\, \Delta p_p(t-\tau)\mathcal{G}(\tau) \tag{3.166}$$

where the $\mathcal{G}$-function is defined by

$$\mathcal{G}(t) = \mathcal{L}^{-1}\left[\frac{\bar{g}_{vp}}{\bar{g}_{pp} + sC_v(\bar{g}_{pp}\bar{g}_{vv} - \bar{g}_{pv}\bar{g}_{vp})}\right]. \tag{3.167}$$

If $C_v = 0$ then Equation (3.167) reduces to Equation (3.103). It is also interesting to observe that the $\mathcal{G}$-function given by Equation (3.167) is independent of the packer module wellbore storage C_p.

3.8.3.3. Pressure-pressure convolution of multiprobe IPTTs

In this section, we provide the p-p convolution for analyzing pressure transient data acquired from multiprobe IPTTs. One of the main configurations of the multiprobe formation testers in a vertical well, particularly for permeability estimation, is a three-probe configuration consisting of a sink probe, a horizontal probe located diametrically opposite to the sink, and a vertical observation probe at z_{o1} from the sink probe. As mentioned previously, in some cases, a second vertical probe at z_{o2} below or above the sink probe is used (Figure 3.20). By withdrawing formation fluid at the sink, four pressure transients (sink, horizontal, vertical 1 and vertical 2 probes) can be recorded at four different locations along the wellbore. Depending on the location of the probes and the top and bottom formation boundaries, and degree of anisotropy, one may observe three distinct flow regimes; a spherical (and/or hemispherical) flow around the sink probe at early times, a transition period and an infinite-acting radial flow regime (Goode & Thambynayagam, 1992).

For the four-probe configuration (Figure 3.20), in principle there are 12 possible p-p convolution options. These are: $\{s, h\}$, $\{s, v1\}$, $\{s, v2\}$, $\{h, s\}$, $\{h, v1\}$, $\{h, v2\}$, $\{v1, s\}$, $\{v1, h\}$, $\{v1, v2\}$, $\{v2, s\}$, $\{v2, h\}$, $\{v2, v1\}$, where s, h, $v1$ and $v2$ donate the sink, horizontal, vertical 1 and vertical 2 probes. For instance, $\{h, s\}$ the p-p convolution from Equations (3.146) and (3.147) for $C_h = C_v = 0$

$$\Delta p_h(t) = \int_0^t \Delta p_s(t-\tau)\, \mathcal{G}_{h-s}(\tau)\, \mathrm{d}\tau, \tag{3.168}$$

where the $\mathcal{G}_{h-s}$-function is given as

$$\mathcal{G}_{h-s} = \mathcal{L}^{-1}\left[\frac{\bar{g}_h(s)}{\bar{g}_s(s)}\right], \tag{3.169}$$

where $\bar{g}_s$ and $\bar{g}_h$ denote the sink-probe and horizontal-probe impulse responses, and $\bar{g}_h = \bar{g}_{hs}$ and $\bar{g}_s = \bar{g}_{ss}$ for simplicity. The pressure change for each probe is defined as: $\Delta p_h(t) = p_{oh} - p_h(t)$ for the horizontal probe, $\Delta p_s(t) = p_{os} - p_s(t)$ for the sink probe, and $\Delta p_v(t) = p_{ov} - p_v(t)$ for the vertical probe.

Although there are twelve different p-p convolution options for the four-probe configuration, the most useful ones are: $\{h, s\}$, $\{v1, s\}$, $\{v1, h\}$, $\{v2, s\}$, $\{v2, h\}$, $\{v2, v1\}$ because, as explained before, a measurable pressure change becomes observable at the location closer to sink/source. For instance, for some of the four-probe configurations, the distance from the sink to horizontal probe is πr_w, where r_w is the wellbore radius, which is about 1 ft for most wells, to the vertical probe 1 is $z_{o1} = 6.4$ ft, and to the vertical probe 2 is $z_{o2} = 14.4$ ft. For most formations, a measurable pressure change becomes observable first at the horizontal probe, and then at the vertical 1 and finally at the vertical 2 if we produce for long enough and $k_h > k_v$.

Similar to Equation (3.168), p-p convolution between the horizontal probe and vertical probes ($\{v1, h\}$ and $\{v2, h\}$) for $C_h = C_{v1} = C_{v2} = 0$ can be written as

$$\Delta p_{vj}(t) = \int_0^t \Delta p_h(t-\tau)\, \mathcal{G}_{vj-h}(\tau)\, \mathrm{d}\tau, \tag{3.170}$$

where the $\mathcal{G}_{vj-h}$-function is defined as

$$\mathcal{G}_{vj-h}(t) = \mathcal{L}^{-1}\left[\frac{\bar{g}_{vj}(s)}{\bar{g}_h(s)}\right] \tag{3.171}$$

for $j = 1$ (vertical 1) and 2 (vertical 2). Here, $\bar{g}_{vj}$ and $\bar{g}_h$ denote the vertical j-probe and horizontal-probe impulse responses, respectively. For a spherical flow geometry, these responses (Goode & Thambynayagam, 1992; Kuchuk, 1996) are given as

$$\bar{g}_h(s) = \frac{\Omega_h \mu}{8\pi r_w \sqrt{k_h k_v}} \exp\left(-\frac{2r_w}{\Omega_h}\sqrt{\frac{s\phi c_t \mu}{k_h}}\right) \tag{3.172}$$

and

$$\bar{g}_{vj}(s) = \frac{\Omega_{vj}\,\mu}{4\pi z_{oj} k_h} \exp\left(-\frac{z_{oj}}{\Omega_{vj}}\sqrt{\frac{s\phi c_t \mu}{k_v}}\right). \tag{3.173}$$

The inverse Laplace transforms of Equations (3.172) and (3.173) can be written, respectively, as

$$g_h(t) = \frac{\mu\sqrt{\mu\phi c_t}}{8\pi k_h \sqrt{k_v}\sqrt{\pi t^3}} \exp\left(-\frac{\phi c_t \mu r_w^2}{\Omega_h^2 k_h t}\right) \tag{3.174}$$

and

$$g_{vj}(t) = \frac{\mu\sqrt{\mu\phi c_t}}{8\pi k_h \sqrt{k_v}\sqrt{\pi t^3}} \exp\left(-\frac{\phi c_t \mu z_{oj}^2}{\Omega_{vj}^2 k_v t}\right) \tag{3.175}$$

for $j = 1$ and 2. In Equation (3.172) to Equation (3.175), Ω_h and Ω_{vj} are the steady-state shape factors for the horizontal and vertical observation probes, respectively. From the results of Goode and Thambynayagam (1992), Ω_h is not a function of anisotropy, and $\Omega_h = 0.5117$, but Ω_{vj} is a function of anisotropy and is given as

$$\Omega_{vj} = 2 - 1.955\xi_{vj} + 3.267\xi_{vj}^2 - 4.876\xi_{vj}^3 + 2.564\xi_{vj}^4, \tag{3.176}$$

where

$$\xi_{vj} = \frac{\sqrt{k_h/k_v}\,\frac{z_{oj}}{r_w}}{1 + \sqrt{k_h/k_v}\,\frac{z_{oj}}{r_w}} \tag{3.177}$$

for $j = 1$ and 2.

Assuming an unbounded formation in the vertical direction, and using Equations (3.172) and (3.173) in Equation (3.171) and inverting the resulting equations yields

$$\mathcal{G}_{vj-h}(t) = \frac{\Omega_{vj}}{\Omega_h}\frac{r_w}{z_{oj}}\frac{\sqrt{\phi c_t \mu}}{\sqrt{k_h}}\frac{1}{\sqrt{\pi t^3}}\left(\frac{z_{oj}}{\Omega_{vj}} - \frac{2r_w}{\Omega_h}\sqrt{\frac{k_v}{k_h}}\right) \times \exp\left[-\frac{\phi c_t \mu}{4k_v t}\left(\frac{z_{oj}}{\Omega_{vj}} - \frac{2r_w}{\Omega_h}\sqrt{\frac{k_v}{k_h}}\right)^2\right]. \quad (3.178)$$

The $t\mathcal{G}_{vj-h}$-function is simply obtained by multiplying both sides of Equation (3.178) by time t and, hence, it follows that

$$t\mathcal{G}_{vj-h}(t) = \frac{\Omega_{vj}}{\Omega_h}\frac{r_w}{z_{oj}}\frac{\sqrt{\phi c_t \mu}}{\sqrt{k_h}}\frac{1}{\sqrt{\pi t}}\left(\frac{z_{oj}}{\Omega_{vj}} - \frac{2r_w}{\Omega_h}\sqrt{\frac{k_v}{k_h}}\right) \times \exp\left[-\frac{\phi c_t \mu}{4k_v t}\left(\frac{z_{oj}}{\Omega_{vj}} - \frac{2r_w}{\Omega_h}\sqrt{\frac{k_v}{k_h}}\right)^2\right]. \quad (3.179)$$

Equations (3.178) and (3.179) assume that $\Omega_h z_{oj}/\left(2r_w\Omega_{vj}\right)\sqrt{k_h/k_v} > 1$ and provide a good approximation for all practical values of anisotropy ratio, k_h/k_v, for all times. The time at which $\mathcal{G}_{vj-h}$ and $t\mathcal{G}_{vj-h}$ functions are maximum (Figures 3.22 and 3.23) can be derived from Equations (3.177) and (3.178), respectively, as

$$t_{\max} = \frac{1}{6}\frac{\phi c_t \mu}{k_v}\left(\frac{z_{oj}}{\Omega_v} - \frac{2r_w}{\Omega_h}\sqrt{\frac{k_v}{k_h}}\right)^2 \quad (3.180)$$

and

$$t_{\max} = \frac{1}{2}\frac{\phi c_t \mu}{k_v}\left(\frac{z_{oj}}{\Omega_v} - \frac{2r_w}{\Omega_h}\sqrt{\frac{k_v}{k_h}}\right)^2. \quad (3.181)$$

Equations (3.180) and (3.181) indicate if the maximums of $\mathcal{G}_{vj-h}$ and $t\mathcal{G}_{vj-h}$ are observed, then Equations (3.180) and (3.181) can be used to estimate the vertical diffusivity of the formation if anisotropy ratio k_h/k_v is known.

The $\mathcal{G}_{vj-h}$ and $t\mathcal{G}_{vj-h}$ functions given by Equations (3.178) and (3.179) can be correlated in terms of $k_h/(\phi c_t \mu)$, $k_v/(\phi c_t \mu)$, and k_v/k_h (or its inverse) because z_{oj}, r_w, and Ω_h are known geometrical parameters. Note that Ω_{vj}, for $j = 1$ or 2, is a function of anisotropy (Equations (3.176) and

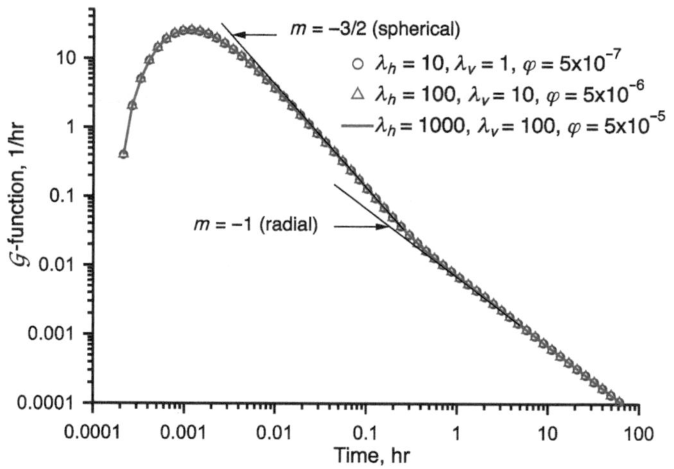

Figure 3.22 Correlation of $\mathcal{G}_{vj-h}$ function for $k_h/k_v = 10, h = 50$ ft, and $z_w = 25$ ft for a vertical well in a single-layer.

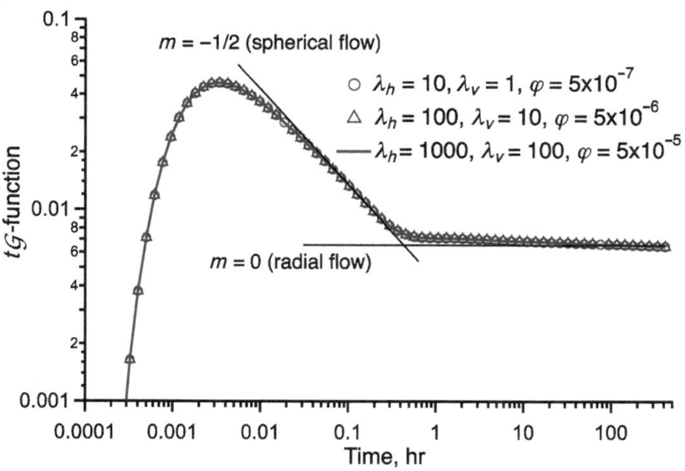

Figure 3.23 Correlation of $t\mathcal{G}_{vj-h}$ function for $k_h/k_v = 10, h = 50$ ft, and $z_w = 25$ ft for a vertical well in a single-layer.

(3.177)). In essence, this means that parameter estimation based on Equation (3.180) could be used to determine the individual values of k_h/μ and k_v/μ only if ϕc_t is known. This result may be considered as one of the weaknesses associated with parameter estimation using $\{v-h\}p$-p convolution because determination of k_h/μ and k_v/μ requires that ϕc_t be known *a priori*. Recall that for the p-r convolution, it is possible to uniquely determine ϕc_t in addition to k_h/μ and k_v/μ from spherical flow analyses of horizontal and vertical observation probe pressures (Goode & Thambynayagam, 1992; Onur, Hegeman, & Kuchuk, 2004b).

For sufficiently large values of time, Equations (3.178) and (3.179) can be further approximated, respectively, by

$$\mathcal{G}_{vj-h}(t) = \frac{\Omega_{vj}}{\Omega_h}\frac{r_w}{z_{oj}}\frac{\sqrt{\phi c_t \mu}}{\sqrt{k_h}}\left(\frac{z_{oj}}{\Omega_{vj}} - \frac{2r_w}{\Omega_h}\sqrt{\frac{k_v}{k_h}}\right)\frac{1}{\sqrt{\pi t^3}} \tag{3.182}$$

and

$$t\mathcal{G}_{vj-h}(t) = \frac{\Omega_{vj}}{\Omega_h}\frac{r_w}{z_{oj}}\frac{\sqrt{\phi c_t \mu}}{\sqrt{k_h}}\left(\frac{z_{oj}}{\Omega_{vj}} - \frac{2r_w}{\Omega_h}\sqrt{\frac{k_v}{k_h}}\right)\frac{1}{\sqrt{\pi t}}. \tag{3.183}$$

Equation (3.182) indicates that a log-log plot of $\mathcal{G}_{vj-h}$ vs. t would yield a straight line with a slope of $-3/2$. Thus, a straight line of $-3/2$ slope on a log-log plot of $\mathcal{G}_{vj-h}$ vs. time indicates a spherical flow regime. Similarly, Equation (3.183) indicates that during the spherical flow regime, a log-log plot of $t\mathcal{G}_{vj-h}$ vs. t would yield a straight line with a slope of $-1/2$, which is the same slope exhibited during the spherical flow regime when the logarithmic derivatives of the unit-rate impulse response functions (g_h and g_{vj}) for the horizontal and vertical probes are plotted as a function of time. Note that multiplying both sides of the impulse response functions given by Equation (3.174) (for the horizontal probe) and Equation (3.175) (for the vertical probe) by time t gives the logarithmic derivatives of the unit-rate drawdown responses.

For the cases during which both vertical and horizontal probes undergo the pseudo-cylindrically radial flow regime associated with the no-flow top and bottom boundaries of the formation, Equation (3.132) to Equation (3.136), and the first equality of Equation (3.136) given for the g and their related responses apply with D_w and D_o replaced by D_h and D_{vj}, respectively. Therefore, during the pseudo-cylindrically radial flow regime, the $\mathcal{G}_{vj-h}$-function displays nearly a straight line of -1 slope, and $t\mathcal{G}_{vj-h}$-function displays a nearly straight line of zero slope (a constant) on a log-log plot. Unfortunately, we do not have the equations for D_h and D_{vj}. However, the results of Onur et al. (2004b) show that for a fixed value of h, the $\mathcal{G}_{vj-h}$ and $t\mathcal{G}_{vj-h}$ functions can be uniquely correlated in terms of $k_h/(\phi c_t \mu)$, $k_v/(\phi c_t \mu)$, and k_h/k_v for *all time ranges* of interest.

Figures 3.22 and 3.23 present log-log plots of the $\mathcal{G}_{vj-h}$ and $t\mathcal{G}_{vj-h}$ functions for three cases where $k_h/(\phi c_t \mu) = 2 \times 10^7$ md.psi/cp, $k_v/(\phi c_t \mu) = 2 \times 10^6$ md.psi/cp, $k_h/k_v = 10$, $h = 50$ ft, $r_w = 0.328$ ft, $z_w = 25$ ft, and $z_{o1} = 2.3$ ft, but the values of $\lambda_h = k_h/\mu$, $\lambda_v = k_v/\mu$, and $\varphi = \phi c_t$ vary from case to case (see Figures 3.22 and 3.23). All cases shown in Figures 3.22 and 3.23 were generated from the Kuchuk (1996) analytical solutions. Note that identical results are obtained for all three cases. We also note that the maximums of the $\mathcal{G}_{vj-h}$ and $t\mathcal{G}_{vj-h}$ functions occur at $t_{max} = 1.3 \times 10^{-3}$ and $t_{max} = 3.6 \times 10^{-3}$ hr, which are almost identical to the ones predicted from Equations (3.180) and (3.181). All these results show that parameter estimation based on Equation (3.170) could be used to determine the individual values of λ_h and λ_v if ϕc_t is known.

As we stated above there are other p-p convolution pairs such as $\{v1, s\}$, $\{v2, s\}$, $\{v2, v1\}$. For instance,

$$\Delta p_{vj}(t) = \int_0^t \Delta p_s(t-\tau)\, \mathcal{G}_{vj-s}(\tau)\, d\tau \tag{3.184}$$

is for $\{v2, s\}$ and $\{v2, v1\}$. Note that in the p-p convolutions given by Equation (3.184) and previous Equation (3.168), the sink probe pressure (Δp_s) data are used. However, sometimes it is difficult to perform the p-p convolutions with the sink probe pressure data because the sink probe pressure measurements can be affected by the continuous cleanup and occasional gas evolution around the sink probe due to a significant pressure drop, causing flowing pressures to be below the bubble point pressure. These formulations are called horizontal-sink ($\{h, s\}$) and vertical-sink ($\{vj, s\}$)p-p convolutions, and their details are presented by Onur et al. (2004b).

In addition, Onur et al. (2004b) have considered a p-p convolution where the second vertical observation probe (vertical probe 2 in Figure 3.20) pressures are expressed as the convolution of the first vertical observation probe; i.e.,

$$\Delta p_{v2}(t) = \int_0^t \Delta p_{v1}(t-\tau)\, \mathcal{G}_{v2-v1}(\tau)\, d\tau. \tag{3.185}$$

The approximate equations for the $\mathcal{G}_{h-s}$, $\mathcal{G}_{vj-s}$ and $\mathcal{G}_{v2-v1}$ functions and the group parameters correlating these functions are given in Onur et al. (2004b). Here, we simply summarize Onur et al. (2004b) results as follows:

1. Individual estimates of k_h/μ, k_v/μ, and S (skin factor) can be determined from the $\{h, s\}p$-p convolution (Equation (3.168)) provided that ϕc_t is known.
2. Only k_v/μ can be determined, but not the individual values of k_h/μ and S from the $\{vj, s\}p$-p convolution (Equation (3.184)) provided that ϕc_t is known.
3. Parameter estimation based the $\{v2, v1\}p$-p convolution (Equation (3.185)) can provide the estimate of k_v/μ if ϕc_t is known, but the value of k_h/μ would be indeterminate.

3.8.3.4. Pressure-pressure convolution of packer-probe IPTTs

Here, we consider p-p convolution formulations of packer-probe wireline formation testers in a vertical well (Figure 3.21). As shown in Figure 3.21, the dual-packer module creates a limited-entry flow geometry. For a limited-entry well producing in a homogeneous transversely isotropic vertically bounded formation (infinite laterally), one may observe three different flow regimes (Kuchuk, 1994; Kuchuk et al., 1994): (1) Early-time radial flow regime due to plane radial flow in the part of the formation adjacent to the open interval, (2) Transitional spherical (and/or hemispherical) flow due to limited-entry completion, and (3) Late-time pseudo-radial flow due to impermeable bed boundaries limiting the reservoir in thickness. The early-time radial flow occurs at very early times, and its duration is usually so short that it cannot be identified in practice, especially when wellbore (or tool) storage and skin effects exist.

For the dual-packer module with the two-probe configuration shown in Figure 3.21, there are six different p-p convolution pairs: $\{p, v1\}$, $\{p, v2\}$, $\{v1, p\}$, $\{v1, v2\}$, $\{v2, p\}$, $\{v2, v1\}$, where p, $v1$ and $v2$ donate the packer, and, vertical 1 and vertical 2 probes. Normally, $\{p, v1\}$, $\{p, v2\}$, and $\{v1, v2\}$ pairs are not useful. For instance, $\{v1, p\}$, $\{v2, p\}$, $\{v2, v1\}p$-p convolution pairs can be written as

$$\Delta p_{vj}(t) = \int_0^t \Delta p_p(t-\tau)\, \mathcal{G}_{vj-p}(\tau)\, \mathrm{d}\tau \qquad (3.186)$$

for $j = 1$ or 2 and

$$\Delta p_{v2}(t) = \int_0^t \Delta p_{v1}(t-\tau)\, \mathcal{G}_{v2-v1}(\tau)\, \mathrm{d}\tau, \qquad (3.187)$$

where Δp_p and Δp_{vj} are the packer and vertical probes pressure data, respectively. If there is only one vertical probe, there is only one useful p-p convolution, which is given by Equation (3.186) with $j = 1$.

Approximate equations for the $\mathcal{G}_{vj-p}$ and $\mathcal{G}_{v2-v1}$ functions for a single-layer system are given by Onur et al. (2004b). For an unbounded formation in the vertical direction and assuming that packer interval and probes undergo the same spherical flow regime, the approximate $\mathcal{G}_{vj-p}$ function is given by Onur et al. (2004b) as

$$\mathcal{G}_{vj-p}(t) = \frac{1}{2\,(1/r_{sw} + S/l_w)} \sqrt{\frac{\phi c_t \mu}{k_v}}\, \frac{1}{\sqrt{\pi t^3}} \exp\left(-\frac{r_{sj}^2 \phi c_t \mu}{4 k_v t}\right), \qquad (3.188)$$

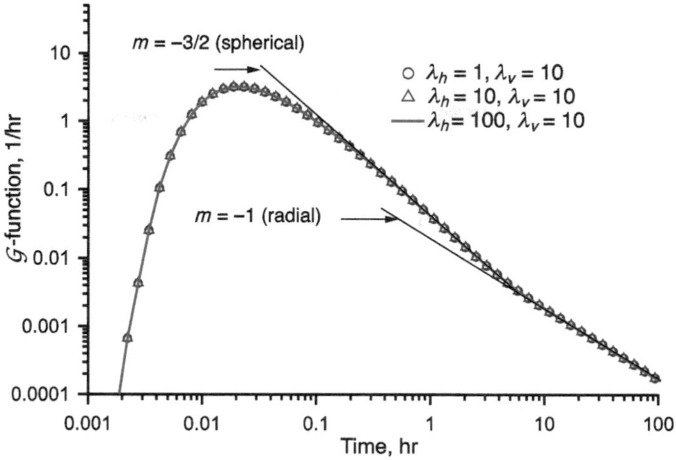

Figure 3.24 Correlation of $\mathcal{G}_{v2-v1}$ functions for three sets of horizontal and vertical mobilities for a vertical well in a single-layer formation.

where r_{sw} and r_{sj} are given by

$$r_{sw} = 2l_w \left[\ln \left\{ \frac{0.5 + 0.5\sqrt{1 + (r_w/l_w)^2 \frac{k_v}{k_h}}}{-0.5 + 0.5\sqrt{1 + (r_w/l_w)^2 \frac{k_v}{k_h}}} \right\} \right]^{-1}, \tag{3.189}$$

$$r_{sj} = \sqrt{\frac{k_v}{k_h} r_w^2 + z_{oj}^2}, \tag{3.190}$$

and S is the skin factor for the unit flow-rate case given as

$$S = \frac{2\pi k_h (2l_w) \Delta p_{u,\text{skin}}}{\mu} \tag{3.191}$$

for $j = 1$ and 2.

For an unbounded formation in the vertical direction and assuming that both vertical probes undergo the same spherical flow regime, the approximate $\mathcal{G}_{v2-v1}$ function is given by Onur et al. (2004b) as

$$\mathcal{G}_{v2-v1}(t) = \frac{r_{s1}(r_{s2} - r_{s1})}{2r_{s2}\sqrt{\pi t^3}} \sqrt{\frac{\phi c_t \mu}{k_v}} \exp\left[-\frac{\phi c_t \mu}{4k_v t} (r_{s2} - r_{s1})^2 \right]. \tag{3.192}$$

Figures 3.24 and 3.25 present $\mathcal{G}_{vj-h}$ and $t\mathcal{G}_{vj-h}$ for various horizontal and vertical mobilities ($\lambda = k/\mu$), and $h = 200$ ft, $\phi c_t = 5 \times 10^{-6}$ psi^{-1},

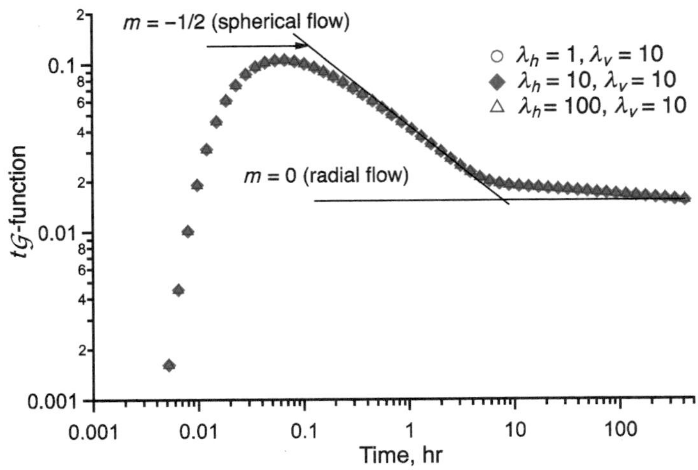

Figure 3.25 Correlation of $t\mathcal{G}_{v2-v1}$ functions for three sets of horizontal and vertical mobilities for a vertical well in a single-layer formation.

$\mu = 1$ cp, $z_w = 100$ ft, $z_{o1} = 6.4$ ft, and $z_{o2} = 14.4$ ft for a vertical well in a single-layer formation. It can be shown that for sufficiently large values of time such that the exponential term in Equations (3.188) and (3.192) becomes unity, log-log plots of $\mathcal{G}_{vj-p}$ and $\mathcal{G}_{v2-v1}$ functions vs. time yield straight lines of $-3/2$ slope, indicating a spherical flow regime. In addition, assuming both packer and vertical probe-j undergo the same late-time radial flow regime due to total formation thickness, Onur et al. (2004b) show that during the radial flow, the $\mathcal{G}_{vj-p}$ and $\mathcal{G}_{v2-v1}$ functions are characterized by a straight line with a slope of nearly -1 on a log-log plot. Finally, we note that the time at which the $\mathcal{G}_{vj-p}$ and $\mathcal{G}_{v2-v1}$ functions reach maximum values, as shown in Figures 3.24 and 3.25 for the $\mathcal{G}_{v2-v1}$ function, is dependent on the vertical permeability, k_v, but independent of the horizontal permeability, k_h.

Regarding parameter estimation based on the vertical probe-packer (v-p) and vertical 2-vertical 1 ($v2$-$v1$) probe p-p convolution equations given by Equations (3.186) and (3.187), Onur et al. (2004b) results show that:

1. Only k_v/μ, but not k_h/μ and skin factor, can be uniquely determined from the analysis of v-p p-p convolution (Equation (3.186)) if ϕc_t is known. One cannot determine individual values of k_h/μ and S because of the term $1/r_{sw} + S/l_w$ in the $\mathcal{G}_{vj-p}$ function given by Equation (3.188). This term introduces a dependency between k_h/μ and S because r_{sw} is a function of k_h/k_v (Equation (3.189)).
2. Parameter estimation based $v2$-$v1p$-p convolution (Equation (3.187)) can provide the estimate of k_v/μ if ϕc_t is known, but the value of k_h/μ would be indeterminate, as clearly confirmed by the numerical results

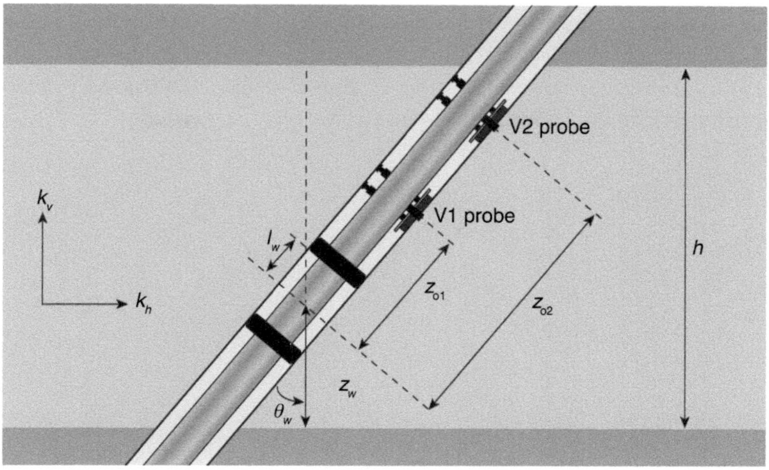

Figure 3.26 Schematic of a packer and two-probe configuration in a slanted well.

shown in Figures 3.24 and 3.25, which present log-log plots of $\mathcal{G}_{v2-v1}$ and $t\mathcal{G}_{v2-v1}$ functions vs. time for three different values of anisotropy ratio ($k_h/k_v = 0.1$, 1, and 10), but k_h/μ varies from case to case. As can be observed in these figures, k_h/μ has no effect at all on the $\mathcal{G}_{v2-v1}$ and $t\mathcal{G}_{v2-v1}$ functions for all time ranges of interest including the late-time radial flow.

The results given above are valid for packer-probe tests conducted in a vertical well. For packer-probe IPTTs conducted in a horizontal well or a slanted well (Figure 3.26), the information content of $\mathcal{G}_{vj-p}$ and $\mathcal{G}_{v2-v1}$ functions is different from that obtained from $\mathcal{G}_{vj-p}$ and $\mathcal{G}_{v2-v1}$ functions for a vertical well. Although we do not have approximate equations for the $\mathcal{G}_{vj-p}$ and $\mathcal{G}_{v2-v1}$ functions for the slanted well case, our numerical results indicate that, unlike the vertical well case, the vj-p convolution (Equation (3.177)) for a horizontal well or a slanted well with inclination angles (measured from vertical) different from 0°(vertical well) will provide unique estimates of the individual values of k_h/μ, k_v/μ and skin factor, and the $v2$-$v1$ convolution (Equation (3.178)) will provide unique estimates of both k_h/μ and k_v/μ, provided that ϕc_t is known and that both spherical and late-radial flow regimes are observed. Of course, we expect that as the slant angle θ_w approaches 0°, the information content of the $\mathcal{G}_{vj-p}$ and $\mathcal{G}_{v2-v1}$ functions for packer-probe tests in a slanted well (Figure 3.26) become similar to those obtained from the $\mathcal{G}_{vj-p}$ and $\mathcal{G}_{v2-v1}$ functions for a vertical well.

Onur et al. (2004b) have also studied the information content of various $\mathcal{G}$ functions for both multiprobe and packer-probe IPTTs conducted in a vertical well in multilayer cross-flow systems. The information content of each $\mathcal{G}$-function in multilayer systems is complex and different and, hence, each $\mathcal{G}$ function is sensitive to different parameter groups. Onur et al. (2004b) recommend simultaneous matching of pressure data sets based on various p-p convolutions for parameter estimation.

CHAPTER 4

Deconvolution

Contents

4.1. Introduction

In recent years, deconvolution techniques have been used increasingly for pressure transient test interpretation. Deconvolution is simply solving the convolution integral (the Volterra integral equation of the first kind) for the convolution kernel (see Chapter 3). Basically deconvolution enables us to reconstruct (compute) an equivalent (extrapolated) constant-rate drawdown response for all production and shut-in periods combined. Although not strictly speaking, the deconvolved constant-rate drawdown response can be thought of as an extrapolated response. For instance, if we produce a well for 100 hr, during which the production rate is accurately measured,

Developments in Petroleum Science, Volume 57
ISSN 0376-7361, DOI: 10.1016/S0376-7361(10)05710-9

and perform a subsequent 24-hr buildup test, during which the downhole pressure is measured, in principle, using deconvolution, we can compute a 124-hr equivalent constant-rate deconvolved drawdown response from the 24-hr buildup pressure data and the production rate data of the drawdown period. This is a rough rule of thumb and not accurate or reliable for every pressure transient test. For example, if the flow-rate data are not accurate and/or if the resolution of the buildup pressure data is low, even a reliable deconvolution algorithm may produce unacceptable behavior and a purely imaginary equivalent constant-rate deconvolved drawdown response. In fact, using a deconvolution algorithm as a black box is highly risky. Therefore, one of the objectives of this chapter is to explain the limitations of deconvolution and to show the influence of various parameters on the deconvolution algorithm performance.

It should be stated, at the outset, that deconvolution (solving the Volterra integral equation) is a highly ill-posed problem in the Hadamard (1953) sense (he investigated the well-posedness problems for differential equations) if there is any noise in the input and output data (see Chapter 3). Therefore, many different regularization techniques (such as the well-known Tikhonov (1963) regularization) have been used in optimization techniques for deconvolution.

To obtain the constant-rate deconvolved behavior (or constant pressure) and its derivative is very important for system identification in pressure transient formation and well testing. Although deconvolution of pressure and rate was not commonly used for reservoir engineering problems until the 1980s, there were previously a few papers on the subject in the petroleum engineering literature. The first three (Coats, Rapoport, McCord, & Drews, 1964; Hutchinson & Sikora, 1959; Katz, Tek, & Jones, 1962) were about determining the influence function directly from field data for aquifers. Jargon and van Poollen (1965) were perhaps the first ones to use a deconvolution technique to compute the constant-rate pressure behavior (the influence function) from transient pressure and rate data. Bostic, Agarwal, and Carter (1980) used another technique to obtain the influence function from variable rate and pressure history. Pascal (1981) also used a deconvolution technique to obtain a constant-rate solution from variable-rate (measured at the surface) and pressure measurements of a drawdown test. Although these techniques for determining influence functions or constant-rate response are practically deconvolution, before 1983 (Kuchuk & Ayestaran, 1983; Stewart, Meunier, & Wittmann, 1983) the word deconvolution was not used in the petroleum engineering literature.

In Chapter 3, we have presented the operational properties and applications of convolution integrals. In this chapter, we present deconvolution methods for pressure transient test interpretation. In pressure transient testing, deconvolution is defined as determining the influence function or unit response behavior of a system (an equivalent constant-rate response) from measured transient pressure and flow-rate data.

Deconvolution is a signal processing method in which the effect of the time-dependent input signal is filtered from the output signal (see Figure 3.1 in Chapter 3). It also simply solves the Volterra convolution integral equation given by Equation (3.1) of Chapter 3 as

$$\psi(t) = f(t) \star \kappa(t) = \int_0^t \mathrm{d}\tau f(\tau)\kappa(t-\tau) \tag{4.1}$$

and in the Laplace domain

$$\bar{\psi}(s) = \bar{f}(s)\bar{\kappa}(s) \tag{4.2}$$

for the convolution kernel $\kappa(t)$ (unit-impulse response) for a given $\psi(t)$ and $f(t)$ data set. The solution of Equation (4.1), the kernel $\kappa(t)$, can be written explicitly using Equation (4.2) and the property of the inverse Laplace transform as

$$\kappa(t) = \int_0^t \mathrm{d}\tau \psi(t-\tau)\mathcal{F}(\tau), \tag{4.3}$$

where

$$\mathcal{F}(t) = \mathcal{L}^{-1}\left[\frac{1}{\bar{f}(s)}\right], \tag{4.4}$$

$\mathcal{L}^{-1}$ denotes the inverse Laplace transform operator, and $\bar{f}(s)$ is the Laplace transform of $f(t)$. As stated in Chapter 3, the kernel κ is the solution of the pressure diffusivity equation in transient testing. The deconvolution is fundamental and is particularly applicable to system identification, where the kernel κ is called the deconvolved unit-impulse response after the deconvolution operation.

Although a generic convolution integral given by Equation (4.2) will be used very often in this chapter for convenience, for pressure transient testing, we normally use the one from Equation (3.51) given in Chapter 3, which can be rewritten as

$$p_m(t) = p_o - \int_0^t q_m(\tau)\, g(t-\tau)\, \mathrm{d}\tau, \tag{4.5}$$

where p_o is the initial reservoir pressure, p_m and q_m are the measured wellbore pressure and flow rate, and g is the unit-impulse response including the skin and wellbore storage effects and is given by Equation (3.52).

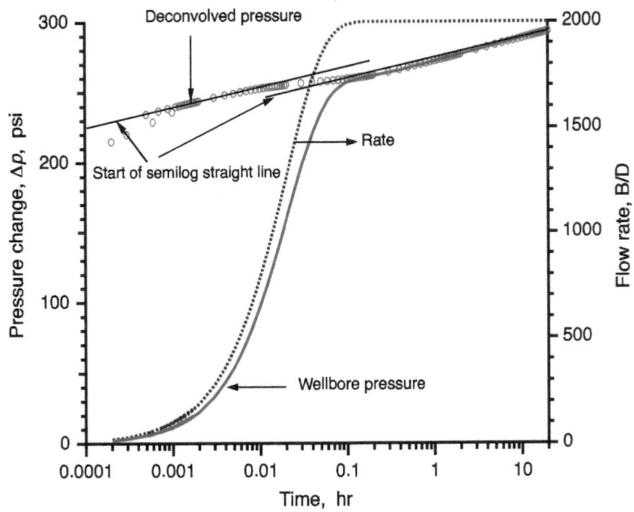

Figure 4.1 Deconvolved and wellbore pressures, and flow rate.

In this chapter, the subscript d denotes all deconvolved quantities. For instance, g_d denotes the deconvolved unit-impulse response. The integral of $g_d(t)$ with respect to time t is called the deconvolved constant unit-rate drawdown pressure change $p_u(t)$ (see Chapter 3 for the definition). $\Delta p_d = qp_u(t)$ denotes the deconvolved constant-rate drawdown pressure change, where q is the reference rate. p_d denotes the deconvolved constant-rate drawdown pressure. The term deconvolved will always refer to the constant-rate or unit-rate drawdown pressure response, or the unit-rate impulse response. Therefore, we will often omit the words constant-rate or unit-rate drawdown pressure response and simply call them deconvolved response or pressure changes, etc.

Figures 4.1 and 4.2 present the deconvolved and wellbore pressure changes, and their logarithmic derivatives for a cylindrical source solution in an infinite acting dual-porosity reservoir. In the same plot, the wellbore flow rate is also shown. As can be seen from Figure 4.1, the deconvolved constant-rate drawdown pressure change exhibits two radial-flow parallel semi-log straight lines, which are well-known characteristics of dual-porosity reservoirs. The wellbore pressure change is dominated by the wellbore storage effect and exhibits only the second radial-flow semi-log straight line, indicating a single-porosity reservoir. These features of the deconvolved and wellbore pressure changes are seen much more clearly in the derivative plots given in Figure 4.2. For an illustration of deconvolution system identification, we have used a simple deconvolution algorithm, which will be presented below, without any noise in pressure and flow-rate measurements for this example.

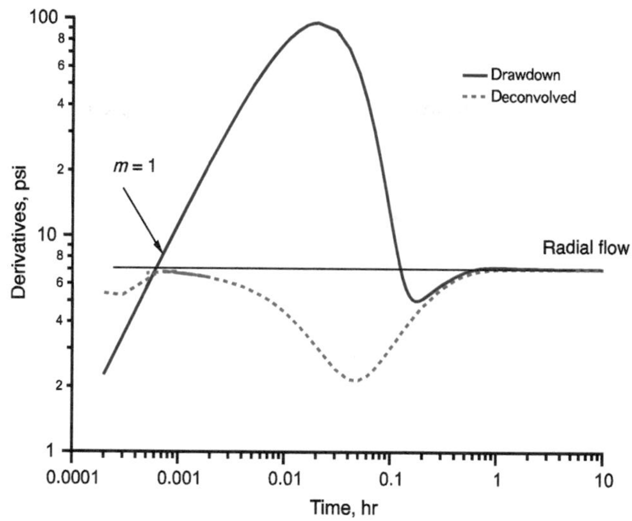

Figure 4.2 Derivatives of the deconvolved and wellbore pressures given in Figure 4.1.

4.2. Analytical Deconvolutions

A number of direct deconvolutions can be performed using Equation (4.3) if the forcing function $f(t)$ (the time-dependent boundary condition, which can be measured flow rate or pressure at the wellbore) can be approximated by suitable functions as in the convolution case.

Power Series:

$$f(t) = \beta t^k, \tag{4.6}$$

where β is a positive constant and $k > -1$. By using the Laplace transform of $f(t)$ given by Equations (4.6) and (4.2), we obtain

$$\bar{\psi}(s) = \beta\bar{\kappa}(s)s^{-(k+1)}\Gamma(k+1), \tag{4.7}$$

from which the kernel κ can be written as

$$\kappa(t) = \frac{1}{\beta\Gamma(k+1)}\mathcal{L}^{-1}\left[\bar{\psi}(s)s^{k+1}\right]. \tag{4.8}$$

Special cases:

$$\kappa(t) = \frac{1}{\beta n!}\frac{\mathrm{d}^{n+1}}{\mathrm{d}t^{n+1}}\psi(t), \quad \text{if } k = n, n \text{ is an integer}$$

$$= \frac{1}{\beta}\frac{\mathrm{d}^2\psi(t)}{\mathrm{d}t^2} \quad \text{for } n = 1 \quad [f(t) = \beta t],$$
$$= \frac{2}{\pi\beta}\int_0^t \mathrm{d}\tau \frac{\psi''(t-\tau)}{\sqrt{\tau}} \quad \text{for } k = 1/2 \quad [f(t) = \beta\sqrt{t}],$$
$$= \frac{1}{\pi\beta}\int_0^t \mathrm{d}\tau \frac{\psi'(t-\tau)}{\sqrt{\tau}} \quad \text{for } k = -1/2 \quad [f(t) = \beta/\sqrt{t}], \tag{4.9}$$

where $\psi(t)$ and all its derivatives up to nth are assumed to be zero when $t = 0$, and the prime $'$ denotes the derivative of ψ with respect to t.
Exponential case:

$$f(t) = 1 - \mathrm{e}^{-\beta t} \tag{4.10}$$

where β is a positive constant. By using the Laplace transform of $f(t)$ given by Equations (4.10) and (4.2), we obtain

$$\bar{\psi}(s) = \frac{\beta}{s(\beta + s)}\bar{\kappa}(s) \tag{4.11}$$

from which the kernel κ can be written as

$$\kappa(t) = \mathcal{L}^{-1}\left[\bar{\psi}(s)\left(s + \frac{s^2}{\beta}\right)\right] = \psi' + \frac{1}{\beta}\psi''. \tag{4.12}$$

In oilfield units, Equation (4.12) in terms of the deconvolved pressure change Δp_d can be written as

$$\Delta p_d(t) = \frac{1}{\alpha}\frac{\mathrm{d}\Delta p_w(t)}{\mathrm{d}t} + \Delta p_w(t), \tag{4.13}$$

where α is a positive constant and its unit is 1/hr, whereas β in Equation (4.10) is dimensionless and given as $\beta = \frac{\alpha\phi c_t\mu}{0.0002637k}$ (Kuchuk, 1990a; Kuchuk & Ayestaran, 1985). It is quite simple to compute Δp_d from the measured downhole pressure p_w, its derivative, and α, which is obtained from the measured downhole rate if the wellbore flow rate varies exponentially (Kuchuk, 1990a).
Gladfelter deconvolution:
Gladfelter, Tracy, and Wilsey (1955) stated that the reciprocal productivity index (rate-normalized pressure, see Chapter 3), $\left(J_w = \frac{\Delta p_{ws}}{\Delta q_m}\right)$, is a linear function of the logarithm of time for buildup tests, where Δp_{ws} is the change in shut-in pressure and Δq_m is the change in afterflow rate. The Gladfelter et al. (1955) method, which we call *Gladfelter deconvolution*,

has been used in well testing since 1955 for the interpretation of measured downhole pressure and flow rate. Over the years a number of studies (Gladfelter et al., 1955; Kuchuk, 1990b; Ramey, 1976a; Winestock & Colpitts, 1965) have appeared on this subject. Simply, the Gladfelter deconvolution can be defined as determining the deconvolved rate-normalized pressure J_{dw} (it is J_w but deconvolved) at the sandface including skin from measured downhole pressure and flow rate. Here we used the term deconvolved rate-normalized pressure in order to differentiate it from that obtained from the logarithmic convolution (see Chapter 3). Once J_{dw} is obtained, the conventional interpretation methods can be used for determination of a well–reservoir system and its parameters. However, Kuchuk (1990b) shows that the Gladfelter deconvolution works for only a few simple flow geometries and a few specific flow-rate schedules, such as linearly increasing rate. It does not work at all for general rate variations or for fractured wells and reservoirs, etc.

4.3. Discrete Numerical Deconvolution without Measurement Noise

Many direct deconvolution algorithms have been presented in the petroleum literature using different operational techniques. These are just a few: Blasingame, Johnston, Lee, and Raghavan (1989), Bostic et al. (1980), Bourgeois and Horne (1993), Jargon and van Poollen (1965), Kuchuk and Ayestaran (1985), Mendes, Tygel, and Correa (1989), Roumboutsos and Stewart (1988), and Thompson and Reynolds (1986). These algorithms present better integration and interpolation methods for discretization of the convolution integral, except the methods given by Bourgeois and Horne (1993), Mendes et al. (1989), Onur and Reynolds (1998), and Roumboutsos and Stewart (1988), in which Laplace-domain deconvolution techniques have been used. Some details of these deconvolution algorithms and their differences can be found in the paper by von Schroeter, Hollaender, and Gringarten (2004). All these methods usually have serious stability problems when $\psi(t)$ and/or $f(t)$ data have measurement errors, even very small, say less then 1%. In the following paragraphs, we present one simple algorithm to illustrate the deconvolution process and the stability problem.

To reveal some basic nature of deconvolution let us rewrite the discrete convolution given by Equation (3.23) as

$$\psi_{n+1} = \sum_{i=0}^{n} \mathcal{P}_{n+1-i}[f_{i+1} - f_i], \quad n = 0, 1, \ldots, N-1, \tag{4.14}$$

where the discrete time intervals are $0 = t_0 < t_1 < t_2 < \cdots < t_n < t_{n+1} = t$,

N is the total number of data points, and $\mathcal{P}$ is the definite integral of κ, and is normally the constant-rate pressure behavior of the system.

A very basic deconvolution algorithm can be written from Equation (4.14) as

$$\mathcal{P}_{n+1} = \frac{1}{f_1}\left\{\psi_{n+1} - \sum_{i=1}^{n} \mathcal{P}_{n+1-i}[f_{i+1} - f_i]\right\},$$
$$n = 0, 1, \ldots, N-1. \tag{4.15}$$

Let us write a few terms of Equation (4.15) as

$$\begin{aligned}
\mathcal{P}_1 &= \frac{1}{f_1}\psi_1, \quad n = 0,\\
\mathcal{P}_2 &= \frac{1}{f_1}[\psi_2 - \mathcal{P}_1(f_2 - f_1)], \quad n = 1,\\
\mathcal{P}_3 &= \frac{1}{f_1}\{\psi_3 - [\mathcal{P}_1(f_3 - f_2) + \mathcal{P}_2(f_2 - f_1)]\}, \quad n = 2.
\end{aligned} \tag{4.16}$$

One may immediately realize the first stability problem in the deconvolution algorithm given by Equation (4.15) because each computed $\mathcal{P}_n$ will be affected by the error in the first data point f_1.

In general, measurements are subject to some errors (measurement noise). There are many different sources of noise that can be encountered in wellbore pressure and flow-rate measurements. Errors in flow-rate measurements, in general, are usually larger than errors in pressure measurements. Suppose that f (flow rate) and ψ (pressure) with measurement errors (noise) can be expressed as

$$\hat{f}(t) = f(t) + \epsilon(t) \tag{4.17}$$

and

$$\hat{\psi}(t) = \psi(t) + \epsilon(t). \tag{4.18}$$

For convenience it is assumed that both measurements have the same noise function defined as ϵ. Substitution of Equations (4.17) and (4.18) into Equation (4.15) gives

$$\mathcal{P}_{n+1} = \frac{1}{f_1 + \epsilon_1}\left\{(\psi_{n+1} + \epsilon_{n+1}) - \sum_{i=1}^{n} \mathcal{P}_{n+1-i}[f_{i+1} - f_i] - \sum_{i=1}^{n} \mathcal{P}_{n+1-i}[\epsilon_{i+1} - \epsilon_i]\right\}, \quad n = 0, 1, \ldots, N-1. \tag{4.19}$$

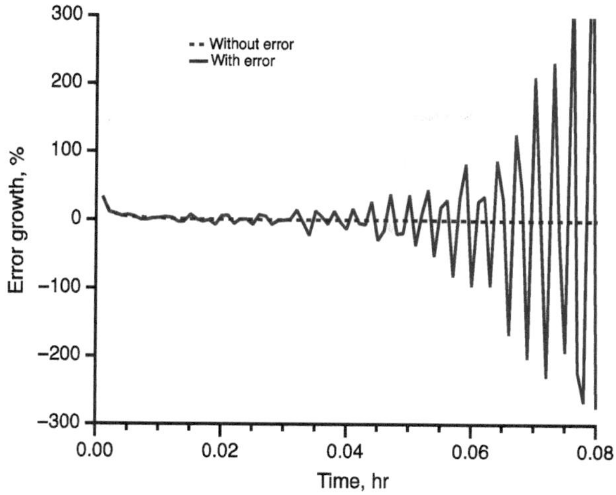

Figure 4.3 The effect of ±1% random noise in flow-rate measurements on a direct deconvolution of pressure and rate data.

After a slight manipulation, Equation (4.19) becomes

$$\mathcal{P}_{n+1} = \frac{1}{f_1 + \epsilon_1}\left\{\psi_{n+1} - \sum_{i=1}^{n} \mathcal{P}_{n+1-i}[f_{i+1} - f_i]\right\} + \Upsilon_{n+1}, \quad (4.20)$$

where

$$\Upsilon_{n+1} = \frac{\epsilon_{n+1}}{f_1 + \epsilon_1} - \frac{1}{f_1 + \epsilon_1}\sum_{i=1}^{n} \mathcal{P}_{n+1-i}[\epsilon_{i+1} - \epsilon_i] \quad (4.21)$$

is the quantity due to errors in the flow-rate measurements. Notice that each computed $\mathcal{P}$ in Equation (4.20) is affected by ϵ_1, which is the error in the first flow-rate measurement. Thus $\mathcal{P}$ will be under or over estimated, depending on whether or not the first measured flow rate is smaller or larger than the actual flow rate. The summation term given in Equation (4.21), which is the summation of $\mathcal{P}$ times the difference of two consecutive error terms, will rapidly grow as n (time) increases.

Figure 4.3 presents the percentage error growth computed from Equation (4.21) using the wellbore pressure computed from Equation (4.5) with a line-source unit-impulse response and an exponential flow rate given by Equation (4.10). As can be seen from Figure 4.3, the percentage error in the deconvolved pressure without flow-rate measurement errors is less than ±0.01%. On the other hand, the percentage error in the deconvolved pressure with flow-rate measurement errors, where a randomly distributed noise with magnitudes between 0 to ±1% was added to the rate data, grows rapidly after 0.04 hr (2.4 min). It reaches ±300% at 0.08 hr, almost an exponential growth. In reality, a 1% error in flow-rate measurements is small and it is very difficult to achieve in downhole wellbore conditions.

4.4. Deconvolution with Constraints

As we know, all measurements, no matter how carefully obtained, are subject to some errors. At the wellbore, usually errors in pressure measurements are smaller than errors in flow-rate measurements. In any event, the direct deconvolution methods mentioned above give unstable solutions if measurements have a small amount of noise (errors) associated with them.

In order to minimize the effects of measurement noise on deconvolution, various constraints have been used by Baygun, Kuchuk, and Arikan (1997), Coats et al. (1964), Gajdica, Wattenbarger, and Startzman (1988), Katz et al. (1962), and Kuchuk, Carter, and Ayestaran (1990a) on the behavior of the convolution kernel κ. In these deconvolution methods, it was assumed that the errors in pressure measurements are negligible, while the flow-rate data contain some measurement errors. The details of the constrained deconvolution were given by Kuchuk et al. (1990a), will not be repeated here, but we will briefly describe the basics of constrained deconvolution.

In general, the constraints for the convolution kernel κ can be expressed as

$$\mathcal{P}(t) = \int_0^t \mathrm{d}\tau \kappa(\tau) \geq 0 \quad \text{for } t \geq 0, \tag{4.22}$$

where $\mathcal{P}(t)$ is a constant-rate solution,

$$\kappa(t) \geq 0 \quad \text{for } t \geq 0, \tag{4.23}$$

$$\kappa'(t) \leq 0 \quad \text{for } t \geq 0, \tag{4.24}$$

and

$$\kappa''(t) \geq 0 \quad \text{for } t \geq 0 \tag{4.25}$$

provided that the real time is greater than a few seconds and the diffusivity constant $k/\phi\mu c_t$ is not very small. The general solution of the diffusivity equation for the second kind of internal boundary condition with non-periodic initial and outer boundary conditions will satisfy the conditions given in Equation (4.22) to Equation (4.25) for single-phase flow of a slightly compressible fluid with constant compressibility and viscosity.

It would, of course, be better to find the smallest perturbation to $f(t)$ (in flow-rate measurements) for which we can solve the convolution integral given by Equation (4.1) exactly. This can be formulated as follows

$$\hat{f}(t) = f(t) + \epsilon(t), \tag{4.26}$$

where $\epsilon(t)$ is the noise term. Substituting Equation (4.26) in Equation (4.1)

$$\psi(t_n) = \int_0^{t_n} [f(\tau) + \epsilon(\tau)]\kappa\,(t_n - \tau)\;\mathrm{d}\tau, \tag{4.27}$$

where $t_n = t$. After further manipulations, the deconvolution with constraints can be written as

$$\text{minimize} \sum_{n=1}^{N} E_n^2(\kappa), \tag{4.28}$$

where

$$E_n(\kappa) = \psi(t_n) - \int_0^{t_n} f(\tau)\kappa\,(t_n - \tau)\;\mathrm{d}\tau, \tag{4.29}$$

and N is the total number of measurements. Equation (4.28) with constraints given above can be written in matrix notation as

$$\text{minimize } E(\mathbf{x}) = \|\mathbf{b} - \mathbf{x}\mathbf{A}\|_2^2 \tag{4.30}$$

$$\text{subject to } \|\mathbf{x}\|_2^2 \leq \mathbf{c}_\kappa, \tag{4.31}$$

where E is the objective function, the vector $\mathbf{b}$ is given by

$$\mathbf{b} = [\psi_1,\ \psi_2,\ \psi_3,\ \ldots, \psi_N]^T, \tag{4.32}$$

the vector $\mathbf{x}$ is given by

$$\mathbf{x} = [\mathcal{P}_1,\ \mathcal{P}_2,\ \mathcal{P}_3,\ \ldots, \mathcal{P}_N]^T, \tag{4.33}$$

the coefficient matrix $\mathbf{A}$ is defined by

$$\mathbf{A} = \begin{bmatrix} f_1 & 0 & 0 & 0 & \ldots & 0 \\ (f_2 - f_1) & f_1 & 0 & 0 & \ldots & 0 \\ (f_3 - f_2) & (f_2 - f_1) & f_1 & 0 & \ldots & 0 \\ \vdots & & & & \ldots & \\ (f_N - f_{N-1}) & (f_{N-1} - f_{N-2}) & (f_{N-3} - f_{N-4}) & \ldots & \ldots & f_1 \end{bmatrix}, \tag{4.34}$$

which is a lower triangular $N \times N$ matrix, T denotes the transpose of a vector, and $\mathbf{c}_\kappa$ is the constraint vector determined from Equation (4.22) to Equation (4.25). Kuchuk et al. (1990a) solved this minimization problem by an active set method for solving a quadratic (or linear least squares) problem subject to simple bounds given by Equation (4.31) on $\mathbf{x}$.

Up to this point, we have used a generic form of the convolution integral given by Equation (4.1), in terms of ψ, f, and κ to avoid dimensional and dimensionless variables in pressure, flow rate, and the unit-impulse response. In the following section we will present a nonlinear least-squares algorithm for deconvolution of transient pressure and flow-rate data.

4.5. Nonlinear Least-Squares Pressure-Rate Deconvolution

The measured pressure in the wellbore (most preferable just above the producing zone and in the tool for all formation testers) can be rewritten from Equation (4.5) as

$$p_m(t) = p_o - \int_0^t q_m(\tau)\, g(t-\tau)\, \mathrm{d}\tau, \tag{4.35}$$

where p_m and q_m are the measured wellbore pressure and flow rate, and g is the unit-impulse response including the skin and wellbore storage effects, and is given by Equation (3.52). Let us assume that N_p is the number of measured p_m and q_m data points. As discussed above we can solve Equation (4.35) for g as a minimization problem given by Equation (4.30) with constraints given by Equation (4.22) to Equation (4.25). However even with constraints, the solution becomes very oscillatory and unacceptable if errors in flow-rate data are more than a few percent (Kuchuk et al., 1990a).

A well-known technique for improving solutions of ill-posed problems is to introduce a regularization term or smoothing term in the least-squares objective function (Levenberg, 1944; Marquardt, 1963; Tikhonov, 1963). Using a regularization method (Levenberg, 1944; Marquardt, 1963) for solving Equation (4.35), Equation (4.30) can be written (Carter & Kuchuk, 1990; Zajic & Kuchuk, unpublished) as

$$E(\mathbf{g}) = \left\| \boldsymbol{\Delta p}_m - \mathbf{Ag} \right\|_2^2 + \lambda \left\| \mathbf{g} \right\|_r^2, \tag{4.36}$$

where E is the objective function, $\boldsymbol{\Delta p}_m$ the vector of pressure measurements with dimension N_p, where $\Delta p = p_o - p_m$, it is assumed that $\mathbf{g}$ is an N_p-dimensional vector with elements g (deconvolved unit-impulse response), λ is a regularization parameter and $\| \cdot \|_r$ is a norm. Let us assume that the dimension of matrix **A** depends on the discretization technique of the convolution integral given by Equation (4.35). For instance if we approximate g and q by piecewise linear functions with equispaced knots at $\{t_i\}$, $t_i = i\Delta t$ in Equation (4.35) for $\tau \in \{t_{i-1}, t_i\}$ and $k \in \{i, N_p\}$ as

$$g(t_k - \tau) = g_{k-i} + \frac{t_{i-1} - \tau}{\Delta t}(g_{k-i} - g_{k-i+1}) \tag{4.37}$$

and

$$q(\tau) = q_{i-1} + \frac{\tau - t_{i-1}}{\Delta t}(q_i - q_{i-1}). \tag{4.38}$$

Substituting g and q given just above in Equation (4.35) yields

$$\begin{aligned}\Delta p(t_k) &= \int_0^{t_k} q(\tau)g(t_k - \tau)\mathrm{d}\tau \\ &= \sum_{i=1}^{k}\int_{t_{i-1}}^{t_i} q(\tau)g(t_k - \tau)\mathrm{d}\tau \\ &= \sum_{i=1}^{k}\int_{t_{i-1}}^{t_i}\left[q_{i-1} + \frac{\tau - t_{i-1}}{\Delta t}(q_i - q_{i-1})\right] \\ &\quad \times\left[g_{n-i} + \frac{t_{i-1} - \tau}{\Delta t}(g_{k-i} - g_{k-i+1})\right]\mathrm{d}\tau.\end{aligned} \tag{4.39}$$

Performing integration and collecting terms yield the matrix **A** in Equation (4.36) as

$$\mathbf{A} = \begin{bmatrix} \frac{1}{6}q_1 + \frac{1}{3}q_0 & \frac{1}{3}q_1 + \frac{1}{6}q_0 & 0 & 0 & \cdots & 0 \\ \frac{1}{6}q_2 + \frac{1}{3}q_1 & \frac{1}{3}(q_2 + q_1 + q_0) & \frac{1}{3}q_1 + \frac{1}{6}q_0 & 0 & \cdots & 0 \\ \frac{1}{6}q_3 + \frac{1}{3}q_2 & \frac{1}{3}(q_3 + q_2 + q_1) & \frac{1}{3}(q_2 + q_1 + q_0) & \frac{1}{3}q_1 + \frac{1}{6}q_0 & \cdots & 0 \\ \frac{1}{6}q_4 + \frac{1}{3}q_3 & \frac{1}{3}(q_4 + q_3 + q_2) & \frac{1}{3}(q_3 + q_2 + q_1) & \frac{1}{3}(q_2 + q_1 + q_0) & \cdots & 0 \\ \cdots & \cdots & \cdots & \cdots & \cdots & \cdots \\ \cdots & \cdots & \cdots & \frac{1}{3}(q_{n-2} + q_{n-3} + q_{n-4}) & \cdots & \frac{1}{3}q_1 + \frac{1}{6}q_0 \end{bmatrix}, \tag{4.40}$$

where **A** is $N_p \times (N_p + 1)$ matrix, and now the unknown **g** is an $(N_p + 1)$

dimensional vector. Notice that unlike the matrix **A** given by Equation (3.26), the above matrix **A** is not a lower triangular matrix. Furthermore,

$$\boldsymbol{\Delta p}_m = \mathbf{Ag} \tag{4.41}$$

given in Equation (4.36) is an underdetermined linear system.

Our linear system can easily be modified by using larger elements to support the approximation to our solution $g(t)$. Rather than taking the nodes to be Δt apart, let us assume that they are $l\Delta t$ apart for some $l > 1$ so that $N_g = N_p/l + 1$ is the number of nodes on g. Note that we still assume that the measured data are collected at intervals of Δt, and that our discretization for q is given by Equation (4.38). Redefining

$$g_i \equiv g_{il} \tag{4.42}$$

and taking g_{il+j} to be an interpolant between g_{il} and $g_{(i+1)l}$ for $0 < j < l$ yields an overdetermined system of linear equations. The new coefficient matrix $\mathbf{A} \in \Re^{N_p \times N_n}$ has columns that are linear combinations of the original columns given in Equation (4.40).

The last term in Equation (4.36) *penalizes* nonsmooth solutions of the nonlinear-least squares. Small values of λ provide little smoothing, while excessively large values of λ force the computed solution to be perfectly smooth with respect to the norm $\|\cdot\|_r$. Typically, the norm selected is

$$\|g\|_r^2 \equiv \frac{1}{L}\sum_{i=1}^{L}\left[g_i^{[r]}\right]^2, \tag{4.43}$$

which is the discrete approximation to

$$\|g\|_r^2 \equiv \frac{1}{t_n}\int_0^{t_n}\left[g^{[r]}(t)\right]^2 \mathrm{d}t, \tag{4.44}$$

where $g^{[r]}$ is the rth order time derivative of g, or the corresponding discrete divided difference term for g: i.e., for $r = 1$ we have

$$g_i^{[1]} = \frac{g_i - g_{i+1}}{l\Delta t}, \tag{4.45}$$

and for $r = 2$ we have

$$g_i^{[2]} = \frac{g_{i-1} - 2g_i + g_{i+1}}{(l\Delta t)^2}. \tag{4.46}$$

Together, the choice of λ and r determines the amount and type of smoothing. For example, selecting $r = 4$ and a very large λ forces the computed solution to be very smooth, while a choice of $r = 1$ would penalize any non-constant solution. The choice $r = 0$ is popular because of its simplicity. In general, adequate solutions ($\mathbf{g}$) can be computed with moderate levels of measurement errors provided a good choice can be found for λ. The choice of a *good* value for λ is a particularly difficult problem. Among the many references on this subject are Groetsch (1984), Kress (1989), Morozov (1984), Wahba (1977), and Zajic and Kuchuk (unpublished).

Another constrained least-squares deconvolution method was presented Baygun et al. (1997) to reconstruct the logarithmic pressure derivative defined as $g_d = \frac{\mathrm{d}p_u}{\mathrm{d}\ln(t)} = tg(t)$, where g is the kernel of the convolution integral, and also known as the Bourdet derivative (Bourdet, Whittle, Douglas, & Pirard, 1983). The constraints for this case are the lower bounds on the first few lags of the normalized autocorrelation coefficients of g_d. Furthermore, the Baygun et al. (1997) algorithm was also applied to the deconvolution of the $\mathcal{G}$-function from the p-p convolution of WFT pressure measurements at different spatial locations in the formation.

The constraints of the Baygun et al. (1997) algorithm require some smoothness to g_d. The Cauchy-Schwarz inequality (Luenberger, 1969), given as

$$\frac{\sum\limits_{n=1}^{N_p-1} g_d(n)g_d(n+1)}{\left\{\sum\limits_{n=1}^{N_p-m} [g_d(n)]^2\right\}^{1/2} \left\{\sum\limits_{n=2}^{N_p} [g_d(n)]^2\right\}^{1/2}} \geq \theta_1, \tag{4.47}$$

is imposed to control the smoothness by varying the value of θ_1 between -1 and 1, where N_p is the number of measured p_m and q_m data points and the number of nodes for g_d. Furthermore, the smoothness constraints given by Equation (4.47) from the adjacent values of $g_d(n)$ and $g_d(n+1)$ can be expanded to $g_d(n)$ and $g_d(n+m)$, where $1 \leq m \leq M_1$, M_1 is an integer, and $M_1 \ll N_p$, as

$$\frac{\sum\limits_{n=1}^{N_p-m} g_d(n)g_d(n+m)}{\|g_d(n)\|_2^2} \geq \theta_m \quad \text{for } 1 \leq m \leq M_1, \tag{4.48}$$

where as in θ_1, $-1 < \theta_m < 1$. Equation (4.48) is an *autocorrelation constraint* and in addition, the energy constraint on g_d given as

$$\sum_{n=1}^{N_p} [g_d(n+1) - g_d(n)]^2 \leq \alpha, \tag{4.49}$$

where $0 \leq \alpha$, is applied to the minimization problem given as

$$E(\mathbf{g_d}) = \left\| \mathbf{\Delta p}_m - \mathbf{A}(\mathbf{g_d}) \right\|_2^2, \tag{4.50}$$

where **A** is now different from that given previously because now Equation (4.50) is solved for $\mathbf{g_d}$, the vector of the solution g_d, subject to constraints given by Equations (4.48) and (4.49). This constrained least-squares minimization problem can be solved using a variety of techniques; for details see Baygun et al. (1997).

Most of the earlier constrained least-squares deconvolution algorithms mentioned above, such as Coats et al. (1964), did not mention anything regarding acceptable noise levels in the data. The constrained least-squares deconvolution algorithms of Kuchuk et al. (1990a) and Baygun et al. (1997) worked well with error levels up to a few percent on the rate measurements; and the algorithm with regularization by Zajic and Kuchuk (unpublished) performed better than both of these. It should be stated that the errors in Δp_m were assumed to be small, and it was therefore not included in the least-squares minimization in all previous deconvolution techniques.

In recent years, there have been new techniques on deconvolution presented by von Schroeter et al. (2004), and subsequently extended or revised by other authors (Levitan, 2005; Levitan, Crawford, & Hardwick, 2006; Onur et al., 2008; Pimonov, Ayan, Onur, & Kuchuk, 2009a; Pimonov, Onur, & Kuchuk, 2009b), which appear to be to more robust and reliable in implementation. The principle of all recent deconvolution techniques is based on the technique of von Schroeter et al. (2004). The important characteristics of the technique of von Schroeter et al. (2004) compared to the prior deconvolution techniques are:

1. Using the total least squares (TLS) method, which is a generalization of the earlier least squares (LS) methods, for instance, Kuchuk et al. (1990a);
2. Deconvolving pressure and rate data in terms of the logarithmic pressure derivative $g_d = tg(t)$, where $g(t)$ is the unit-impulse function;
3. Estimation of the reservoir initial pressure if it is unknown;
4. Transforming Equation (4.35) into a nonlinear one to ensure positivity of g_d (the solution); and
5. Including buildup test data or any portion of test data in the deconvolution.

The first two items have been known in the literature. Particularly, the total least squares method (TLS) has been widely used in engineering, physics, and mathematics, for instance see Golub and Van Loan (1993), De Moor (1993), and an unit-impulse response estimation by discrete deconvolution by Van Huffel (2003). The deconvolution of g_d was introduced by Baygun et al. (1997). The innovation of von Schroeter et al. (2004) in deconvolution of pressure and rate data is given by the last three items. Therefore, we will present the technique of von Schroeter et al. (2004) briefly in the following paragraphs. Let us rewrite Equation (4.35) as

$$p_m(t) = p_o - \int_0^t q_m(t - \tau)\, g(\tau)\, \mathrm{d}\tau, \tag{4.51}$$

where both wellbore pressure and flow-rate measurements contain errors (noise) that are defined as $\epsilon_p(t)$ for pressure and $\epsilon_q(t)$ for rate. Let $\xi = \ln(\tau)$ and $z_d(\xi) = \ln[\tau g(\tau)]$, where this is, not strictly speaking, the logarithmic derivative of the constant-rate response of the system (deconvolved pressure). Substituting these two terms in Equation (4.51) yields

$$p_m(t) = p_o - \int_{-\infty}^{\ln t} q_m\left(t - \mathrm{e}^{\xi}\right) \mathrm{e}^{z_d(\xi)} \mathrm{d}\xi. \tag{4.52}$$

This change of the integration variable and the transformation ensure that the deconvolved influence function $g_d(t) = \mathrm{e}^{z_d(\xi)}$ remains positive, however it comes at the cost of losing the linearity of the convolution equation and thus will require iterative solution techniques. Furthermore, in order to account for uncertainties in measured rate, the objective function has to be minimized including flow rate during the deconvolution process to obtain a better match with the measured pressure. This naturally calls for penalizing the difference between measured and reconstructed (computed) rates in the objective function.

Finally, the solution of Equation (4.52) is expected to exhibit a certain degree of smoothness to satisfy the conditions given in Equation (4.22) to Equation (4.25) for single-phase flow of a slightly compressible fluid with constant compressibility and viscosity. In other words, the deconvolved response should not have rapid fluctuations. Therefore, a smoothness constraint is applied on the deconvolved response by means of adding a weighted measure of response curvature to the objective function.

The combination of all these aspects for solving Equation (4.52) was formulated by von Schroeter et al. (2004) as a nonlinear total least squares minimization problem and can be written (Kuchuk, 2009b) as

$$\begin{aligned} E(p_o, \mathbf{z_d}, \mathbf{q}_c) = \frac{1}{2} \Big\{ &\left\| p_o\mathbf{e} - \mathbf{p}_m - \mathbf{C}(\mathbf{z_d})\mathbf{q}_c \right\|_2^2 + \nu \left\| \mathbf{q}_c - \mathbf{q}_m \right\|_2^2 \\ &+ \lambda \left\| \mathbf{D}\mathbf{z_d} - \mathbf{k} \right\|_2^2 \Big\}, \end{aligned} \tag{4.53}$$

where E is the objective function; p_o is the unknown initial pressure;

$\mathbf{z_d}$ is the N_g dimensional vector with the elements $z_d(t_i) = \ln[t_i g(t_i)]$; $i = 0, 1, 2, \ldots, N_g$ (the number of nodes of z_d); $\mathbf{e}$ is an N_p dimensional unit vector; N_p is number of measured pressure points; $\mathbf{p}_m$ is the vector of pressure with dimension N_p; $\mathbf{q}_m$ and $\mathbf{q}_c$ are the vectors of measured and computed flow rate with dimension N_q; $\mathbf{C}$ is the $N_p \times N_q$ dimensional matrix with elements defined as $\mathbf{C}_{ij}(z) = \int_{-\infty}^{\ln t} \theta_j [t_i - \exp(\xi)] \exp[z_d(\xi)] d\xi$; θ is a binary operator given by Equation (3.30); $\mathbf{D}$ is the $N_g - 1 \times N_g$ dimensional constant matrix in the curvature measure (von Schroeter et al., 2004); and $\mathbf{k}$ is the $N_g - 1$ dimensional vector in the curvature measure as given by $\mathbf{k}^T = (1, 0, \ldots, 0)$. Notice that a factor of 1/2 is added to the right-hand side of Equation (4.53) for convenience; i.e., to indicate that errors in pressure and rate are of Gaussian distributions. Basically it has no effect on the values of p_o, $\mathbf{z_d}$, and $\mathbf{q}_c$ that minimize Equation (4.53), though the constant 1/2 is not present in the objective function originally presented by von Schroeter et al. (2004).

The elements of the curvature measure matrix $\mathbf{D}$ for the first row ($i = 1$) and the other rows ($i = 2, 3, \ldots, N_g - 1$) for logarithmically uniformly spaced nodes are given by von Schroeter et al. (2004) as

$$d_{1,j} = \begin{cases} -\dfrac{1}{\Delta\xi} & \text{if } j = 1 \\ \dfrac{1}{\Delta\xi} & \text{if } j = 2 \\ 0 & \text{if } j = 3, 4, \ldots, N_g \end{cases} \tag{4.54}$$

and

$$d_{i,j} = \begin{cases} \dfrac{1}{\Delta\xi} & \text{if } j = i - 1 \\ -\dfrac{2}{\Delta\xi} & \text{if } j = i \\ \dfrac{1}{\Delta\xi} & \text{if } j = i + 1 \\ 0 & \text{otherwise.} \end{cases} \tag{4.55}$$

It should be noticed that the last term in the right-hand side of Equation (4.53) represents the curvature of the solution, namely, on the sum of squares of the angles between the piecewise linear segments joined at logarithmically uniformly spaced nodes of ξ_l values generated from

$$\xi_l = \xi_1 + \frac{l - 1}{N_g - 1} \left(\xi_{N_g} - \xi_1\right) \quad \text{for } l = 1, 2, \ldots, N_g \tag{4.56}$$

with a uniform node increment such that $\Delta\xi = \xi_l - \xi_{l-1}$ for $l = 1, 2, \ldots, N_g - 1$. The first node ξ_1 is usually chosen as less than or equal to the minimum elapsed time of any point in the pressure signal. The techniques of von Schroeter et al. (2004) assume ξ_1 by default, e.g., the minimum time $t_{\min} = 10^{-3}$ hr. However, if $t_{\min} = 10^{-3}$ hr is greater than the time of the first point in the pressure signal, then its value can be automatically changed to this minimal time. The last node ξ_{N_g} is chosen as the total test duration.

Although different interpolation schemes could be used for the deconvolved response z_d and the flow rate, for simplicity, as used by von Schroeter et al. (2004), the values of $z_d(\xi)$ are interpolated linearly between values $(z_d)_k = z_d(\tau_k)$ using

$$z_d(\xi) = \varphi_k + \xi \nu_k, \quad \text{for } \xi_{k-1} \leq \xi \leq \xi_k, k = 1, 2, \ldots, N_g, \tag{4.57}$$

where $\nu_1 = 1$ to model the unit-slope wellbore storage period and at the first node. For the other nodes,

$$\nu_k = \frac{(z_d)_k - (z_d)_{k-1}}{\xi_k - \xi_{k-1}}, \quad \text{for } k = 2, 3, \ldots, N_g, \tag{4.58}$$

and

$$\varphi_k = (z_d)_k - \nu_k \xi_k, \quad \text{for } k = 1, 2, \ldots, N_g. \tag{4.59}$$

The algorithm considers that flow rates $q_j = q(t_j)$, $j = 1, \ldots, M$, are reported as stepwise constants over time. In this case, if q_j is the rate over a time interval $a_j \leq t \leq b_j$, then the rate interpolation scheme for $q(t)$ can be written as

$$q(t) = \sum_{j=1}^{M} q_j \theta_j(t), \tag{4.60}$$

where θ is a binary operator such that

$$\theta_j(t) = \begin{cases} 1, & \text{if } a_j \leq t \leq b_j \\ 0, & \text{otherwise.} \end{cases} \tag{4.61}$$

It should be pointed out that M denotes the total number of flow-rate steps for the entire test, and can be equal to or greater than N_q, which denotes the number of unknown flow-rate steps.

The terms λ and ν in the right-hand side of Equation (4.53) are the relative error weight factor and regularization parameter, respectively, and given by von Schroeter et al. (2004) as

$$\nu = \frac{N_q}{N_p} \frac{||\Delta \mathbf{p}_m||_2^2}{||\mathbf{q}_m||_2^2} \tag{4.62}$$

and

$$\lambda = \frac{1}{N_p} \frac{||\Delta \mathbf{p}_m||_2^2}{||\mathbf{D}||_F^2}, \tag{4.63}$$

where $||\mathbf{D}||_F$ denotes the Frobenius operator of the matrix $\mathbf{D}$ and is given by

$$||\mathbf{D}||_F = \sqrt{\sum_{i=1}^{N_g-1} \sum_{j=1}^{N_g} |d_{i,j}|^2}. \tag{4.64}$$

The objective function given by Equation (4.53) considers three error measure terms that can be seen as the pressure match, rate match, and curvature measure. ν and λ are weighting factors that are used to enforce either a better fit on the measured pressure or flow-rates, or to ensure a smoother response. Increasing ν would force a better match on the measured flow-rates while λ would force the response to be smoother.

Inputs for Equation (4.53) can be categorized into two groups: (1) Measured data (pressure and flow-rates) and (2) Weighting factors used to obtain a better fit on the pressure or rates or to obtain a smooth deconvolved derivative response. Unknowns to be solved for are the transformed deconvolved derivative $\mathbf{z}_d$, computed (reconstructed) flow rate $\mathbf{q}_c$, and the initial reservoir pressure p_o. Other parameters are intermediate matrices or vectors. When implementing an algorithm to solve the minimization problem given by Equation (4.53), the initial pressure p_o or rates can either be estimated as part of the deconvolution process or are assumed to be correct and fixed during the deconvolution process.

The above deconvolution formulation raises a few questions:

1. The discretization problem in the vicinity of $-\infty$, which is the lower limit of the integral given in Equation (4.52), for the values of z_d,
2. How the first node should be modeled, particularly if there is negligible or no wellbore storage effect at the early times, and
3. How to set the deconvolution algorithm parameters (ν, λ, and $\mathbf{k}$).

With these concerns, Levitan (2005) proposed a slightly different formulation of the objective function, by incorporating some estimates of noise levels or uncertainty in pressure and rate measurements to replace the weighting factors by some physical parameters, and by replacing the constraint on the first node by taking the first node out of the formulation. The convolution integral given by Equation (4.52) with the modification of Levitan (2005) can be written as

$$p_m(t) = p_o - q_m(t)p_u(t_1) - \int_{\xi_1}^{\ln t} q_m\left(t - e^{\xi}\right)e^{z_d(\xi)}d\xi, \qquad (4.65)$$

then the objective function of Levitan (2005) (Onur et al., 2008) can be expressed as

$$E(p_o, p_u(t_1), \mathbf{z}_d, \mathbf{q}_c) = \frac{1}{2}\left\{\left\|\frac{p_o\mathbf{e} - \mathbf{p}_m - p_u(t_1)\tilde{\mathbf{q}}_c - \tilde{\mathbf{C}}(\mathbf{z}_d)\mathbf{q}_c}{\varsigma_p}\right\|_2^2 + \left\|\frac{\mathbf{q}_c - \mathbf{q}_m}{\varsigma_q}\right\|_2^2 + \left\|\frac{\mathbf{D}\mathbf{z}_d}{\varsigma_c}\right\|_2^2\right\}, \qquad (4.66)$$

where ς_p, ς_q, and ς_c denote *a priori* error bounds on the pressure data, rate data, and acceptable curvature between points. $\tilde{\mathbf{q}}_c$ denotes the flow-rates at the time of each pressure measurement. If the flow rate is not measured at each pressure point, then they are interpolated from $\mathbf{q}_c$. The matrix $\tilde{\mathbf{C}}$ in Equation (4.66) is low slightly different from that is given in Equation (4.53) and given as $\tilde{\mathbf{C}}_{ij}(z) = \int_{\ln t}^{\xi_1} \theta_j[t_i - \exp(\xi)]\exp[z_d(\xi)]d\xi$, where θ is a binary operator given by Equation (3.30).

The technique of Levitan (2005) given by Equation (4.66) has some advantages because some of the minimization parameters have a better physical sense, but it does require some *a priori* knowledge of measurement uncertainties in pressure and rate data. The first node unit-rate pressure $p_u(t_1)$ in this formulation becomes an independent parameter that can be optimized without any constraint on the slope of the response before the first node.

This of course has direct implications on the solution algorithm and in fact the techniques of von Schroeter et al. (2004) and Levitan (2005) differ in how the problem is solved from an algorithmic standpoint, but more importantly, it has an impact the stability of the solution. At the expense of possibly getting a negative unit-rate pressure change at the first node in some cases (Onur et al., 2008), the solution of Levitan (2005) does not make any assumption regarding the slope of the response before the first node. The use of physical numbers as inputs for expected measurements errors definitely adds to this implementation by allowing the interpreter performing the deconvolution to use meaningful weight factors to the error measure, but the ς parameters from Levitan (2005) can be easily related to the ν and λ parameters by von Schroeter et al. (2004). A comparison of algorithms by von Schroeter et al. (2004) and Levitan (2005) has been given by Onur et al. (2008), in which the differences between these two algorithms and their respective benefits were presented in detail. It is also important to point out that *a priori* estimates of error bounds are not straightforward to determine. Pressure measurement uncertainty can easily be related to the resolution of the pressure gauge used, but in reality the downhole wellbore environment has other sources of noise that can affect measured data. Furthermore, uncertainties in flow-rate measurements are very difficult to quantify.

In general, it can be stated that all pressure and rate measurements are unlikely to exhibit the same level of uncertainty throughout a data set, and thus two approaches should be considered:

1. Using a consistent quality subset data chosen from all available data sets, for instance considering only buildup pressure data for deconvolution purposes, as proposed by Levitan (2005), where a single measurement error bound should apply, although this might not be true of flow-rate data; and
2. Performing a more thorough assessment of data quality and applying a different uncertainty error bound to each subset of data sets.

The deconvolution algorithms presented by Pimonov et al. (2009a; 2009b) have incorporated the above two points by adding a measure of confidence in individual measurements, where the first algorithm is based on the method of von Schroeter et al. (2004), Equation (4.52), and the second one is based on the method of Levitan, Equation (4.65). The objective function given by Pimonov et al. (2009b), based on the encoding of Equation (4.52), can be expressed as

$$E(p_o, \mathbf{z}_d, \mathbf{q}_c) = \frac{1}{2}\left\{\sum_{i=1}^{N_p}(w_p)_i^2\left[\frac{p_o - (p_m)_i - \Delta p_c(t_i)}{\sigma_p}\right]^2\right.$$

$$\left. + \sum_{j=1}^{N_q}(w_q)_j^2\left[\frac{(q_m)_j - (q_c)_j}{\sigma_q}\right]^2 + \sum_{k=1}^{N_g-1}(w_c)_k^2\left[\frac{(\varrho_c)_k}{\sigma_c}\right]^2\right\}, \quad (4.67)$$

where Δp_c is to be computed from Equation (4.52) and q_c is the computed (reconstructed) flow rate. The curvature constraints $(\varrho_c)_k$ in Equation (4.67) is based on the the sum of the squares of the angles between the piecewise linear segments joined at nodes (von Schroeter et al., 2004) and computed from

$$(\varrho_c)_1 = \frac{(z_d)_2 - (z_d)_1}{\Delta\xi} - 1 \quad (4.68)$$

and

$$(\varrho_c)_1 = \frac{(z_d)_{k+1} - 2(z_d)_k + (z_d)_{k-1}}{\Delta\xi}$$
$$\text{for } k = 2, 3, \ldots, N_g - 1, \quad (4.69)$$

where $\Delta\xi = \xi_2 - \xi_1$ is the node increment of equally spaced grid points ξ_k.

The parameter σ_p denotes the standard deviation of error in the measured pressure, and $0 \leq (w_p)_i \leq 1$ denotes the weight associated with the pressure measurements $(p_m)_i$. The value of $(w_p)_i$ depends on a degree of trust in the quality of measurement for pressure point $(p_m)_i$. If the pressure $(p_m)_i$ is measured with the highest accuracy then the value $(w_p)_i = 1$. Similarly, if the pressure point $(p_m)_i$ is measured with the low accuracy then the value $(w_p)_i = 0$. The remaining $(w_p)_i$ are distributed between 0 and 1 inclusive. In much the same way, σ_q denotes the standard deviation of error in measured (or allocated) rate step $(q_m)_j$ and $0 \leq (w_q)_j \leq 1$ denotes the weight associated with the flow-rate step $(q_m)_j$. Weights for rates are chosen in accordance with the same rule as for pressure points. The parameter $(\varrho_c)_k$ in Equation (4.67) is the curvature constraint, and σ_c denotes the standard deviation of the curvature constraint. This value should provide a small degree of regularization and at the same time not overconstrain the problem and not create a significant bias. Most of our deconvolution applications to real and synthetic tests indicate that the value $\sigma_c = 0.05$ works well. However, it sometimes needs to be changed for deconvolution of some pressure-rate data.

Equation (4.67) can also be expressed by using a weighted Euclidean norm as

$$E(p_o, \mathbf{z_d}, \mathbf{q}_c) = \frac{1}{2}\left\{\left\|\mathbf{W}_p[p_o\mathbf{e} - \mathbf{p}_m - \mathbf{C}(\mathbf{z}_d)\mathbf{q}_c]\right\|_2^2 + \left\|\mathbf{W}_q(\mathbf{q}_c - \mathbf{q}_m)\right\|_2^2 \right.$$
$$\left. + \|\mathbf{W}_c(\mathbf{D}\mathbf{z}_d - \mathbf{k})\|_2^2\right\}, \tag{4.70}$$

where $\mathbf{W}_p$, $\mathbf{W}_q$, and $\mathbf{W}_c$ are the diagonal weighting matrices of appropriate sizes for pressure, rate, and curvature, respectively. Specifically, $\mathbf{W}_p$ is an $N_p \times N_p$ diagonal weighting matrix of pressure measurements given as

$$\mathbf{W}_p = \frac{1}{\sigma_p}\begin{bmatrix} (w_p)_1 & 0 & \cdots & 0 \\ 0 & (w_p)_2 & \cdots & 0 \\ \cdots & \cdots & \cdots & \cdots \\ 0 & \cdots & 0 & (w_p)_{N_p} \end{bmatrix}, \tag{4.71}$$

$\mathbf{W}_q$ is an $N_q \times N_q$ diagonal weighting matrix of flow-rate measurements given as

$$\mathbf{W}_q = \frac{1}{\sigma_q}\begin{bmatrix} (w_q)_1 & 0 & \cdots & 0 \\ 0 & (w_q)_2 & \cdots & 0 \\ \cdots & \cdots & \cdots & \cdots \\ 0 & \cdots & 0 & (w_q)_{N_q} \end{bmatrix}, \tag{4.72}$$

and finally $\mathbf{W}_c$ is an $(N_g - 1) \times (N_g - 1)$ diagonal weighting matrix of curvature constraints given as

$$\mathbf{W}_c = \frac{1}{\sigma_c}\begin{bmatrix} (w_c)_1 & 0 & \cdots & 0 \\ 0 & (w_c)_2 & \cdots & 0 \\ \cdots & \cdots & \cdots & \cdots \\ 0 & \cdots & 0 & (w_c)_{N_g-1} \end{bmatrix}. \tag{4.73}$$

In the deconvolution applications considered here, unless stated otherwise, all weight factors of curvature constraints are equal to unity, i.e., $(w_c)_k = 1$ for $k = 1, 2, \ldots, N_g - 1$, and hence Equation (4.73) can be expressed as $\mathbf{W}_c = \frac{1}{\sigma_c}\mathbf{I}$, where $\mathbf{I}$ is an $(N_g - 1) \times (N_g - 1)$ identity matrix.

If we incorporate a measure of confidence in individual measurements for different data types in the technique of Pimonov et al. (2009a; 2009b), based on the convolution formula of Levitan (2005) (see Equation (4.65)), to eliminate the unit-slope assumption at and before the first node and introduce $p_u(t_1)$ (see Equation (4.65)), then the appropriate objective function becomes

$$E(p_o, p_u(t_1), \mathbf{z}_d, \mathbf{q}_c) = \frac{1}{2}\left\{\sum_{i=1}^{N_p}(w_p)_i^2\left[\frac{p_o - (p_m)_i - \Delta p_c(t_i)}{\sigma_p}\right]^2 + \sum_{j=1}^{N_q}(w_q)_j^2\left[\frac{(q_m)_j - (q_c)_j}{\sigma_q}\right]^2 + \sum_{k=2}^{N_g-1}(w_c)_k^2\left[\frac{(\varrho_c)_k}{\sigma_c}\right]^2\right\}, \tag{4.74}$$

where Δp_c is to be computed from Equation (4.65). Also note that now there are $N_p - 2$ curvature constraints (see the last summation in the right-hand side of Equation (4.74)). Equation (4.74) can be expressed by using the Euclidean norm as

$$E(p_o, p_u(t_1), \mathbf{z}_d, \mathbf{q}_c) = \frac{1}{2}\left\{\left\|\mathbf{W}_p[p_o\mathbf{e} - \mathbf{p}_m - p_u(t_1)\tilde{\mathbf{q}}_c - \tilde{\mathbf{C}}(\mathbf{z}_d)\mathbf{q}_c]\right\|_2^2 + \left\|\mathbf{W}_q(\mathbf{q}_c - \mathbf{q}_m)\right\|_2^2 + \left\|\mathbf{W}_c(\tilde{\mathbf{D}}\mathbf{z}_d)\right\|_2^2\right\}, \tag{4.75}$$

where $\tilde{\mathbf{C}}$ given above, an $N_p \times N_q$ matrix function of the vector $\mathbf{z}_d$, denotes the convolution operation for the convolution equation of Levitan (2005) (see Equation (4.65)), and $\tilde{\mathbf{D}}$ is an $(N_g - 2) \times N_g$ matrix for the curvature measure, which can be explicitly written for uniformly spaced nodes z_d as

$$\tilde{\mathbf{D}} = \frac{1}{\Delta\xi}\begin{bmatrix} 1 & -2 & 1 & 0 & \cdots & \cdots & 0 \\ 0 & 1 & -2 & 1 & 0 & \cdots & 0 \\ 0 & 0 & \ddots & \ddots & \ddots & \cdots & 0 \\ \vdots & \vdots & \ddots & \ddots & \ddots & & \vdots \\ 0 & \ldots & \ldots & 0 & 1 & -2 & 1 \end{bmatrix}. \tag{4.76}$$

$\mathbf{W}_p$ and $\mathbf{W}_q$ are given by Equations (4.71) and (4.72), and $\mathbf{W}_c$, given by Equation (4.73), becomes

$$\mathbf{W}_c = \frac{1}{\sigma_c}\begin{bmatrix} (w_c)_1 & 0 & \cdots & 0 \\ 0 & (w_c)_2 & \cdots & 0 \\ \cdots & \cdots & \cdots & \cdots \\ 0 & \cdots & 0 & (w_c)_{N_g-2} \end{bmatrix}. \tag{4.77}$$

As stated above, $(w_c)_k = 1$ for $k = 1, 2, \ldots, N_g - 2$, and $\mathbf{W}_c = \frac{1}{\sigma_c}\mathbf{I}$, where $\mathbf{I}$ is an $(N_g - 2) \times (N_g - 2)$ identity matrix.

In general, weight factors w (assuming values between zero and unity) denote the confidence level associated with each measurement, treated independently from the measurement intrinsic error bound. These factors are defined independently for each data type (pressure, rate, etc.) but also individual points or subsets of points in a given data type. Using factors w, for instance, one can assign a lower weight or higher error bound (i.e., small values of w) for drawdown pressure measurements (because drawdown measurements are normally expected to exhibit higher noise levels than buildup measurements). From different implementations of deconvolution algorithms, the selection of error bounds are generally quite transparent. Using a pressure measurement error bound of 0.01 psi for quartz gauges has been proposed by Levitan (2005), Pimonov et al. (2009a), and Pimonov et al. (2009b); as stated above it could be much higher in a downhole wellbore environment (discussed in Chapter 1). Flow-rate error bounds have been proposed in the range of 0.1-1 B/D. While this range is reasonable for WFTs, it is very optimistic for the performance of flow measurement devices used for well testing. A reasonable range for well testing is about 10-100 B/D depending on fluid type, wellbore deviation, wellbore conditions, tools used, the magnitude of the flow rate, etc. Generally, the number of pressure data points is higher than the number of rate data points, particularly when the rate is measured by the surface production-test tank by at least 100 orders of magnitude. However, when production logging tools and/or new generation multiphase flow-rate devices are used, the number of data points for pressure and rate could be the same or at least of a similar order of magnitude. The pressure sampling frequency for most commercial well testing pressure gauges is about one second (but it could range from 0.3 to 10 seconds), while flow rates are measured at best hourly at surface production-test tanks. The sampling frequency of production logging and multiphase flow-rate devices can be as low as one second.

For most well tests, surface production-test tanks are still used commonly, having far fewer flow-rate measurement points compared to pressure measurements. This creates a favorable stability condition for deconvolution, in which the system remains over-determined. Having much more pressure data points than rate data points creates a bias in the objective function, giving more importance to matching the pressure measurements. In other words, as the error function contains far more pressure measurements than rate measurements then more weight is given to the pressure match in nonlinear deconvolution. This in turn requires stringent error bounds on the rate data. In practice, this could be accounted for by including a normalization of the weighting coefficients depending on the number of pressure and rate measurements. It must be emphasized that, while the errors in pressure measurements can generally be considered as random noise, errors in rate measurements can be more consistent and would thus introduce a bias in the deconvolved response. Having flow-rate measurements systematically in error by, say +5%, would simply result in a shift of the deconvolved response by −5%, in much the same way that rate measurement errors in a single drawdown case would shift the conventional multirate derivative by such an amount.

With respect to the error bound on the response curvature, or more precisely an acceptable amount of curvature, various authors, as reported by Pimonov et al. (2009a; 2009b), have found that a value of 0.05 seems to provide acceptable levels of response smoothness. This value obviously depends on the number of nodes in comparison to the number of pressure points and of rate measurements, as well as on the actual reservoir response, but appears to be a good generic starting point. Except in specific cases, the use of variable weights on the curvature constraint is clearly not of obvious interest and is of lesser use than its application to pressure and rate data.

Different formulations of deconvolution clearly provide different degrees of control and have their own advantages and disadvantages in terms of stability or quality of the response. However, as shown by Onur et al. (2008), the results from different formulations tend not to be very different. We therefore use the deconvolution algorithms of Pimonov et al. (2009a; 2009b) for deconvolving pressure-rate and pressure-pressure data sets. Next, the focus will be on the practical implementation of deconvolution algorithms as well as their application to formation and well testing pressure transient data.

4.6. PRACTICALITIES OF DECONVOLUTION

In most applications, performing deconvolution for formation and well testing pressure transient data sets is not a fully-automated operation and requires a careful selection of control parameters. When we are performing deconvolution using a program or software, the following specified series of actions should be considered for a successful application of deconvolution for pressure and rate transient data:

- Data selection;
  - Considering pressure data for the entire test duration,
  - Considering pressure data recorded only during buildups,
  - Considering only consistent buildups,
  - Considering, independently, several buildups,
  - Simplifying the rate history before the pressure data subset to be used, and
  - Excluding some production periods prior to a long shut-in after which the reservoir pressure distribution has become uniform.
- Deconvolution parameters selection;
  - Applying some default error bounds,
  - Refining default parameters, and
  - Using constant error bound values and weighting factors or making those specific to data subsets.
- Use of constraints or fixed inputs;
  - Setting the initial pressure at a fixed value, and
  - Setting the flow-rates at fixed values.

Most of above decisions to be made are neither obvious nor necessarily straightforward. However some practical points presented by Levitan et al. (2006) and others can be used as guidelines towards defining a workflow for pressure-rate and pressure-pressure deconvolution implementations.

4.6.1. Data selection

In order to perform deconvolution for a given data set, the linearity of the convolution integral given by Equation (4.5) must be maintained. This implies that the kernel function $g(t)$ remains the same and thus that its parameters such as thickness of formation, permeability, porosity, skin, etc. remain constant (independent of time and pressure) during the test. There are possible instances where the linearity of the convolution integral could be violated (this will be called *inconsistent data or inconsistency*):

- During clean-up sequences, where the skin factor caused by mud filtrate invasion and completion fluids decreases as the formation is flushed by the production rate;

- Over long periods of production the wellbore skin factor may change due to calcium carbonate or silica scaling, asphaltene precipitation, fines movements, etc;
- Workover or stimulation (fracturing or perforation) operations over time may change the effective thickness of flowing interval and/or formation of damage skin;
- Wellbore storage coefficients for the drawdown and buildup tests may not be the same for the deconvolution period, particularly if a downhole shut-in tool is used for the buildup tests and/or downhole flow rates are not measured;
- The fluid compressibility and viscosity may change due to its pressure dependence: such as gas, gas condensate, and light oil. Using the real-gas potential (Al-Hussainy, Ramey, & Crawford, 1966), the kernel function $g(t)$ in Equation (4.5) can be linearized;
- Change in the kernel function parameters due to formation compaction;
- In the presence of multiphase flow in the vicinity of the wellbore due to gas evolution as the flowing wellbore pressure goes below the bubble point pressure;
- Moving fluid banks or aquifer water influx in the reservoir during the deconvolution period;
- etc. ...

In fact, it would be very optimistic to assume that the superposition principle holds in earnest for the entire duration of a given pressure transient test, particularly for very long production tests.

However, while pressure-rate deconvolution requires a complete production rate history, only a subset of the pressure data can be used to obtain a deconvolved response corresponding to the entire duration of a test. This is a key feature of pressure-rate deconvolution that can easily be explained from an algorithmic standpoint because pressure is only used as a series of measurements to be matched to a given model, even though that model depends on the complete production history.

What might be less straightforward to identify is how, for instance, a single buildup can bear information about the entire flow sequence that would allow for the determination of a deconvolved response over a time interval potentially much longer than that of this buildup.

We will illustrate this with a synthetic example called Example A, as shown in Figure 4.4, where the model is a limited-entry vertical well in an infinite reservoir. It is well known that the buildup response shown in Figure 4.4 is strongly affected by the final rate (2nd DD), but also is affected to some extent by the earlier production period (1st DD). The deconvolved response can therefore be obtained from 0 to t_3 (see Figure 4.4), representing the entire test duration by using only a subset of the pressure data.

For test Example A, Figure 4.5 presents the buildup (represented by BU), model (drawdown for the entire test duration), and deconvolved derivatives.

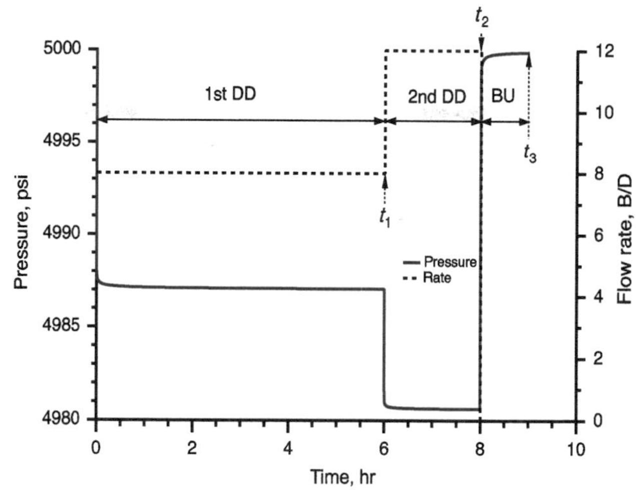

Figure 4.4 Pressure and rate history used for the synthetic Example A.

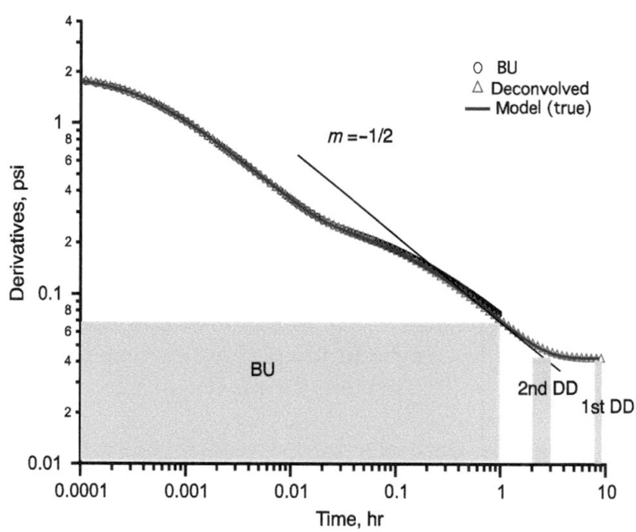

Figure 4.5 The true and buildup (BU), and deconvolved derivatives corresponding to a 3-rate flow sequence: first drawdown, second high-rate drawdown, and short buildup for Example A.

Overlaid on this graph are the time periods represented, via superposition, in the data. This representation (Figure 4.5) clearly highlights one of the key benefits of the logarithmic formulation of the convolution integral presented in Equation (4.52). As the deconvolved response is presented at a logarithmic scale, even significant gaps in terms of available response in linear terms become small at such a scale. In other terms, most of the response interval can be defined based on the pressure response from a single buildup, and in the short intervals where information is missing the smoothness constraint imposed in terms of curvature does ensure that a smooth and interpretable response is obtained, faithful to the slow nature of diffusive events.

Therefore, the deconvolution process makes it possible to use only a few selected consistent data subsets to obtain a deconvolved response. This could be the case of a single buildup, as shown above, or of only a selection of several high-quality buildups, or of the last drawdown sequence followed by the ensuing buildup. Thus, while an entire test is unlikely to exhibit a consistent response following the convolution integral, using only parts of the pressure data satisfying this condition makes deconvolution an attractive technique. While this property (obtaining reservoir information for the entire duration of the test) makes deconvolution a practical tool for interpretation, it should not be inferred that by extension of this property a good deconvolved response could always be obtained from an extremely short buildup. The information content and quality of the data are still critical to make the deconvolution workable and provide representative results. Even in the case presented in Figure 4.5, significant constraints needed to be applied to obtain a good result, such as imposing both flow rates and initial pressure as well as forcing an extremely high quality match of the pressure data that was not significantly affected by any noise.

For this reason, selecting which flow periods to consider for the deconvolution process from available pressure data is a critical first step in ensuring that representative deconvolution results are obtained. The simplest logical way to approach that challenge is obvious: using as much consistent data as possible. Levitan (2005) suggested using only buildup portions of the test, which is understandable as buildup test data usually tend to have significantly higher data quality (much less noise and the rate approaches zero rapidly) than drawdown data. In addition, we can expect the noise levels for buildups to remain similar, making the process of selecting a single error bound far more realistic.

In order to assess the data selection process in deconvolution, we use a synthetic multirate test (Example B) that consists of multiple drawdowns with three buildup periods for a fully penetrated vertical well. The formation and fluid properties for this example are: $h = 100$ ft, $k = 30$ md, $\phi = 0.20$, $\mu = 1$ cp, $c_t = 10^{-5}$ psi^{-1}, $r_w = 0.3$ ft, $p_o = 5000$ psi. The reservoir model assumes having two parallel no-flow faults at 300 ft and 1200 ft away from the well. It should be stated that only the closest boundary (300 ft) can be identified from individual buildup tests. As shown in Figure 4.6, in which both pressure and flow-rate data are presented, this test consists of three main periods: an initial short flow period followed by a buildup, an assumed clean-up period (multiple drawdowns or production periods) during which the clean-up occurs gradually (the skin value decreases) and finally a flow-after-flow test followed by a long buildup. The reservoir model considered here

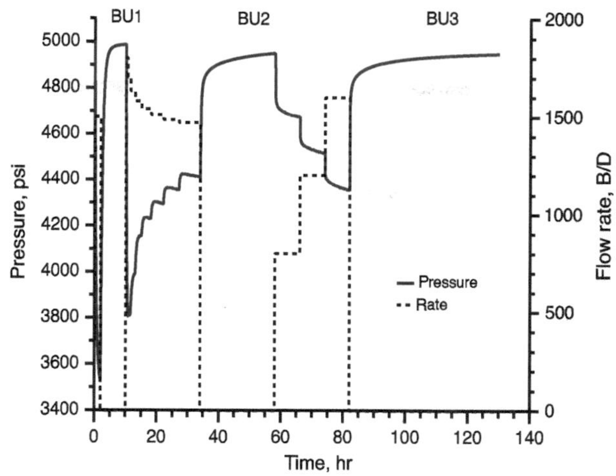

Figure 4.6 Pressure and rate history used for the synthetic Example B.

assumes a skin factor of 15 during the first drawdown, gradually decreasing to zero for the last two buildups. In the model, a wellbore storage coefficient 0.02 B/psi is used for both drawdown and buildup tests (surface shut-in is assumed), except for the last (third) buildup test for which it is assumed that a downhole shut-in is used with a 0.001 B/psi wellbore storage coefficient. Random noises are added to both drawdown (± 2 psi) and buildup (± 0.1 psi) data. These contaminated pressures and flow rates are called measured (input) pressures and flow rates as if they were measured.

In this example, the linearity of the convolution equation Equation (4.5) is deliberately violated (inconsistency in data) because of the change in skin and wellbore storage values during the test. As stated above, for the first two buildup tests it is assumed that surface shut-in is used while for the last one downhole shut-in is used. Furthermore, it is assumed that the filtrate and completion fluid invasion created a significant skin factor (15) for the first drawdown, which gradually decreases during the production periods. For these reasons, the pressure derivatives of the three buildup periods are different, as shown in Figure 4.7.

Due to changes in the skin and wellbore storage of the model among the three build-up periods, only the last buildup (BU3) will be deconvolved initially. A very good match is obtained for the pressure history, where the flow rates are essentially left unaltered by the deconvolution process. The deconvolved pressure derivative is presented in Figure 4.8 and matches very well with the derivative of BU3, but does not match the derivative of the true model, as shown in the figure. In fact, the deconvolved derivative only maintains the trend of the BU3 derivative until the end of the BU3, and then goes up slightly. Both BU3 and deconvolved derivatives shown in Figure 4.8 exhibit a signature of a single no-flow fault, where the second infinite-acting

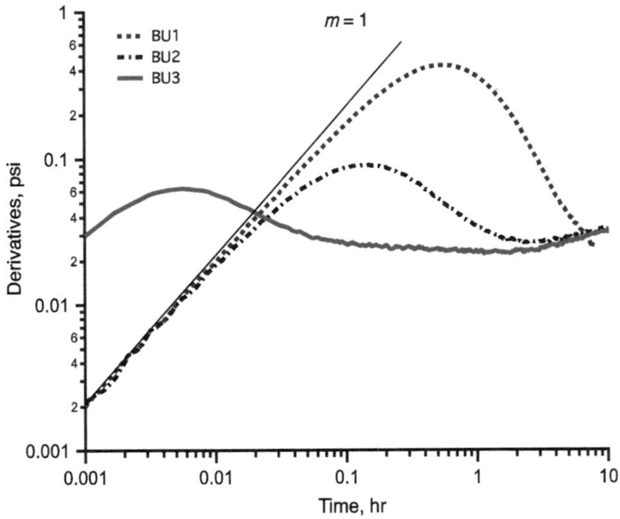

Figure 4.7 Comparison of the conventional derivatives of the three buildup tests of Example B (rate-normalized to the unit-rate).

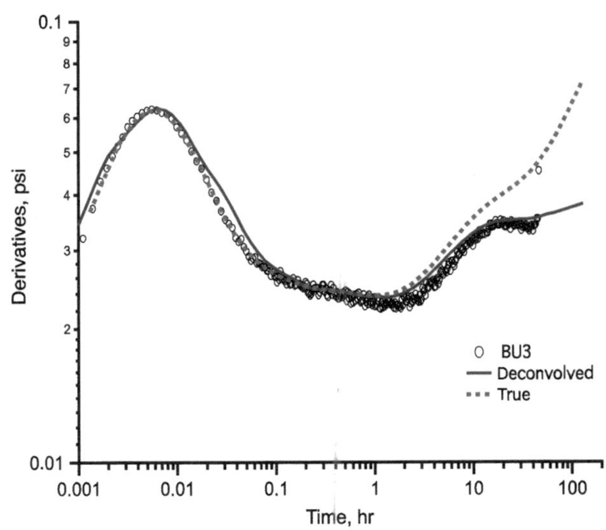

Figure 4.8 The true and buildup (BU) derivatives, and the deconvolved derivative for which only the last buildup data are used without added constraints.

flow regime is apparently due to a half-infinite system. The semilog slope (derivatives towards the end of BU3) becomes almost twice that of the first infinite-acting period. Without any other data, one may conclude from both BU3 and deconvolved derivatives that we have an infinite reservoir with a

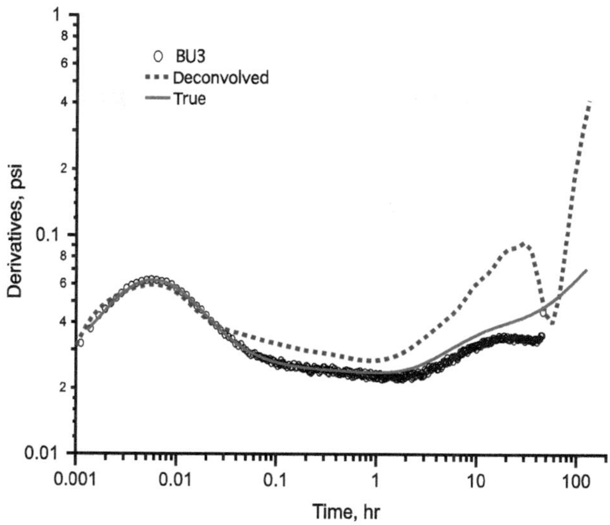

Figure 4.9 The true and buildup (BU3) derivatives, and the deconvolved derivative for which only the last two buildups' data are used without added constraints.

single no-flow fault. As can be seen from the derivative of the true model (Figure 4.8), these are distorted signatures due to superposition effects but not the true reservoir behavior.

The behavior discussed in the above paragraph, whereby the deconvolved derivative only extends the conventional derivative, is an artifact of the curvature constraint used for this deconvolution when limited data are used. In order to minimize the response curvature, the algorithm tends to extend the response without reflecting slope changes that might be present in the true response. Hence, this may be a key limitation of using a single buildup period without additional constraints.

In order to include more data in deconvolution, such as the data pertinent to the last two buildups (BU2 and BU3) can lead to instabilities in the derivative response because the linearity of the convolution integral Equation (4.5) was violated by a change of the wellbore storage in the model between those two periods. This is illustrated in Figure 4.9, where the deconvolved derivative obtained by using both BU2 and BU3 pressure data exhibits an oscillatory behavior at the late times, even creating a slight distortion of the first infinite-acting flow regime.

The deconvolved derivative is greatly improved by using a part of the last flow period along with the last buildup (BU3), as shown in Figure 4.10, although it is still slightly different from the true model behavior. In this case, the deconvolution process requires assigning different error bounds to drawdown and buildup pressure data because each data set has considerably different noise characteristics and levels.

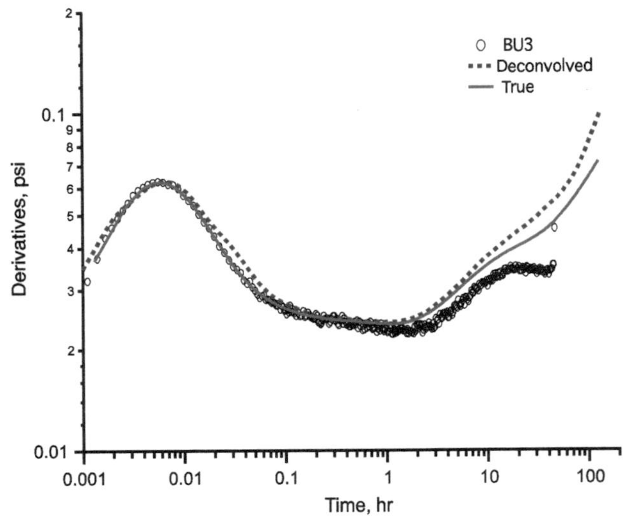

Figure 4.10 The true and buildup (BU3) derivatives, and the deconvolved derivative for which only the last flow period and buildup data are used without added constraints.

The above three cases shown in Figures 4.8–4.10 clearly indicate that it is important to determine what subsets of data among production and buildup periods should be used for deconvolution. Although it is assumed that flow rates are known exactly (without measurement noise) in these three deconvolution cases, deconvolution of a single buildup period without using constraints could be still uncertain and does fail if the data are inconsistent. As shown in Figure 4.10, combining consistent parts of drawdown data with a buildup period can considerably improve deconvolution results. However, in reality it may be difficult to define consistent and inconsistent parts of production and buildup periods from the field data.

4.6.2. Flow-rate estimation from deconvolution

As mentioned previously, the recent deconvolution algorithms of von Schroeter et al. (2004), Levitan (2005), Pimonov et al. (2009a; 2009b) also account for uncertainty (based on normal errors with zero mean and specified variance) in flow-rate data and hence one could also estimate or attempt to "correct" the flow-rate history simultaneously along with $z_d(\xi) = \ln[\tau g(\tau)]$ and initial pressure by using these deconvolution algorithms. This is actually one of the significant advantages of the recent deconvolution algorithms over the earlier ones. However, we emphasize that the proper use of these algorithms requires an understanding of their assumptions, and also an understanding of the deconvolution problem itself. Deconvolution is an ill-posed problem, particularly when estimating flow

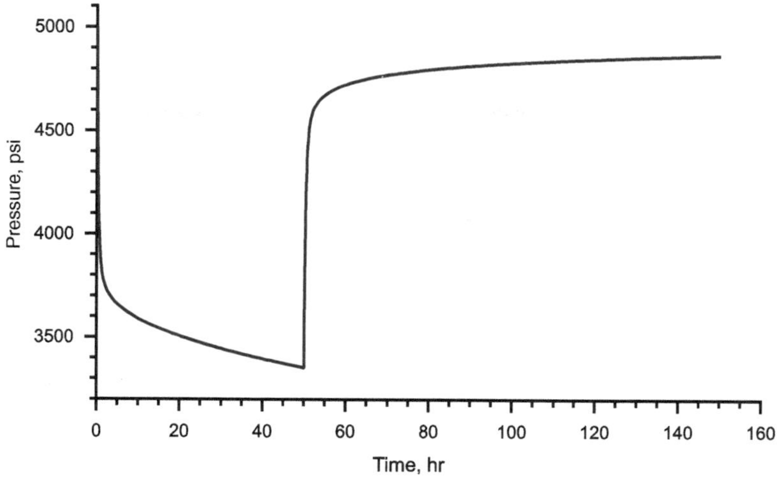

Figure 4.11 A drawdown and a subsequent buildup test for a fully penetrating well located in the middle of two parallel faults in an infinite reservoir.

rate by these deconvolution algorithms. The adjusted rates obtained on the basis of pressure data alone may not necessarily be correct, because deconvolution adjusts the rates for the sake of "honoring" measured pressures. Assuming that both rates and the impulse response (kernel of the convolution) are unknown, there is an infinite number of combinations of deconvolved responses and updated rates that could provide a proper match of the pressure data, but only one such pair represents the true behavior of the well/reservoir considered. Therefore, we do not view deconvolution in general as a method to correct the rates from pressure data alone for possible measurement errors. In fact, as to be shown with a synthetic test application given below, we could easily show that there is no deconvolution method or algorithm that can estimate the rate history from the pressure data, or "correct" the measured rate, even in the case of consistent data sets *unless we know the spread of the error (or expected error margins) in rate data.* Here, we will give a simple example to illustrate this point. Suppose that we have a constant rate drawdown followed by a buildup, as shown in Figure 4.11, for a fully penetrating vertical well located in the middle of two parallel faults. The true, but unknown production rate is 1000 B/D. Furthermore, let us assume that pressure/rate data are consistent with the deconvolution model given by Equation (4.5). Further suppose that we use the entire drawdown and buildup pressure data in deconvolution with the drawdown rate and unit-rate (or z_d) impulse response as unknown.

Figure 4.12 presents pressure changes (both drawdown and buildup periods) for the true value and deconvolved (reconstructed) values for three different production rates: $q_m = 800$, 1000, and 1200 B/D. As can be seen

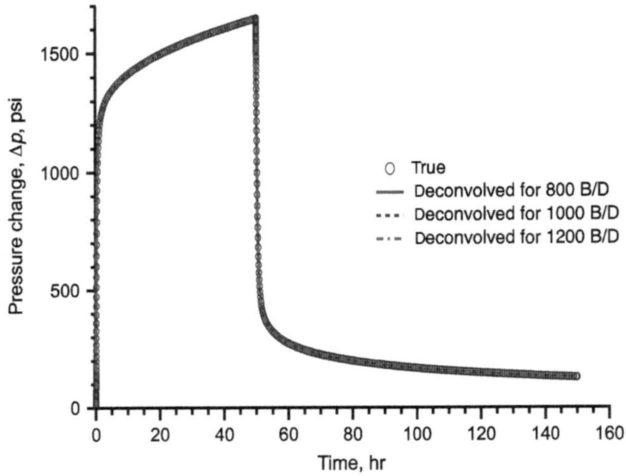

Figure 4.12 Match of pressure changes of the true value with the computed (reconstructed) deconvolved values for three different production rates: $q_m = 800, 1000$, and 1200 B/D.

from this figure, the matches are perfect, i.e., all pressure changes are almost identical.

Figure 4.13 presents the unit-impulse responses $q_m = 800$, 1000, and 1200 B/D. As can be observed from this figure, the unit-impulse responses are vertically displaced depending on the input value of drawdown flow rate. The reason is that we do not have a unique solution for the unit-impulse response (z_d) and drawdown rate, and hence any value of drawdown flow rate with the estimated z_d (based on this flow rate) will reproduce the same pressure history. Therefore, it can be stated that the deconvolution methods (Levitan, 2005; Pimonov et al., 2009a; 2009b; von Schroeter et al., 2004) do not estimate the true (unknown) drawdown rate if an initial guess used for the rate is different from the true value of 1000 B/D because the algorithms will (and should) give back the same initial input rate without providing any correction. It should be pointed out that the input rate should be matched with the final rate estimated from deconvolution for consistency but not estimation of the unknown rate history.

4.6.3. Deconvolution parameters selection

In the previous section, no consideration was given to the selection of algorithmic deconvolution parameters, i.e., error bounds for the various types of data (pressure, rate, and curvature constraints) and weight factors. In this section, the deconvolution algorithm given by Equation (4.70) will be the basis for the discussion given below. First, the impact of the curvature constraint on deconvolution will be explored using the above synthetic multirate test (Example B) given in Figure 4.6. As in Figure 4.10, the

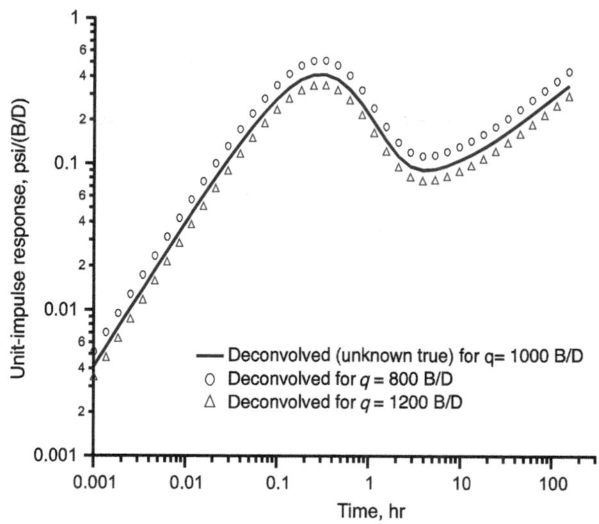

Figure 4.13 Reconstructed (estimated) unit-impulse responses with an initial guess of flow rate $q_m = 800$, 1000, and 1200 B/D.

consistent last drawdown data, along with the last buildup data, will be deconvolved while increasing and decreasing the curvature error bound σ_c by a factor of 30 in Equation (4.70). A decrease in σ_c would lead to a smoother derivative response, while an increase in σ_c would reduce the constraint on the curvature and thus tend to yield a noisier derivative response. This is illustrated in Figure 4.14, which clearly shows that relaxing the curvature constraint too much (increasing, large σ_c) yields a noisier deconvolved derivative, which is essentially non-interpretable at late times. On the other hand, a small σ_c may produce overly an smoothed deconvolution derivative, which can lead to incorrect interpretation of the reservoir signature. For the case where the curvature error bound σ_c is divided by 30 from the base case, the effect of the no-flow fault located at 300 ft away from the well becomes smoothed out, as shown in Figure 4.14.

The qualities of history matches of pressure and rate obtained from deconvolution are a good measure of whether a deconvolved response is over-smoothed. In general, it is assumed that noise in the pressure data is random or at least not biased in most cases. Therefore, the pressure mismatch would be expected to be mostly distributed around a zero mean with no particular bias. Figure 4.15 presents the pressure and rate mismatch obtained from the deconvolution for the case with a lower curvature error bound shown in Figure 4.14 (error bound 1/30 curve). As shown in Figure 4.15, the pressure mismatch bears a low-frequency bias to the response over the entire data sequence (long-term oscillations). Clearly, the computed pressure response differs from the input one in not following a randomly

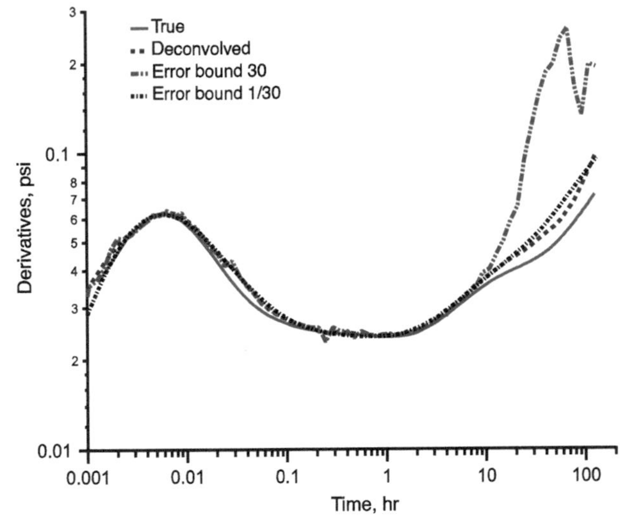

Figure 4.14 The true and deconvolved derivatives and the impact of curvature error bounds on derivatives for Example B.

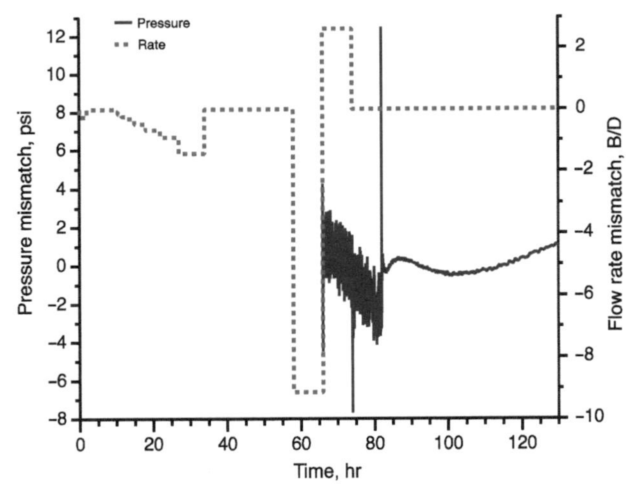

Figure 4.15 Pressure and rate mismatch [computed-input (measured)] for the case with a low curvature error bound presented in Figure 4.14.

pattern, but actually showing differences in the slope of pressure changes. This indicates an over-smoothing of the deconvolved response due to the application of a low curvature error bound. This can be seen in general from the pressure match itself but becomes clearer when comparing just the differences. Ideally, the optimal curvature weight or error bound should

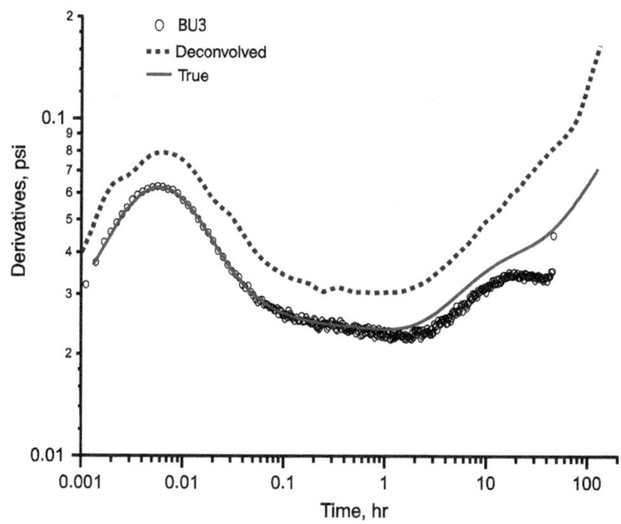

Figure 4.16 The true and buildup (BU3) derivatives, and the deconvolved derivative with a 100-B/D rate error bound.

be the one providing an interpretable response but still leaving some minor fluctuations in the response.

The selection of pressure and rate error bounds might appear straightforward because those bounds can sometimes be estimated from the specifications of the pressure gauge and flow-rate device (both will be called tools) that are used for the test. However, as discussed previously, these specifications should be treated as lower bounds. Actual bounds sometimes can be greater than the tool specifications. Furthermore, tool specifications may not always be representative or even be available. Generally, pressure tool specifications are more reliable and available for measurement noise characterization, but flow-rate measurements bear more uncertainties depending on the measurement technology and measurement environment. Furthermore, the measurement errors tend to be not randomly distributed but are often biased. Still working with Example B, noise has been added to the rate (range of ±10%) and a deconvolution has been applied considering different constraints on the rate error bounds [0.1, 3 and 100 B/D (the rate with the error (uncertainty) is called measured (input) rate to be deconvolved with the pressure), from a tight constraint to a weak constraint]. The deconvolution results for these noises in rate measurements are presented in Figures 4.16 through 4.21. It should be pointed out that the deconvolution results given in these figures were obtained by using only pressure data pertinent to the last buildup (BU3) and parts of the last production periods shown in Figure 4.16.

The results given in Figure 4.16 through Figure 4.21 show that the rate error bounds have a significant impact on the deconvolution results.

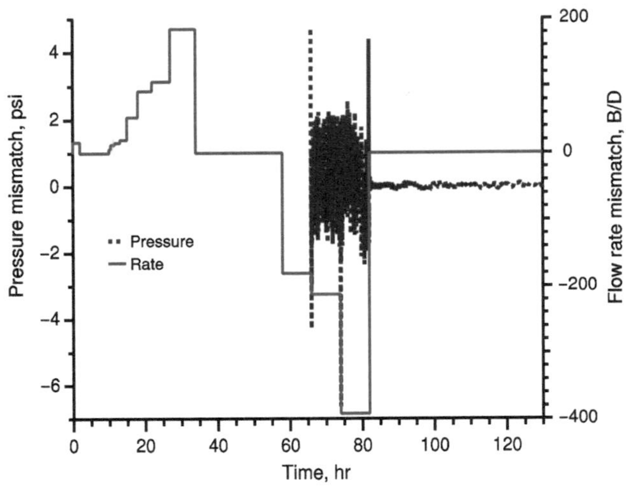

Figure 4.17 Pressure and rate mismatch with a 100-B/D rate error bound.

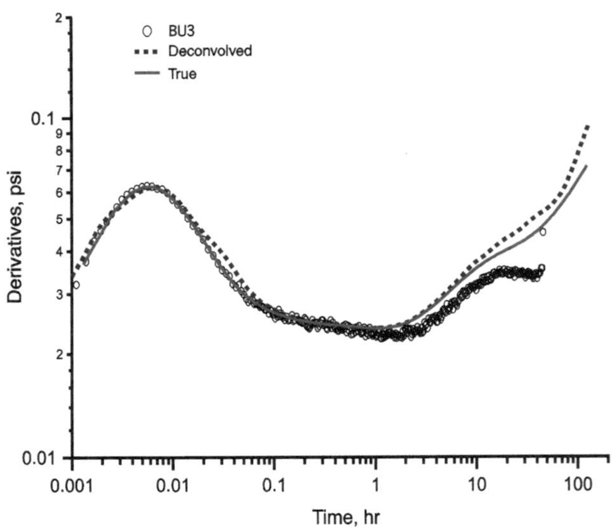

Figure 4.18 The true and buildup (BU3) derivatives, and the deconvolved derivative with a 3-B/D rate error bound.

Enforcing too few constraints, as shown in Figures 4.16 and 4.17, leads to a response that ultimately does not match the true response because the rates are changed so significantly by the deconvolution process that the response ends up shifted and slightly deformed compared to the true one. The pressure match is not necessarily improved by leaving much freedom on the rates as seen by the pressure mismatch. The magnitude of the rate mismatch is a clear indication of using rate constraints that are too lax. On the other hand,

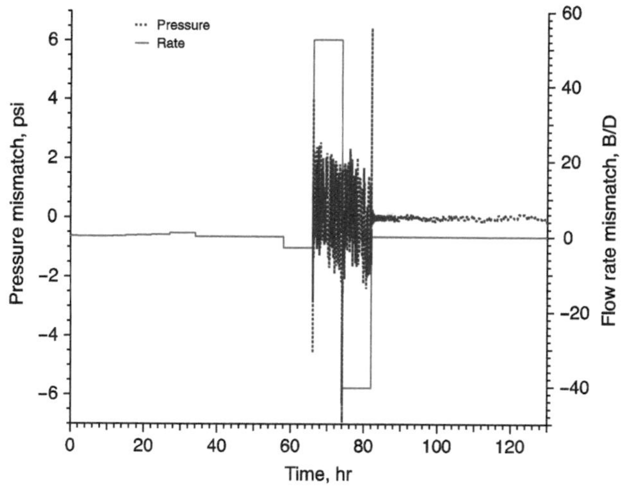

Figure 4.19 Pressure and rate mismatch with a 3-B/D rate error bound.

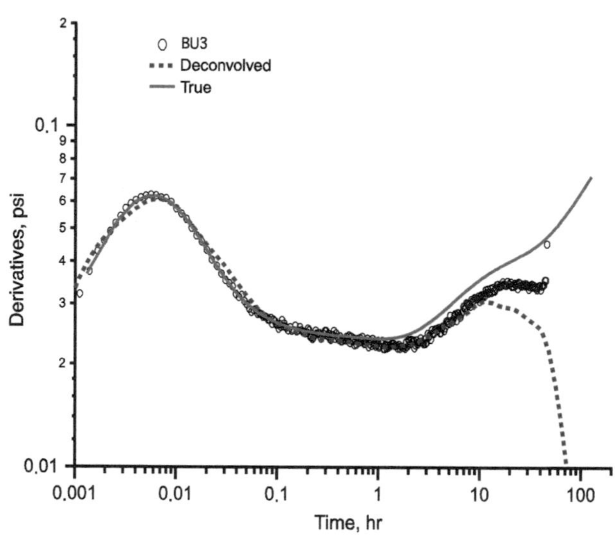

Figure 4.20 The true and buildup (BU3) derivatives, and the deconvolved derivative with a 0.1-B/D rate error bound.

on imposing very strong rate constraints, as shown in Figures 4.20 and 4.21, its impact is also very significant on the deconvolved response, but in this case this translates into a clear non-random signature on the pressure mismatch. In extreme cases, the rate and pressure mismatch results imply improper settings of the constraints. Therefore, the rate and pressure mismatch can be used as guides to adjust the deconvolution constraints.

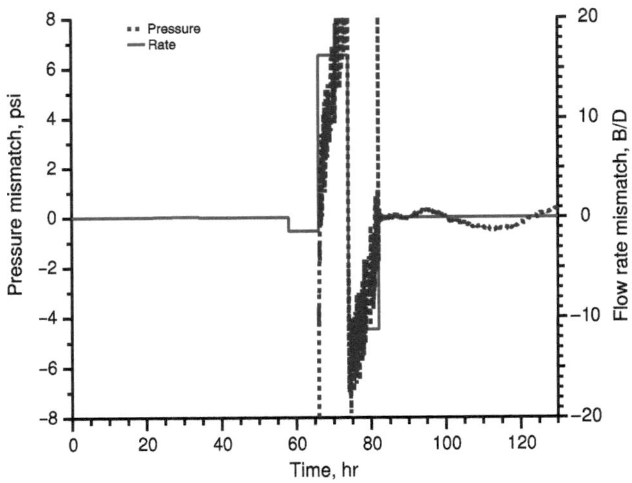

Figure 4.21 Pressure and rate mismatch with a 0.1-B/D rate error bound.

The median case shown in Figures 4.18 and 4.19 limits both those deviations and yields a deconvolved response that is much closer to the true response. An interesting related question may arise in the value of using computed (updated or modified) rate estimates obtained from deconvolution: Do they represent a good match to the true rates that can be considered as more accurate than the measured (input) rates? If such was the case, updated rates obtained from the deconvolution process could then be used for conventional pressure transient analysis based on single flow-period superposition plots and multirate derivatives.

As shown in Figure 4.22, the computed (updated) rates appear to offer a better match to the true values than the input (measured) rate contaminated with noises. This can often be true in reality, but updated rates have to be used with care. First, updated rates are significantly improved only for time periods that are reasonably close to the times for which the pressure data are being used for the deconvolution. This is expected as the information content in the pressure data is mostly dominated by recent production, with earlier production creating comparatively small pressure changes during the last build-up. As can be seen in Figure 4.22, the computed (updated) flow rates are closer to the true rates during the last flow sequence, but the initial production data are left essentially unchanged as these have a much lower impact on the pressure data used in the deconvolution process.

4.6.3.1. Use of constraints

Defining (characterizing) error bounds thoroughly is an important step for obtaining representative deconvolution results and provides a good balance between the pressure match, rate match, and response curvature.

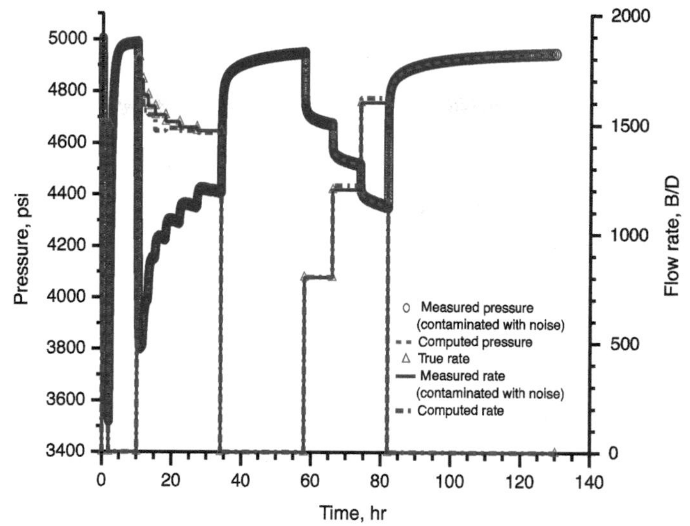

Figure 4.22 Comparisons of the measured (contaminated) and computed pressures and the true, measured (contaminated), and computed rates for Example B.

Furthermore, it is possible that the production rate and/or the initial reservoir pressure are accurately known, therefore they can be set at fixed values during deconvolution. As stated above, for WFT, the production rate and the initial reservoir pressure are measured accurately for most cases, except in very tight formations, where measurement uncertainties are high. On the other hand, the production rate may have high measurement uncertainties for most cases in well testing. The initial reservoir pressure may also be inaccurate in some cases.

For the synthetic examples presented above, deconvolution is performed without updating those rates. The resulting deconvolved response is presented in Figure 4.23 and shows a very good conformance to the true model even though little data have been used (the last flow sequence and the final buildup test). The simplification of the problem allowed the interpretation to focus only on finding a good balance between curvature constraint and pressure match.

Even an error of a few percent in flow-rate data has a very significant impact on deconvolution results, as shown in Figure 4.24. As can be seen in this figure, the true signature of the reservoir is totally distorted and the deconvolved derivative leads to an incorrect model.

Accurately known initial pressure does essentially constrain the end point of the deconvolved response, representing the entire duration of the flow sequence and hence linking the final pressure point to the initial pressure via the rate history. Combining this final point with a good estimate of the deconvolved response over the time intervals where data are used (e.g.,

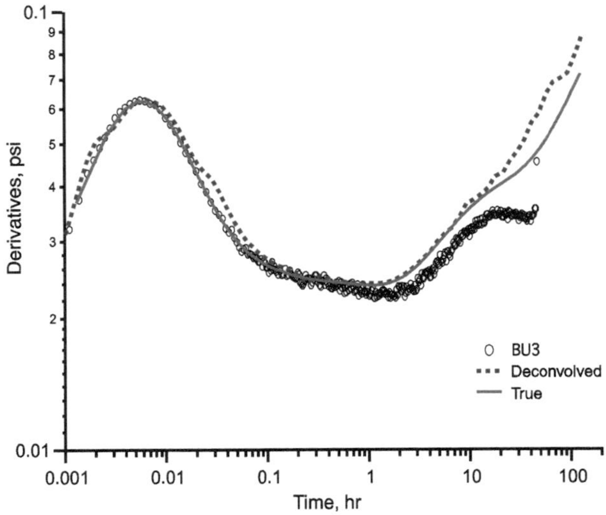

Figure 4.23 The true and buildup (BU3) derivatives, and the deconvolved derivative assuming a fixed and correct flow-rate data for Example B.

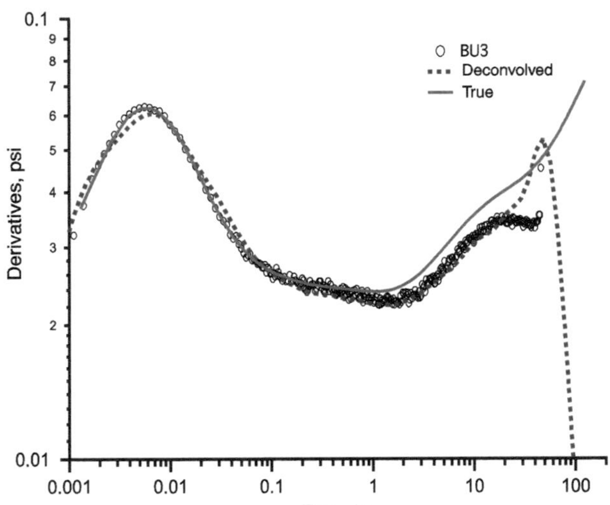

Figure 4.24 The true and buildup (BU3) derivatives, and the deconvolved derivative assuming a fixed but inaccurate flow-rate data for Example B.

extent of the buildup used for deconvolution) and a curvature constraint significantly increases the robustness of the process.

Using again the same example with the rate errors presented above (Example B with ±10% error on the rate measurements), considering only the final buildup for deconvolution and assuming the initial pressure is

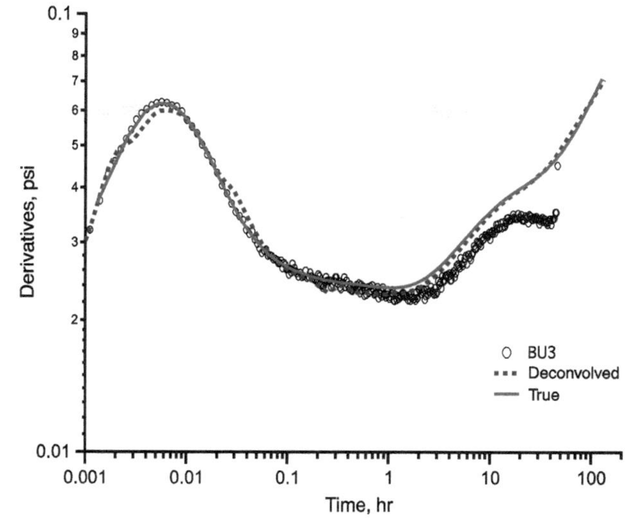

Figure 4.25 The true and buildup (BU3) derivatives, and the deconvolved derivative using a fixed and accurate initial pressure, but inaccurate flow-rate data for Example B.

accurately known, the deconvolved response matches very well with the true reservoir response (Figure 4.25). However, deconvolution will eventually fail when there are high uncertainties in flow-rate measurements.

Using an inaccurate initial pressure, however, can lead to significant errors in deconvolution, as shown in Figure 4.26, where a 10-psi error in the initial pressure is assumed (taken at 4990 psi instead of 5000 psi). As can be seen in Figure 4.26, the deconvolved response deviates significantly from the true reservoir model. This sensitivity of the deconvolution process to the initial pressure has been discussed thoroughly by Levitan (2005).

To illustrate the effects of the initial pressure on deconvolution, we consider two cases in Example B: (1) The initial pressure is fixed at 4990 psi, which is 10 psi lower than the actual one (under-estimated), and (2) at 5010 psi, which is 10 psi higher (over-estimated). With these two fixed initial pressures, both second and last buildup test pressure are deconvolved. The comparison of deconvolved derivatives for these under- and over-estimated initial pressures are presented in Figures 4.27 and 4.28. As shown in these figures, if the initial pressure is under-estimated, the shorter (BU2) deconvolved response (from an earlier flow period) goes to a lower value towards the end of the second test than that from the longer (BU3) deconvolved response. This is reversed towards the end of the second test (BU2) if the initial pressure is over-estimated.

From the differences in the deconvolved derivatives shown in Figures 4.27 and 4.28, it can be concluded that the initial pressure estimate is erroneous, and further whether it is over- or under-estimated. Using

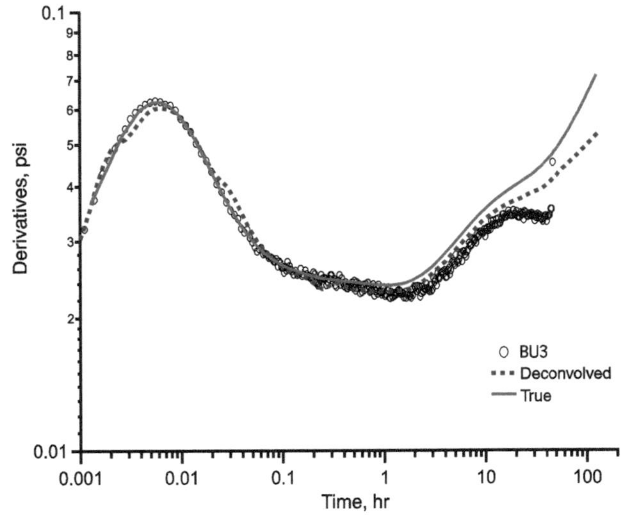

Figure 4.26 The true and buildup (BU3) derivatives, and the deconvolved derivative using a fixed but incorrect initial pressure for Example B.

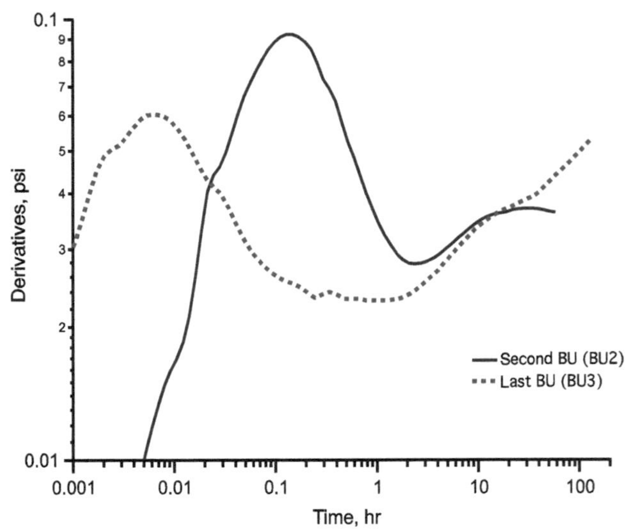

Figure 4.27 Comparison of deconvolved derivatives from two different buildups with an under-estimated initial pressure.

the differences as a measure, representative deconvolved responses can be obtained from two different buildup tests by iterating the initial pressure estimate. Figure 4.29 presents the deconvolved derivatives from two different buildups with the correct initial pressure. The deconvolved derivatives are almost identical and yield the same reservoir signature at the late time. It

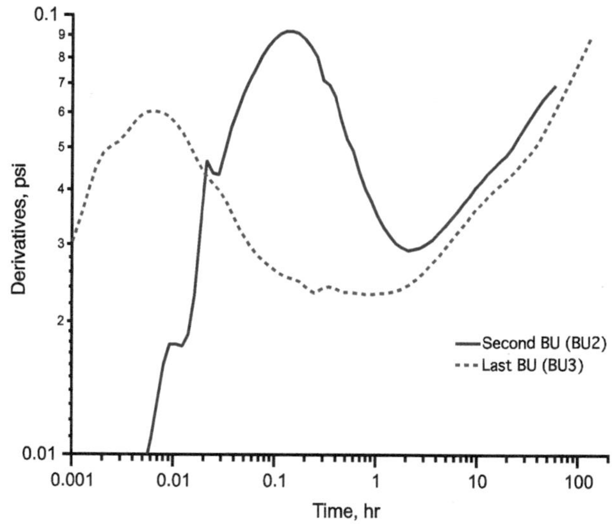

Figure 4.28 Comparison of deconvolved derivatives from two different buildups with an over-estimated initial pressure.

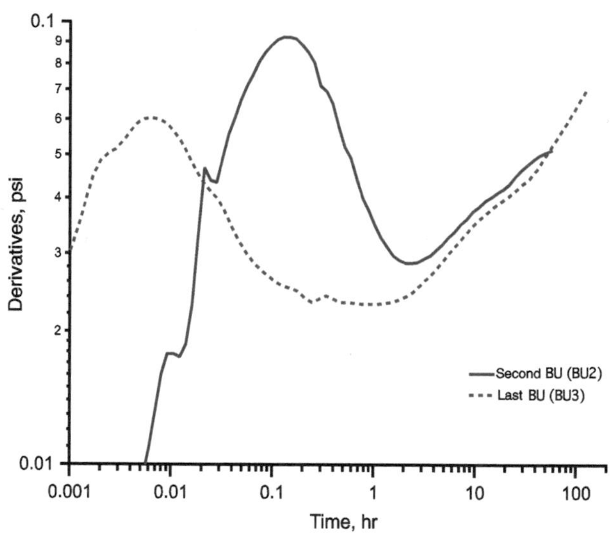

Figure 4.29 Comparison of deconvolved derivatives from two different buildups with the correct initial pressure.

should be pointed out that the initial pressure can be estimated accurately if several buildup tests are available for deconvolution (Levitan, 2005), even though these tests might be inconsistent because of skin of wellbore storage changes. As also can be seen in Figure 4.29, the two deconvolved derivatives differ significantly at early time because the two buildup periods used are inconsistent (namely affected by different wellbore storage values due to the surface and downhole shut-in). Using an accurate initial pressure and fixing it during deconvolution affects the accuracy of deconvolution results significantly. This should be taken seriously at the test design stage to ensure that the data acquired are adequate to apply deconvolution and obtain representative results.

4.7. Pressure-Rate Deconvolution Examples

In this section, various applications of deconvolution are presented for simulated and field pressure transient and wireline formation tests. In these applications, we will use the deconvolution algorithm recently presented by Pimonov et al. (2009a; 2009b) while using the basic ideas of Levitan (2005) and von Schroeter et al. (2004), but a more general weighted least-squares objective function accounting for a measure of confidence in individual measurements, as discussed in detail previously. The simulated examples were designed so that we can illustrate advantages and disadvantages associated with the modern deconvolution techniques presented recently.

4.7.1. Simulated well test example

Here we present a deconvolution example for a buildup test from a hypothetical fully penetrating vertical well in a homogeneous reservoir with a sealing (no-flow) fault. The formation and fluid properties and the production data for this example are: $k = 100$ md, $h = 100$ ft, $\phi = 0.20$, $\mu = 1$ cp, $c_t = 10^{-5}$ psi^{-1}, $r_w = 0.354$ ft, $p_o = 4400$ psi, the production rate at the onset of buildup, $q = 1000$ B/D, $t_p = 2100$ hr, and the distance from the fault to the wellbore is 700 ft. The reservoir is assumed to be an infinite isotropic single-layer reservoir.

Figure 4.30 presents the pressure and flow-rate data for this example. Although the buildup period looks very short in Figure 4.30, its duration is 37 hr. For this example, we only use the buildup pressure and flow-rate data during the drawdown period for deconvolution. Based on our discourse given previously for buildup deconvolution (see 4.6.1), there is a gap of information for 2063 hr (in the time interval from 37 to 2100 hr) for the constant-rate drawdown response to be generated for the entire test duration of 2137 hr by buildup deconvolution. So, one of our objectives in this buildup deconvolution example is to show how successfully the deconvolution algorithm could fill this large gap of information, depending on the different levels of noise in pressure and rate data.

A random noise with zero mean and standard deviation of 10 B/D is added to the flow-rate data, as can be observed in Figure 4.30. This

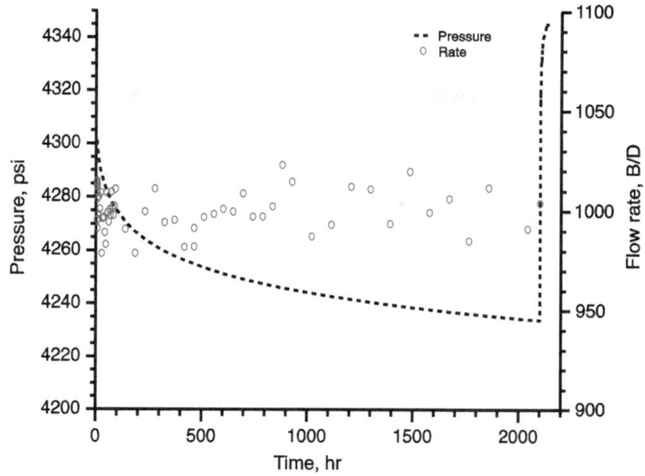

Figure 4.30 Pressure and flow-rate history for the simulated deconvolution example.

is about a 2% maximum error in the rate measurements. Today, the best downhole pressure gauges have resolutions of about 0.002 psi, and the more common gauges used in pressure transient tests have a 0.01-psi resolution. Furthermore, as discussed in Chapter 1, the apparent resolution could be much less than the stated gauge resolution. Therefore, Gaussian noise is added to pressure data. Two levels of random noise with zero mean and standard deviation are considered for pressure measurements: (1) $\sigma = 0.01$ psi (attainable today) and (2) $\sigma = 1$ psi (an extreme case).

Figure 4.31 presents the results of deconvolution obtained for the $\sigma = 0.01$-psi error case with the three different initial reservoir pressures: (1) 4400 psi, true value, (2) 4380 psi (−0.5% error), and (3) 4420 psi (0.5% error) using only the 2100-hr production rate data and the downhole buildup test pressure data. Here the flow rate and the initial pressure were fixed during the deconvolution. The curvature constraint was set to $\sigma_c = 0.05$ in this deconvolution application. The true drawdown and superposition buildup derivatives are also given in the same figure. It should be noted we have used the noisy rate data in generating the deconvolved response and computing the superposition buildup derivative (without any smoothing) shown in Figure 4.31. The infinite-acting period after the wellbore storage effect occurs at about 6 hr, as shown in Figure 4.31. At about 10 hr, the effect of the sealing fault becomes observable, and all derivatives are indistinguishable until about 37 hr and exhibit a 1/4 slope period. The 37-hr marks the end of the Horner buildup derivative. The 1/4 slope period can be interpreted for a few different geological features. The effect of the producing time on the Horner buildup derivative is not observable because of a long production period. If we want to see the doubling of the slope (second

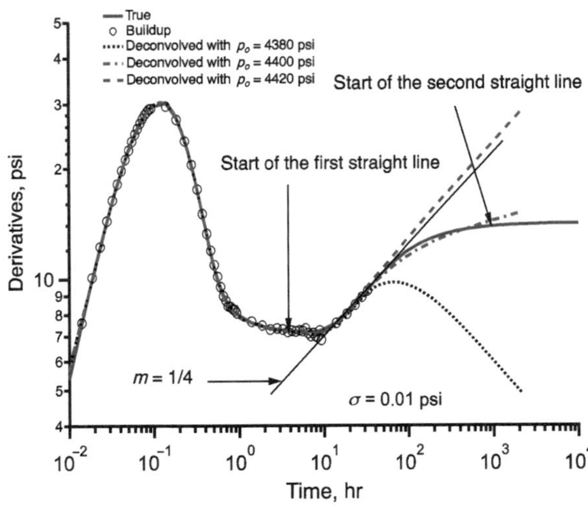

Figure 4.31 Comparison of deconvolution derivatives with the true and buildup derivatives for the simulated example with different initial reservoir pressures.

infinite-acting flow regime) that is definitely an indication of a sealing fault, the buildup test had to be run for about 1000 hr (slightly more than a month). As shown in Figure 4.31, if we have "accurate" buildup pressure and production rate measurements with an accurate initial reservoir pressure, deconvolution provides a 2137-hr drawdown derivative that exhibits the doubling of the slope period reasonably well and, hence the deconvolution algorithm successfully (or correctly) fills the information gap of 2063 hr with the algorithmic parameters used ($\sigma_c = 0.05$ and $\sigma_p = 0.01$ psi, $(w_p)_i = 1$ for all buildup pressures, and $(w_c)_k = 1$ for all curvature constraints, see Equation (4.70)).

However, the effects of an inaccurate initial reservoir pressure are very detrimental for system identification, as shown in Figure 4.31. For the 4380-psi (-0.5% error) initial reservoir pressure case, the deconvolution derivative indicates constant-pressure boundary effects, while the 4420-psi deconvolution derivative perhaps indicates leaky faults or fractures.

Figure 4.32 presents the results of deconvolution for the $\sigma = 1$-psi error case with the three different initial reservoir pressures as in the previous case in comparison to the conventional buildup pressures based on the superposition time function. It should be noted that we have used the same value of the curvature constraint ($\sigma_c = 0.05$) as in the previous example application, but $\sigma_p = 1$ psi. Conventional buildup derivatives based on superposition time function shown in Figure 4.32 were smoothed by using a smoothing factor of $L = 0.4$ [also referred to as the dimensionless differentiation length on the semilog axis (Bourdet, Ayoub, & Pirard, 1989)] because the buildup derivatives without smoothing were highly oscillatory. It

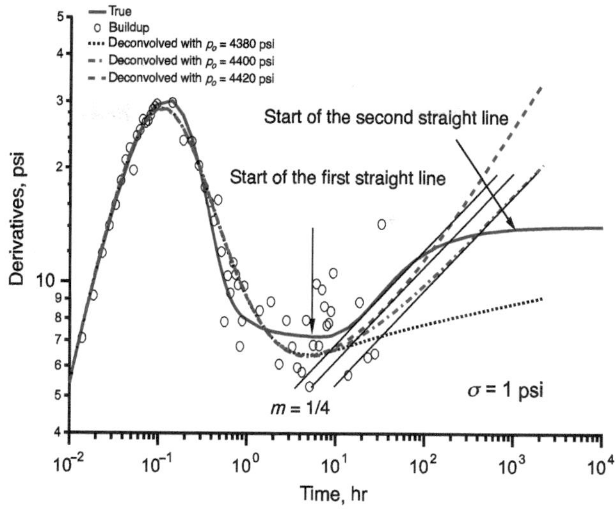

Figure 4.32 Comparison of deconvolution derivatives with the true and buildup derivatives for the simulated example with different initial reservoir pressures.

is clear that with this level of noise in buildup pressures, conventional buildup derivatives even with a high level of smoothing may not be very useful for identifying the correct geological features. For these three cases, the deconvolution derivatives also indicate totally incorrect geological features, although they are quite smooth, particularly, the 4380-psi case indicates almost an infinite-acting flow regime (derivative slightly increases with time) about 1000 hr, as shown in Figure 4.32. It should be emphasized that deconvolution derivatives are valuable system identification tools by adding to the drawdown period to the buildup period to look deeper into the reservoir, but accurate downhole pressure and surface or downhole flow-rate measurements with an accurate initial reservoir pressure are needed. The next field example will show a methodology when an accurate initial reservoir pressure is not available.

4.7.2. Horizontal field test example

This example (Kuchuk, 2009b) consists of an extended drawdown/short buildup well test sequence taken over a 1346-hr period from a horizontal well. The rate measurements were acquired every 2 hours at the surface. As can be observed from Figure 4.33, both pressure and rate decline steadily during the production period that followed a 270-hr buildup test. The initial reservoir pressure is not known for this example. The reservoir model is quite complicated to apply the Horner method to obtain the initial reservoir pressure. Therefore, for this data set, we cannot directly apply the procedure of Levitan (2005) because the buildup data in this example

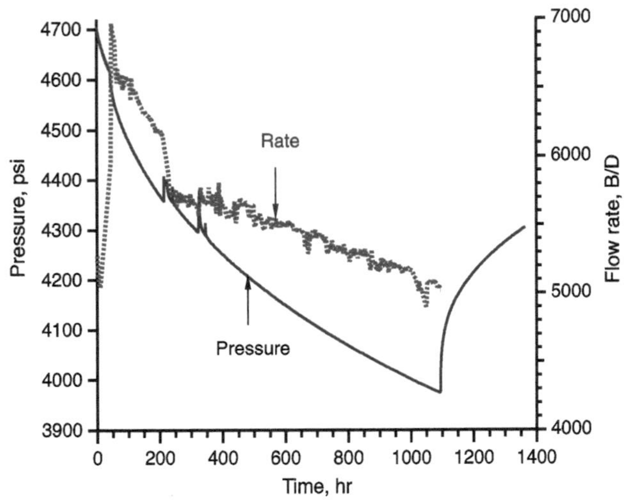

Figure 4.33 Pressure and flow-rate history for the field data deconvolution example.

has insufficient information to accurately determine the initial pressure and unit-rate drawdown pressure response as his deconvolution strategy requires that we determine the initial pressure from at least two buildup periods. As a result, we apply the procedure of von Schroeter et al. (2004); i.e., we apply the deconvolution algorithm to both drawdown and buildup pressure data for the entire test to estimate the initial reservoir pressure and rates simultaneously (assuming some uncertainty in the rate measurements), and the unit-rate derivative response.

The maximum pressure of the entire test is 4702.5 psi, which is much higher than the final pressure of the buildup test. In this example, the error levels in pressure and rate data (σ_p and σ_q) and the appropriate value of curvature constraint σ_c are not known *a priori*. Therefore, we need to make different guesses for the values of σ_p, σ_q and σ_c to be used in the objective function of Equation (4.70), compare reconstructed deconvolved responses, pressure and rate histories for each of these guesses and then select the appropriate set of σ_p, σ_q and σ_c that provides "reasonable" matches of pressure and rate measured. We could use the approaches suggested by von Schroeter et al. (2004) and (Onur et al., 2008) to compute initial guesses of the algorithmic parameters σ_p, σ_q and σ_c. So, we can take the maximum pressure as an initial guess for the initial pressure and consider 40 uniform logarithmically spaced nodes for the z_d response, the initial guesses for ν and λ are computed from Equations (4.62) and (4.63) to be $\nu = 0.00627$ psi D/B and $\lambda = 79.61$ psi^2. Furthermore, the values of the parameters ν and λ could be related to the average variances of pressure (σ_p^2) and rate (σ_q^2) data, and the curvature constraint (σ_c^2) by the following

Table 4.1 Deconvolution parameters used for the horizontal-well field test example

Parameters	ν **(psi D/B)**	λ **(psi^2)**	σ_p^2 **(psi^2)**	σ_q^2 **((B/D)^2)**	σ_c^2 **(unitless)**
Case 1	0.00627	79.61	0.2	31.8	0.0025
Case 2	0.00627	796.1	2.0	318.0	0.0025
Case 3	0.00627	7961.0	20.0	3180.0	0.0025

equations (Onur et al., 2008):

$$\nu = \frac{\sigma_p^2}{\sigma_q^2} \tag{4.78}$$

and

$$\lambda = \frac{\sigma_p^2}{\sigma_c^2}. \tag{4.79}$$

Using the above values of ν and λ, and the initial guesses for $\sigma_c^2 = 0.0025$, $\sigma_p^2 = 0.2$ psi^2 and $\sigma_q^2 = 31.8$ (B/D)^2 are computed from Equations (4.78) and (4.79). Then, fixing the values of $\sigma_c^2 = 0.0025$ and $\nu = 6.265 \times 10^{-3}$ (psi D/B)^2, three different values of λ were computed such that $\lambda = 10^r \times 79.61$, where $r = 0, 1, 2$. Then, we computed three values of σ_p^2 and σ_q^2 for the given values of ν and λ. Table 4.1 lists the values of ν, λ, σ_p^2 and σ_q^2, and σ_c^2 computed for $r = 0$, 1, and 2, which are called Case 1, 2 and 3, respectively, by using this procedure.

For this pressure-rate deconvolution, different values of the standard deviation of measurements and the curvature constraint should be used until a satisfactory solution is obtained, although we must admit that it is somewhat subjective. Because there is no an objective measure for defining what a satisfactory solution is, and it depends on the interpreter's judgment, therefore, the pressure-rate deconvolution results should be interpreted together with conventional derivatives, other specialized plots, history matching, and geoscience data (see Figure 6.3 in Chapter 6 for the iterative nature of the well test interpretation).

Figure 4.34 presents the results of deconvolution obtained from the algorithm of Pimonov et al. (2009b) with the deconvolution parameters Table 4.1 and the initial pressure of 4702.5 psi. As can be seen from Figure 4.34, the derivative for Case 1 undulates visibly, while for Case 2 the derivative undulates very slightly. The Case 3 derivative is very smooth, but it follows closely the buildup (superposition) derivative from 1 to 30 hr during

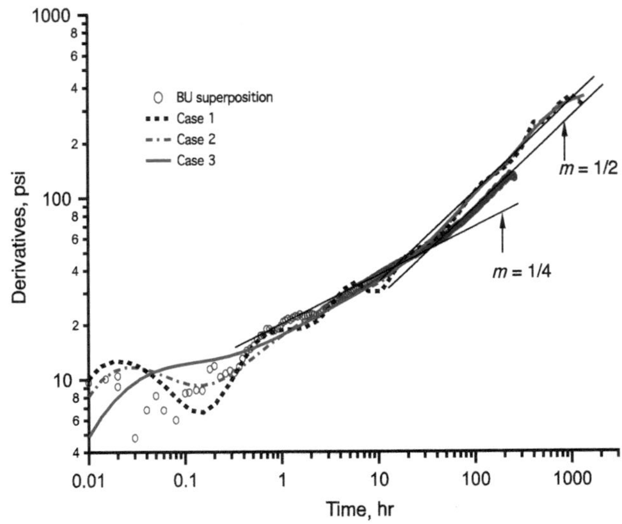

Figure 4.34 Buildup and deconvolution derivatives of the field example for the entire drawdown and buildup durations for three different cases of parameters given in Table 4.1.

which a 1/4 slope period is observed, perhaps due to wellbore-intersecting finite conductivity fractures (observed from the openhole logs). This 1/4 slope period cannot easily be seen from the derivative for Case 1. All derivatives' early times (0.01 to 1 hr) behave differently, perhaps due to the dynamics of the fluid movement in the wellbore during which measurement uncertainties are high (fluid segregation, uneven potential distribution along the wellbore, etc.). However, the Case 2 derivative matches the buildup derivative much more closely than that of Case 1 and 3 from 0.1 to 1 hr.

The derivatives for all three cases behave similarly from 30 to 320 hr, during which a 1/2 slope period is observed, perhaps due to discrete finite and/or infinite conductivity fractures in one major direction laterally. The buildup (superposition) derivative also exhibits an 1/2 slope period 30 to 270 hr (the end of the buildup).

After this analysis of all derivatives, it is safe to say that Case 2 parameters are the optimal values for deconvolution of this test pressure-rate data, although the difference between Case 2 and Case 3 derivatives is small. This is further validated by inspecting the reconstructed pressures and rates by deconvolution for these three cases.

Using the deconvolution algorithm parameters for Case 2, the computed pressure and rate are matched with the corresponding measured data with the estimated initial reservoir pressure 4791.0 psi (the initial guess was 4702.5 psi). Given the uncertainties in selecting deconvolution parameters, this last check is necessary. Figures 4.35 and 4.36 present the pressure and rate matches. The pressure match is excellent—the Root Mean Square (RMS)

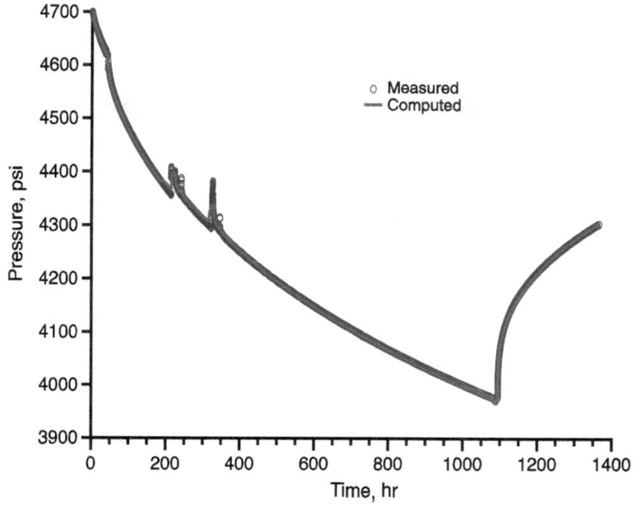

Figure 4.35 Comparison of the computed and measured pressures for the field data deconvolution example.

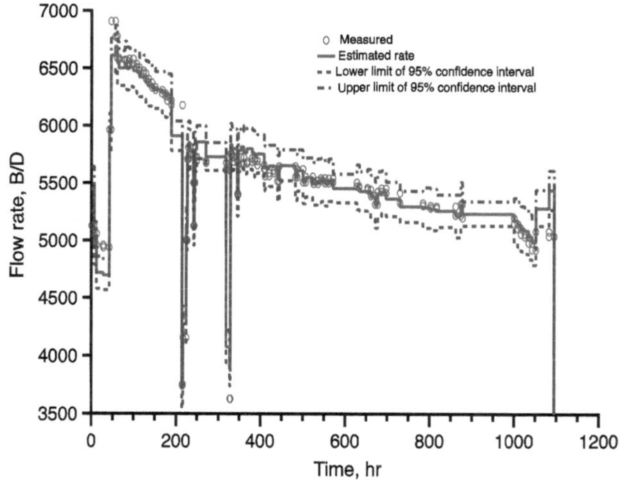

Figure 4.36 Comparison of the computed and measured flow-rate histories, and the computed flow-rate confidence intervals for the field data deconvolution example.

error is 6.91 psi. The rate match seems reasonably good because the rate for each flow period is not changed by more than 10%—the overall RMS error for the rate is 193.69 B/D (Figure 4.36). The 95% confidence intervals for the estimated rates are also presented in Figure 4.36. It should be stated that the measured drawdown pressure significantly reduced the uncertainty

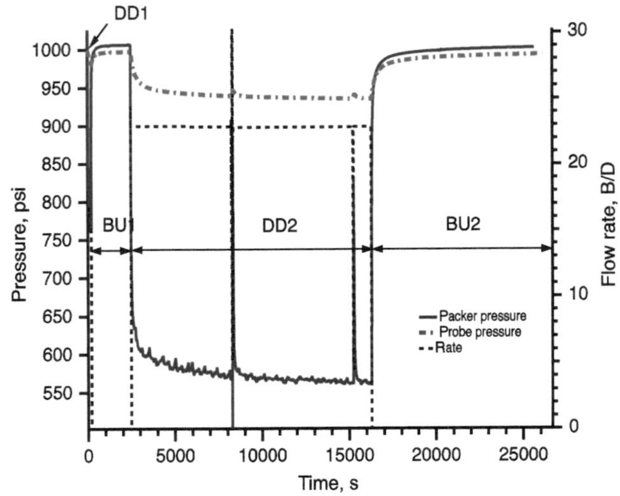

Figure 4.37 Packer and probe pressure, and flow-rate measurements for the field IPTT.

in the deconvolution of this test example, particularly for the estimation of the initial reservoir pressure.

4.7.3. Interval pressure transient test (IPTT) field example

As described in Chapter 3, 3D (spatial r and z and time) IPTTs are frequently performed with multiprobe Wireline Formation Testers (WFT) for the estimation of horizontal and vertical permeabilities and delineation of fracture and fault conductivities using the dual-packer module and observation probes of the multiprobe wireline formation tester. In recent years, deconvolution algorithms have been applied to WFT single-probe, multiprobe, and packer-probe pressure-rate and pressure-pressure data; for example Onur, Ayan, and Kuchuk (2009a), Pimonov et al. (2009a; 2009b), and Wu, Georgi, and Ozkan (2009).

This field IPTT was conducted with the dual-packer module and a single observation probe and consists of three main sequences:

- A pretest, short flow sequence followed by a relatively short buildup,
- A production period for about 4 hr during which fluid samples were acquired, and
- A final buildup period of slightly less than 3 hr.

In total the IPTT lasted about 7 hr, and the packer and probe pressures, along with flow-rate data are presented in Figure 4.37. This is a very long IPTT.

In this test, the final buildup was run long enough so that both packer and probe pressures exhibit the same radial flow regime, as shown in Figure 4.38. In addition, both packer and probe responses exhibit the same spherical

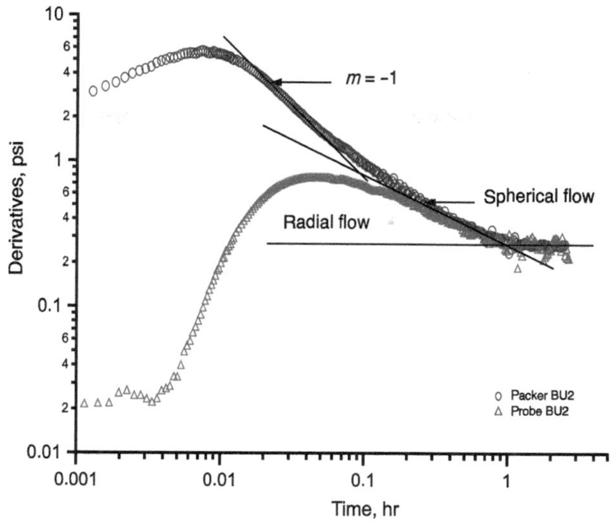

Figure 4.38 Packer and probe pressure derivatives of the last buildup.

flow regime. Therefore, both horizontal and vertical permeabilities can be obtained readily from these radial and spherical flow regimes, for instance see Kuchuk and Onur (2003) and Onur, Hegeman, Gok, and Kuchuk (2009b).

Although this IPTT is a complete test with good data quality, only affected by two pressure spikes during the drawdown due to short production interruptions, it can be interpreted with well-documented techniques (Kuchuk & Onur, 2003). However, although the total test duration is about 7 hr, the last buildup investigates (identifies) the formation for only 3 hr. Therefore, applying deconvolution for the last buildup will provide an equivalent constant-rate drawdown response that investigates the formation for 7 hr. In other words, any formation features from 3 hr to 7 hr will appear in the deconvolved derivative.

For this example, first the packer pressure with flow-rate data will be deconvolved, and then the probe data.

4.7.3.1. Packer data

As discussed above, the deconvolution process is very sensitive to the initial formation (reservoir) pressure. Normally, the pressure at the end of the pretest buildup is representative of the true initial formation pressure if the formation is not very tight. In this case, the buildup pressure is quite stabilized at a value of 1006.8 psi towards the end of the pretest (see Figure 4.37). This value will be assumed as the initial formation pressure. The flow-rate data was measured by the pump and will be assumed to have high accuracy. Hence, the deconvolution will be performed first using only

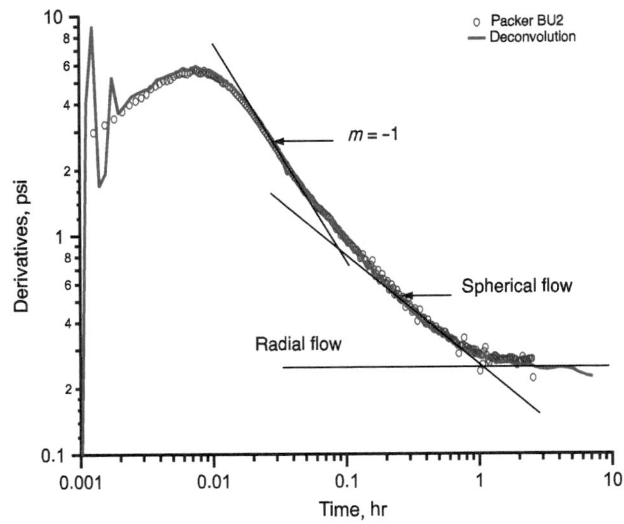

Figure 4.39 The final buildup pressure and deconvolution derivatives.

the final buildup pressure data with the flow rate of the production periods with a known initial pressure.

Figure 4.39 presents both final buildup and deconvolved response derivatives with a RMS of only 0.044 psi highlighting a very good match. As can be seen in the same plot, the final buildup and deconvolved response derivatives are identical until the end of the buildup test. However, the deconvolution derivative goes downwards slowly towards the end of the test, potentially reflecting the presence of formation heterogeneities such as an increase in reservoir permeability and/or thickness at some distance away from the wellbore. This may also be caused by slight uncertainties in the flow-rate data and/or the initial formation pressure.

This extra feature highlighted by deconvolution, and not available from the build-up alone, needs validation because it could have significant implications on the actual formation model. Thus far, all inputs had been set and fixed based on available measurements. To confirm or not those results, a second deconvolution run has been performed, aiming at setting the initial formation pressure on the basis of comparison of the two available buildups BU1 and BU2, as shown in Figure 4.37, and allowing for rate modifications (computing rates during deconvolution). In this case, the initial pressure was determined by comparing the deconvolved responses from the two buildups available (pretest build-up and main build-up). The objective was to ensure that both responses had the same behavior, even though those could be shifted because of possible pretest rate errors. The resulting deconvolved responses from each buildup are presented in Figure 4.40.

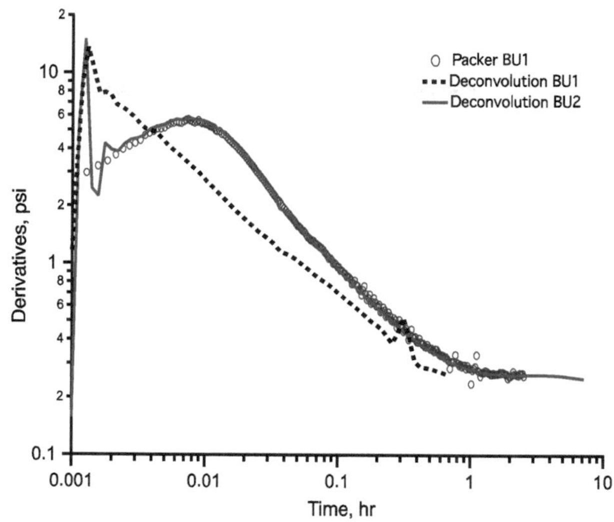

Figure 4.40 Deconvolved responses using a fixed pressure input at 1007.2 psi that is based on the two buildups.

It is noticeable from Figure 4.40 that the slight change in formation pressure input (from 1006.8 to 1007.2 psi) affects somewhat the late-time derivative response, reducing significantly the late-time derivative decrease. This realization would refute the possibility of a mobility change away from the well. The rate changes remained minimal and the computed (reconstructed) pressure is very well matched with the measured data in this case as well. This application clearly shows the large degree of sensitivity of the response to the initial pressure input.

In both cases, the application of deconvolution to refine and confirm the interpretation model is clear. In the first case, deconvolution is used to highlight some features not seen in the conventional single flow-period response, while in the latter it essentially confirms the infinite-acting nature of the formation past the radius of investigation achieved from a single flow period. The limitation here, however, is that there is no immediate way to confirm which model is truly the correct one between these two deconvolution interpretations. The differences are tenuous but the significance of these differences may not be negligible from a reservoir management standpoint. Using both buildup sequences together does not bring any reliable deconvolved response because the two buildups are clearly inconsistent (the near-wellbore formation was not cleaned up properly from the short flow sequence generated during pretest). Furthermore, significant doubts can be cast regarding the validity of the very short build-up pretest to obtain a representative reservoir model.

However, there is an additional feature of formation testers that can bring supplementary value by providing the possibility of comparing the results from the packer pressure response with that from the observation probe. This will be discussed next.

4.7.3.2. Observation probe data

Here, the response from the observation probe will be reviewed and subject to deconvolution to further check whether it allows for a selection of the most adequate reservoir.

It should be highlighted that the observation probe data should not be affected by non-linearities due to changing skin or storage factor. While the formation is being produced through the packer, no flow takes place through the observation probe. Therefore, skin and storage effects are negligible on the observation probe pressure without production noise. As a result, the observation probe data should exhibit a very smooth derivative, hence, a larger span of data should be useable for deconvolution purposes, constraining the response not through fixed input parameters but solely by using a more complete pressure record. This should be taken with some qualifiers however. First, the tool storage at the packed-interval affects the flow-rate inputs and by extension the pressure response at the observation probe. Significant changes in storage value (for instance when changing the fluids withdrawn, from a low-compressibility water filtrate observed during pretest, to formation gas following a more extensive clean-up) would affect the pressure response at an observation probe. Second, as clean-up leads to a change in near-wellbore saturation, those changes could encompass a significant volume between the packed-interval and observation probes, leading to an actual change in formation mobility between those probes, by extension breaking the response linearity.

First, a similar approach to that presented above for the deconvolution of packer pressure data has been used, assuming the flow rates to be known and setting the initial pressure at a fixed value corresponding to the last pressure observed at the observation probe during the pretest of 997.6 psi (Figure 4.37). The deconvolved and conventional derivative responses for the last buildup period (BU2) for the observation probe along with the corresponding ones for the packed interval are presented in Figure 4.41.

The two deconvolved responses (Figure 4.41) are clearly very similar at late times, where both packer and probe investigate similar formation volumes. Some minor discrepancies are observed, but overall the slight decrease in derivative response past the duration of the last buildup appears on both deconvolved responses.

Determining, as previously, the initial pressure from a comparison of the deconvolved responses from both buildups taken individually yields an estimated formation pressure of 997.9 psi, marginally higher than that estimated from the pretest. These results are presented in Figure 4.42.

This seems to confirm the deconvolution results from the packer data, highlighting an excellent match between the two deconvolved responses at

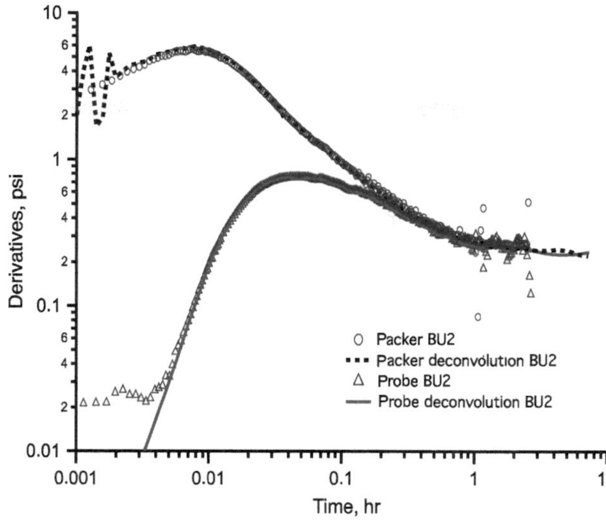

Figure 4.41 Probe and packer deconvolved derivative responses for the last buildup period (BU2) with conventional responses using fixed inputs.

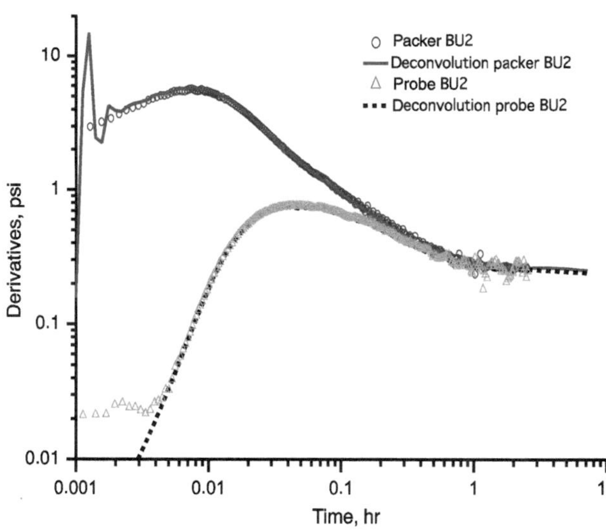

Figure 4.42 Probe and packer deconvolved responses with conventional responses using the fixed initial pressure obtained from buildup comparisons.

late time. Here again, the late-time derivative decrease observed from the deconvolved response based on fixed inputs is not clearly apparent.

To confirm the deconvolution results, a third configuration has been applied, considering both the final buildup pressure along with part of the drawdown pressure during stable flow sequences. The data used consisted

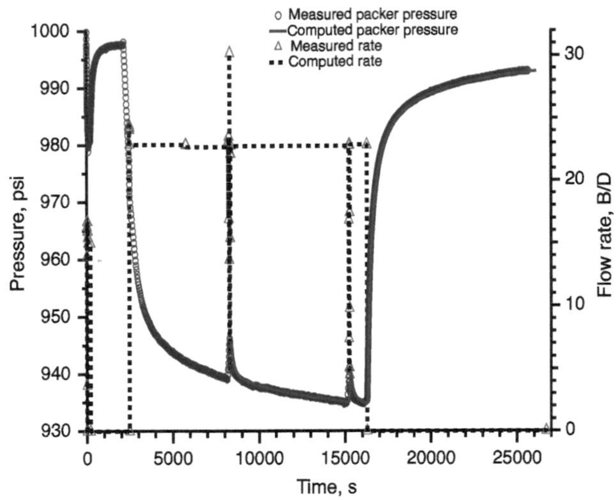

Figure 4.43 Pressure match and updated rates obtained after performing deconvolution on observation probe data using buildup and some drawdown data.

of the final buildup along with the second half of the flow sequence. This should reduce the uncertainty linked both to the use of fixed inputs as well as to that possibly associated with an erroneous initial pressure estimate due to poor pretest data quality. A very good pressure match was obtained, and the flow rates were only marginally updated, as shown in Figure 4.43. The initial pressure obtained from the process is 997.5 psi, i.e. very similar to that from the pretest flow sequence. Moreover, the resulting deconvolved response matches very well with that observed using the packer data with fixed inputs, as shown in Figure 4.44.

All these results tend to confirm the presence of a zone with increasing total mobility away from the wellbore. Both responses are very similar, albeit slightly shifted. This could not have been identified using conventional interpretation techniques. Some variability remains in the deconvolution results, however different approaches have provided consistent results and the combination of packer data with observation probe data has been proven to be able to provide consistent results with little variability.

4.8. Pressure-Pressure (p-p) Deconvolution

In Chapter 3, the details of p-p convolution were given, and p-p deconvolution will be presented here. The objective of p-p deconvolution is to compute (reconstruct) $\mathcal{G}$- and/or $t\mathcal{G}$-functions (defined in Chapter 3) of the system from pressure measurements made at two different spatial

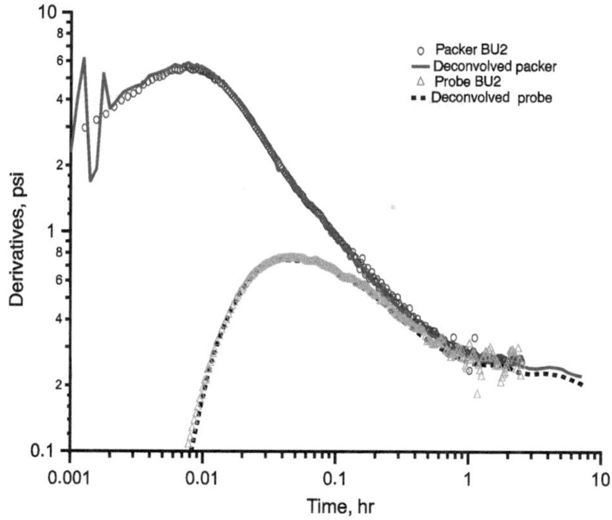

Figure 4.44 Deconvolved responses obtained using the packer pressure data with fixed inputs and using the probe data with a combination of drawdown and buildup data without constraints.

locations in the formation. Let us rewrite Equation (3.98) given in Chapter 3 as

$$p(\mathbf{r}_2, t) = p_{o2} - \int_0^t \Delta p(\mathbf{r}_1, t - \tau)\mathcal{G}(\mathbf{r}_1, \mathbf{r}_2)(\tau)\,\mathrm{d}\tau, \tag{4.80}$$

where $\Delta p(\mathbf{r}_1, t) = p_{o1} - p(\mathbf{r}_1, t)$, the $\mathcal{G}$-function is defined from Equation (3.99) as

$$\mathcal{G}(\mathbf{r}_1, \mathbf{r}_2, t) = \mathcal{L}^{-1}\left[\frac{\bar{g}(\mathbf{r}_2, s)}{\bar{g}(\mathbf{r}_1, s)}\right], \tag{4.81}$$

where $\bar{g}(\mathbf{r}_1, s)$ and $\bar{g}(\mathbf{r}_2, s)$ are the Laplace transforms of impulse responses at the spatial locations $\mathbf{r}_1$ and $\mathbf{r}_2$. Unlike the pressure-rate deconvolution given above, the p-p deconvolution does not require the flow-rate data to be used, as is clear from Equation (4.80). Therefore, it provides another useful system identification tool if the flow-rate data are not measured or are unreliable. As mentioned previously, although recent p-r deconvolution algorithms (Levitan, 2005; Pimonov et al., 2009a; 2009b; von Schroeter et al., 2004) allow jointly estimating flow rate, initial pressure, and the unit-impulse response g of the system, their use requires *a priori* knowledge of the errors in flow-rate data to be available. However, it is often difficult to estimate uniquely the unit-impulse and flow-rate signals from the pressure signal alone (Onur et al., 2008).

Because p-p deconvolution does not require flow-rate data, it becomes quite attractive for interference test applications, particularly for IPTTs for which pressures at different spatial locations along the wellbore are directly and very accurately measured. Computation (reconstruction) of the $t\mathcal{G}$-function from p-p deconvolution based on the v-p convolution equation Equation (3.186) for a packer-probe IPTT was first investigated by Baygun et al. (1997). They solved the deconvolution problem by using a nonlinear least-squares minimization with regularization based on a set of constraints expressed in terms of the $t\mathcal{G}$-function. Baygun et al. (1997) show that their deconvolution algorithm is able to reconstruct the $t\mathcal{G}$-function with pressures containing measurement errors up to 5%, but as the noise level increases, the reconstructed $t\mathcal{G}$-response from their algorithm shows large oscillations.

Recently, Onur et al. (2009a) investigated in detail the applicability of p-p deconvolution for pressure transient formation and well testing. They proposed a p-p deconvolution algorithm originating from the work of von Schroeter et al. (2004) by using the following p-p deconvolution equation given as

$$\begin{aligned} p(\mathbf{r}_2, t) &= p_{o2} - \int_{-\infty}^{\ln t} \Delta p_m(\mathbf{r}_1, t - \mathrm{e}^{\xi})\mathrm{e}^{z_d(\xi)}\mathrm{d}\xi \\ \Delta p(\mathbf{r}_2, t) &= \int_{-\infty}^{\ln t} \Delta p_m(\mathbf{r}_1, t - \mathrm{e}^{\xi})\mathrm{e}^{z_d(\xi)}\mathrm{d}\xi, \end{aligned} \tag{4.82}$$

where $\Delta p(\mathbf{r}_2, t) = p_{o2} - p(\mathbf{r}_2, t)$. Equation (4.82) is obtained from Equation (4.80) by using $z_d(\xi) = \ln(t\mathcal{G}(\mathbf{r}_1, \mathbf{r}_2, t))$ as a function of $\xi = \ln t$. Equation (4.82) is exactly the same as the p-r convolution equation formulated by von Schroeter et al. (2004) if the pressure change $\Delta p_m(\mathbf{r}_1, t)$ in Equation (4.82) is replaced by the flow rate q_m. Therefore, the above deconvolution algorithms can also be used for p-p deconvolution by simple replacing the flow-rate q_m data by the measured pressure change data $\Delta p_m(\mathbf{r}_1, t)$. As for the p-r deconvolution, the deconvolution algorithm of Pimonov et al. (2009a; 2009b) will be used for p-p deconvolution of the examples given below. Then the $t\mathcal{G}$-function is obtained from $t\mathcal{G}(t) = \mathrm{e}^{z_d(\xi)}$, where $z_d(\xi)$ is computed from the deconvolution algorithm. To compute the $\mathcal{G}$-function, we simply divide the $t\mathcal{G}$-function by time. The algorithm requires that the pressure change data measured at the spatial location $\mathbf{r}_1$, i.e., $\Delta p(\mathbf{r}_1, t)$ be represented as stepwise constant, and uses a piecewise linear approximation for $z_d(\xi)$. We simply use a two point averaging scheme for two subsequent pressure points to represent $\Delta p(\mathbf{r}_1, t)$ data as stepwise constant. As will be shown in the example section, this representation produces very satisfactory results for reconstructing the $t\mathcal{G}$-function from Equation (4.82).

The algorithm of Pimonov et al. (2009a; 2009b) has the flexibility to adjust both initial pressure at the gauge location $\mathbf{r}_2$ (p_{o2} in Equation (4.82)) and pressure change data $\Delta p(\mathbf{r}_1, t)$ during reconstruction of the $t\mathcal{G}$-function. The algorithm can be applied to any selected data intervals in a given test sequence as in p-r deconvolution. The latter feature is very useful if Equation (4.82) is used for inconsistent data sets. Hence, the data intervals in a given test should be selected so that Equation (4.82) produces consistent $t\mathcal{G}$-functions for each of the test intervals. For instance, we may select only the buildup periods among the test sequences to apply p-p deconvolution. In cases where p-p deconvolution is applied to individual buildup periods (if the test data contain more than one buildup period), we may be able to simultaneously estimate the $t\mathcal{G}$-function and the correct initial formation pressures at the gauge location p_{o2}, as suggested by Levitan et al. (2006).

4.9. Pressure-Pressure Deconvolution Examples

In this section, we apply p-p deconvolution to three packer-probe IPTTs. The first example is a simulated packer-probe test in a slanted well. The second one is a field example for a packer-probe IPTT conducted in a vertical well. The third one is also a field example for a packer-probe IPTT conducted in a vertical well in a three-layer system.

4.9.1. Simulated slanted well IPTT example

This is a simulated packer-probe test (with a single observation probe) in a slanted well in a single layer system (Figure 3.26). The input model parameters used to simulate the test are given in Table 4.2.

The simulated packer and probe pressures along with the flow-rate history are shown in Figures 4.45 and 4.46. As shown in these figures, the IPTT contains three flow and two buildup (BU) periods. The total number of pressure data points is $N_p = 256$ for a 5040-s (1.4-hr) test duration.

For this example, the packer is set at the wellbore and its pressure given as $p_p(r = r_w, \theta = 0, z = 0, t) = p_p(t)$, where $\{r = r_w, \theta = 0, z = 0, t\}$ is the coordinate and the observation probe pressure $p_v(r = 0, \theta = 0, z = z_o, t) = p_v(t)$. Notice that the probe is set at the wellbore with a distance of z_o from the center of the packer (Figure 3.26). The pressure change for the packer is defined as: $\Delta p_p(t) = p_{op} - p_p(t)$ and for the probe $\Delta p_v(t) = p_{ov} - p_v(t)$. Now let us set $\Delta p_m(\mathbf{r}_1, t) = \Delta p_p(t)$ and $\Delta p(\mathbf{r}_2, t) = \Delta p_v(t)$ in Equation (4.82). The objective is to estimate the $t\mathcal{G}$-function, where $\mathcal{G} = \mathcal{G}_{v-p}$ (see Chapter 3), based on Equation (4.82) using the deconvolution algorithm given by Equation (4.70) with observation probe pressure change $\Delta p_v(t)$ and packer pressure change $\Delta p_p(t)$. The initial formation pressures

Table 4.2 Formation and fluid properties for the slanted well IPTT example

ϕ	fraction	0.2
c_t	psi^{-1}	7.5×10^{-6}
μ	cp	1.0
k_h	md	20
k_v	md	5
S	unitless	2.5
C	B/psi	2×10^{-6}
h	ft	60
z_w	ft	30
z_o	ft	6.4
θ_w	degree	45
r_w	ft	0.354
l_w	ft	1.6
p_{op} (packer)	psi	4200
p_{ov} (probe)	psi	4197

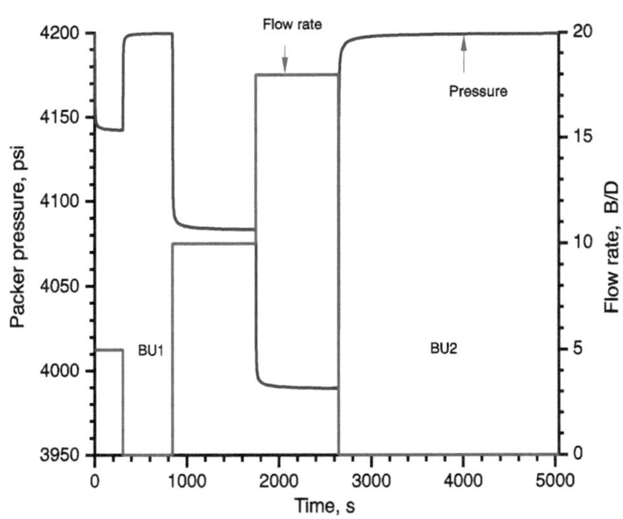

Figure 4.45 Packer pressure and flow-rate data for the simulated IPTT in a slanted well in a single-layer system.

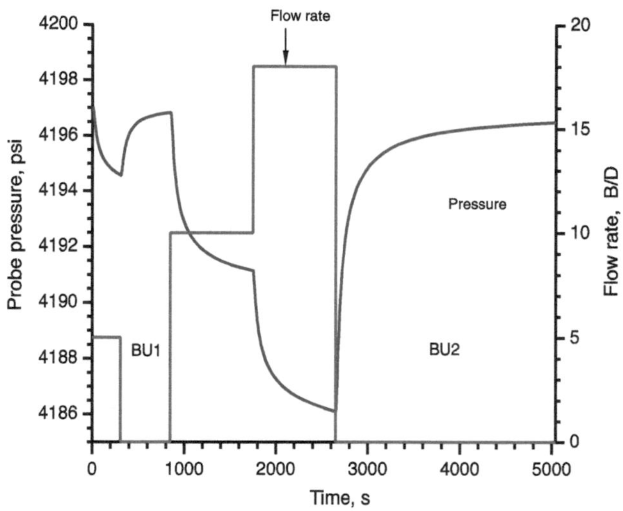

Figure 4.46 Probe pressure and flow-rate data for the simulated IPTT in a slanted well in a single-layer system.

at the packer and probe locations are $p_{op} = 4200$ psi and $p_{ov} = 4197$ psi, respectively.

To investigate the effect of the noise on the reconstructed (deconvolved or computed) $t\mathcal{G}$-functions for this simulated test example, we use simulated packer and probe pressure data sets corrupted by noise, which is treated as an independent and identically distributed normal random error with zero mean and specified standard deviation (σ) computed as

$$\sigma = \frac{\epsilon||\Delta\mathbf{p}||_2}{\sqrt{N}_p}, \tag{4.83}$$

where $\Delta\mathbf{p}$ is the N_p-dimensional vector of true pressure changes (either for packer or probe) generated from the model, where the pressure and flow-rate data are shown in Figures 4.45 and 4.46, and $||\mathbf{x}||_2(=\sqrt{x_1^2 + x_2^2 + \cdots})$ is the l_2-norm. The error level as a fraction of the pressure change is represented by ϵ in Equation (4.83). For example, $\epsilon = 0.01$ corresponds to an error level of 1%. Given the true pressure change data, along with specified ϵ, we generate the N_p-dimensional vectors of normal errors with zero mean, and σ is computed from Equation (4.83). The generated random errors are then added to the true (absolute) pressure data to generate the corrupted (or "measured") packer and probe pressure to be used for p-p deconvolution. This error generation process provides the same standard deviation of errors in both the absolute pressure p and the pressure change Δp $(= p_o - p)$. However, the specified error level (ϵ) for the pressure change data may

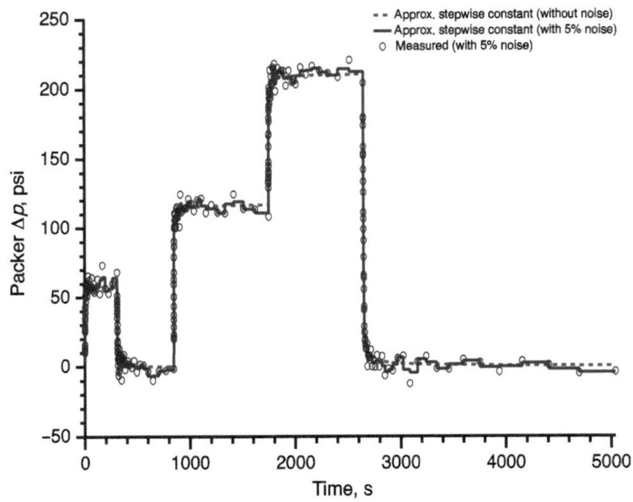

Figure 4.47 Comparison of the measured (simulated data 5% noise) packer pressure change with the pressure change approximated stepwise-constants without noise and with 5% noise.

not correspond to the same error level in the pressure data p because magnitudes of the pressure and pressure change data may differ significantly. The l_2-norms of true probe and packer pressure changes ($||\Delta\mathbf{p}||_2$ used in Equation (4.83)) are 86 and 1667.5 psi, respectively, and $N_p = 256$ in Equation (4.83) for this simulated test example.

In general, observation probe pressure data contain much less noise than packer pressure data; hence we generate random errors from Equation (4.83) with a "small" noise level of 1.0% for the probe data. This level of percentage error in probe pressure change generates a standard deviation of errors in probe pressure data equal to $\sigma = 0.05$ psi. For the packer, we consider three different sets of corrupted packer pressure change data generated from Equation (4.83) with three different levels of noise; 1%, 5%, and 10%. We deliberately chose such high levels of noise for packer pressure change data to test the ability of the deconvolution algorithm to recover the unknown $t\mathcal{G}$ response closely to its true value. The standard deviations of errors in packer pressure change data with 1%, 5% and 10% error are $\sigma = 1.04, 5.21$, and 10.42 psi, respectively.

As mentioned previously, when performing p-p deconvolution based on the algorithm of Pimonov et al. (2009a; 2009b), we approximate the measured packer pressure change ($\Delta p_m = \Delta p_p(t)$ in Equation (4.82)) as stepwise constants. Figure 4.47 shows a comparison of the approximated stepwise constant packer pressure changes, where the packer pressure change data contain a noise level of 5%, with the true value. Noisy (contaminated called measured) packer pressure data are also shown in the same figure. Note that some steps at late build times in both buildup periods are negative due

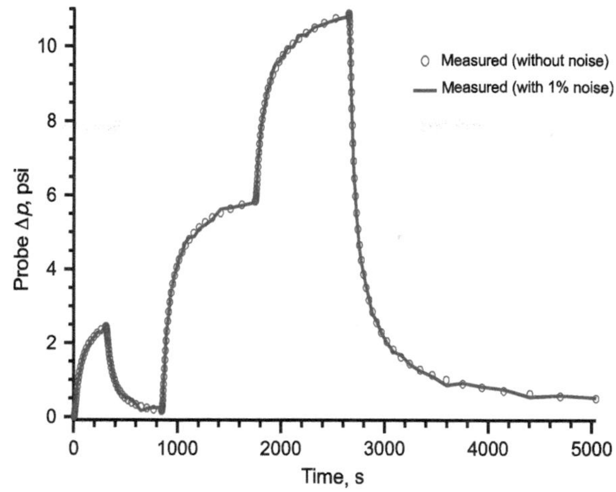

Figure 4.48 Comparison of the measured (simulated) and corrupted with 1% noise probe pressure change.

to noise, which may not be the case in real data. Shown in Figure 4.48 is a comparison of the true probe pressure data with the noisy probe pressure data that will be used in *p–p* deconvolution. For this deconvolution, we will consider only a single noisy set of probe pressure data with a noise level of 1%, but consider three different noisy sets of packer pressure change data with the noise levels of 1%, 5% and 10%.

First, the $t\mathcal{G}$-function is computed (deconvolved) by using the packer and probe pressure change data set [$\Delta p_p(t)$ and $\Delta p_v(t)$] for the entire test duration. We use 40 uniform logarithmically spaced time nodes to reconstruct the $t\mathcal{G}$-function. For this example and others to be given, we use the same input tuning parameters of the deconvolution algorithm; e.g., the same "standard deviation" of the curvature constraints ($\sigma_c = 0.17$) and the same number of nodes (equal to 40). As this is a simulated test, we know the standard deviation of errors in noisy probe pressure data and packer pressure change data and input these values in the deconvolution algorithm. As an initial guess for the initial formation pressure at the probe location, we take the highest probe pressure data in the noisy probe data set, which is equal to 4197.05 psi. It should be pointed out that we assume that the initial formation pressure at the packer location is known so that we can compute the packer pressure change to be used in Equation (4.82). In practice, we will have an error or uncertainty in the initial pressure and, hence, this error will be reflected in the computed values of packer pressure changes. In fact that is one of the reasons why we have deliberately considered different error levels for packer changes for the deconvolution applications to be given for this simulated test, because the error in initial packer pressure will be reflected in the total error in the computed packer pressure changes, and

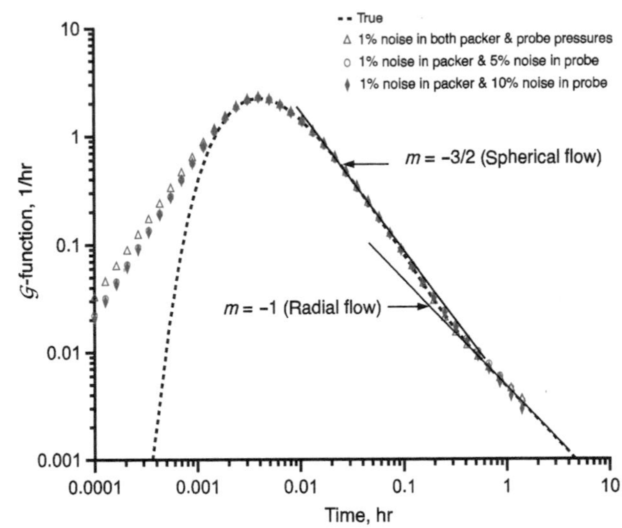

Figure 4.49 Comparison of the true response with the reconstructed $\mathcal{G}$-functions from noisy probe and packer data sets.

we allow updating of packer pressure changes in deconvolution, i.e., treating it as unknown. Perhaps, starting from Equation (4.82) one could develop a p-p deconvolution equation allowing one to treat the initial pressure at the packer location as unknown in addition to the initial pressure of the observation probe.

Figures 4.49 and 4.50 show the reconstructed (deconvolved) $\mathcal{G}$- and $t\mathcal{G}$-functions from the probe data with a 1% noise level and three different sets of noisy packer pressure data (1%, 5%, and 10%) in p-p deconvolution. As is clear from Figures 4.49 and 4.50, the reconstructed $\mathcal{G}$ and $t\mathcal{G}$ from noisy probe pressure and packer pressure change data sets agree very well with the true responses, except at very early times. The difference at early times could be due to the treatment of the $t\mathcal{G}$-function at the first node in the deconvolution algorithm. In all error cases, spherical and radial flow regimes are identifiable, and the deconvolution algorithm yields the correct initial pressure for the probe, which is equal to 4197 psi. Furthermore, the estimated (or reconstructed) packer pressure changes and probe pressures are reasonably close to corresponding true (stepwise constants) ones (Figure 4.51). Figure 4.52 shows the reconstructed packer pressure stepwise-constant changes and probe pressure for the probe data set with 1% error and for the packer data set with 5%. Thus, we have confidence in the reconstructed $\mathcal{G}$- and $t\mathcal{G}$-functions.

One of the useful features of the recent deconvolution algorithms (Levitan, 2005; Pimonov et al., 2009a; 2009b; von Schroeter et al., 2004) is to apply deconvolution to the selected data intervals, e.g., buildup

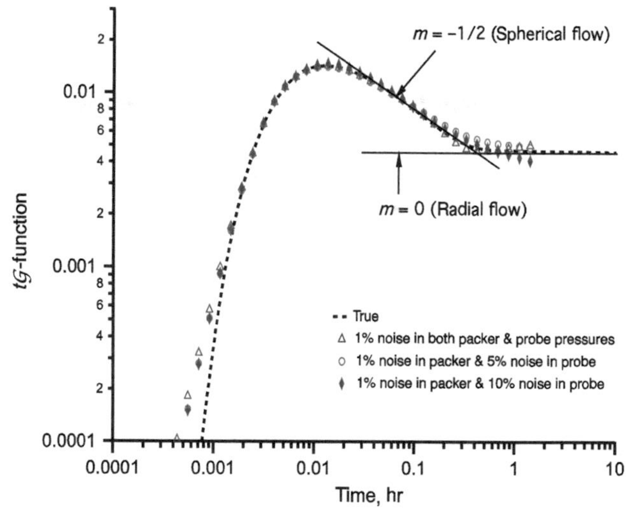

Figure 4.50 Comparison of the true response with the reconstructed $t\mathcal{G}$-functions from noisy probe and packer data sets.

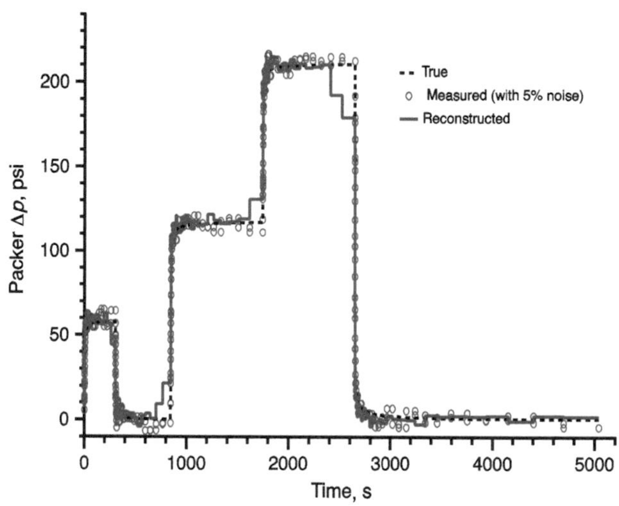

Figure 4.51 Comparison of the reconstructed packer pressure changes by *p-p* deconvolution.

portions, in a given test sequence. This feature is useful, particularly in dealing with inconsistent data sets, because by applying deconvolution to different selected data intervals, we may identify the data intervals that are consistent with the deconvolution model. Applying deconvolution to the entire data set that is inconsistent with the deconvolution model does

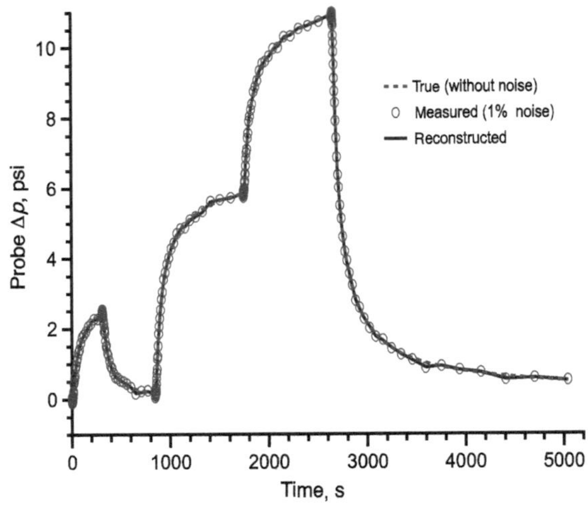

Figure 4.52 Comparison of reconstructed probe pressures by p-p deconvolution.

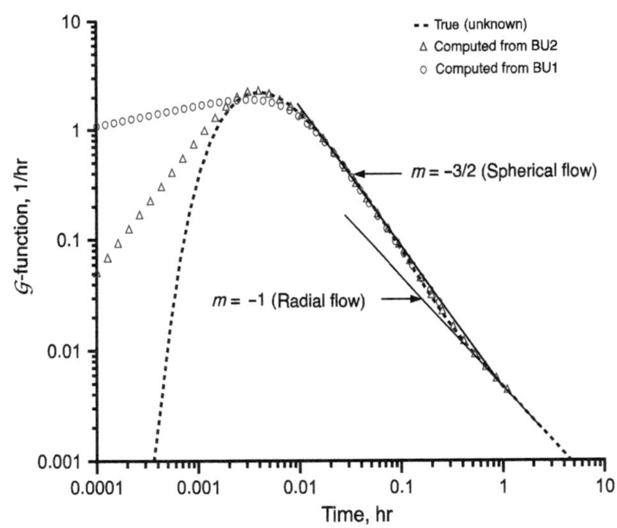

Figure 4.53 Comparison of the true response with reconstructed $\mathcal{G}$-functions from buildup (BU 1 and 2) probe pressures with 1% noise and packer pressure change data with 5% noise.

not produce meaningful results, as shown by Levitan (2005). Although the simulated slanted-well test example considered here is consistent with the p-p deconvolution model of Equation (4.82), we want to show that one can also apply p-p deconvolution to two individual buildup periods of the simulated test (shown as BU1 and BU2 in Figures 4.45 and 4.46) to reconstruct the $\mathcal{G}$- and $t\mathcal{G}$-functions. Figures 4.53 and 4.54 show the

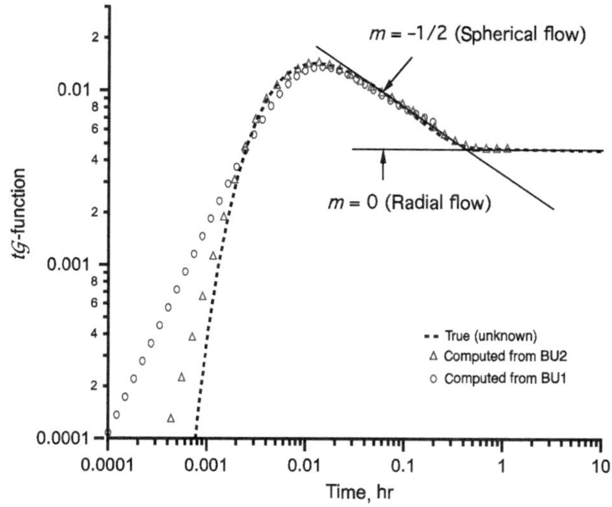

Figure 4.54 Comparison of the true response with the reconstructed $t\mathcal{G}$-functions from buildup (BU 1 and 2) probe pressures with 1% noise and packer pressure change data with 5% noise.

reconstructed $\mathcal{G}$ and $t\mathcal{G}$ by processing probe pressures pertaining to first (BU1) and second (BU2) periods, separately, for the case of a 1% error in the probe pressure and a 5% error in the packer pressure data. In these *p*–*p* deconvolution applications, we do not treat initial pressure at the probe location as unknown, but fix it at its true value of 4197 psi. The entire period of the packer pressure changes, including buildup periods, are jointly estimated along with the $t\mathcal{G}$-function during deconvolution.

As can be seen from Figures 4.53 and 4.54, the deconvolution algorithm reconstructs the $\mathcal{G}$- and $t\mathcal{G}$-functions reasonably well from BU1 and BU2 alone except at very early times. As noted previously, when reconstructing $\mathcal{G}$ and $t\mathcal{G}$ shown in Figures 4.53 and 4.54, we fixed the initial probe pressure at its correct value of 4917 psi in the deconvolution algorithm. This was necessary because the algorithm converges to an incorrect value of initial probe pressure if we treat the initial probe pressure as unknown in addition to $t\mathcal{G}$ and packer pressure changes when reconstructing the $t\mathcal{G}$-function from the buildup period alone. For instance, Figures 4.55 and 4.56 show reconstructed $\mathcal{G}$- and $t\mathcal{G}$-functions (with the initial probe pressure treated as unknown in *p*–*p* deconvolution algorithm) from the probe-pressure data pertaining to BU2 alone. In this application, the deconvolution algorithm converged to the initial probe pressure value equal to 4196.67 psi, which is only 0.33 psi lower than the correct initial probe pressure value of 4197 psi. It is remarkable to notice, from Figures 4.55 and 4.56, a *small* difference in the initial pressure has a profound effect towards end of the reconstructed $\mathcal{G}$ and $t\mathcal{G}$. This application clearly shows there is insufficient information about the

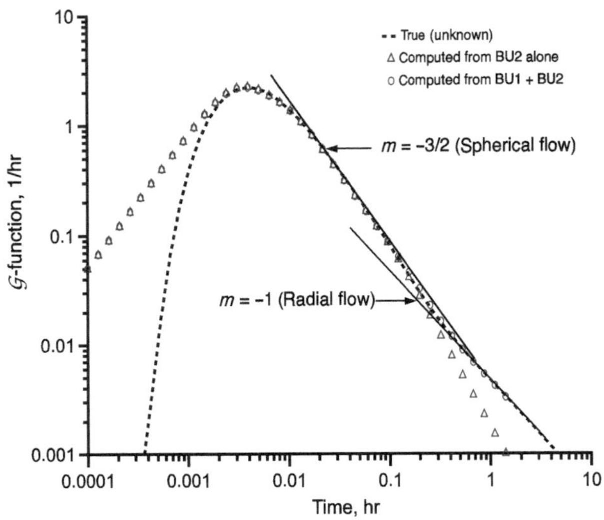

Figure 4.55 Comparison of the true response with the reconstructed $\mathcal{G}$-functions from a single buildup (BU2) and both buildup (BU1 and BU2) periods with an unknown initial probe pressure and 1% noise in probe and 5% noise in packer pressure change data.

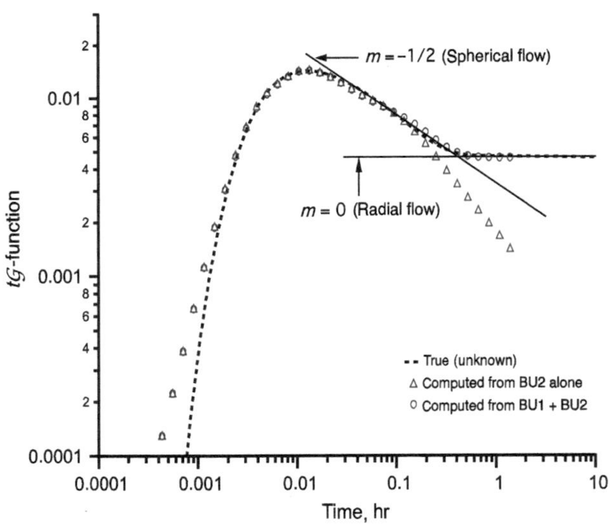

Figure 4.56 Comparison of the true response with the reconstructed $t\mathcal{G}$-functions responses from a single buildup (BU2) and both buildup (BU1 and BU2) periods with an unknown initial probe pressure and 1% noise in probe and 5% noise in packer pressure change data.

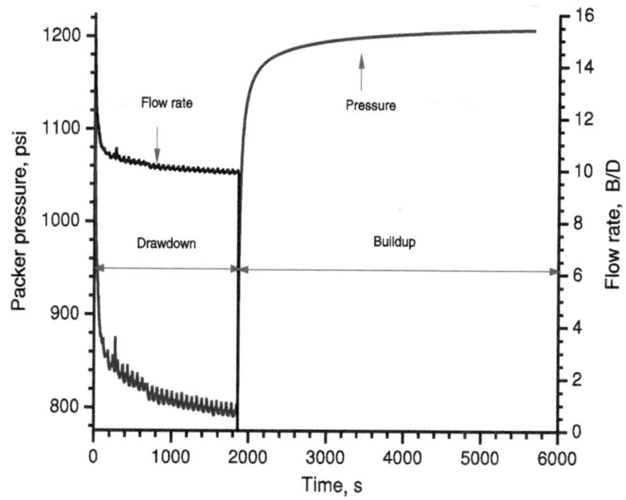

Figure 4.57 Measured packer pressure and flow-rate data for Field Example 1.

initial pressure in the data from a single buildup period. Similar observations have been already made by Levitan et al. (2006) for the p-r deconvolution method. If the probe pressure data from both buildup periods are included, then the initial probe pressure can be treated as unknown. Then, as shown in Figures 4.55 and 4.56, the deconvolution algorithm correctly yields the probe initial pressure as 4197 psi and the reconstructed $\mathcal{G}$ and $t\mathcal{G}$ functions.

4.9.2. Vertical well IPTT field Example 1

The next example is a field packer-probe IPTT conducted at a vertical well in an 11-ft thick water zone, where a single vertical probe was mounted 6.4 ft above the middle of the dual-packer interval. The other pertinent parameters are $z_w = 2$ ft, $l_w = 1.6$ ft, $z_o = 6.4$ ft, $c_t = 8 \times 10^{-6}$ psi^{-1}, $\mu = 1$ cp, $r_w = 0.354$ ft, $\phi = 0.22$, and the tool storage coefficient (C) is about 2.0×10^{-6} B/psi. The test consists of a 1862-second (0.52 hr) production period, during which about 34 liters of water produced, followed by a 3826-second (1.06 hr) buildup period. Pressures at the packer and probe were recorded with high-resolution quartz gauges. The measured pressures at the packer and probe, and the measured flow rate from the pump displacement are shown in Figures 4.57 and 4.58.

The initial pressures measured at probe and packer locations are 1211.4 and 1218.58 psi, respectively. Normally, three pressure measurements per second are acquired by most WFTs at both packer and probes. We therefore filter data by applying different filter criteria for drawdown and buildup periods as well as the packer and probes. In the end, 513 pressure data

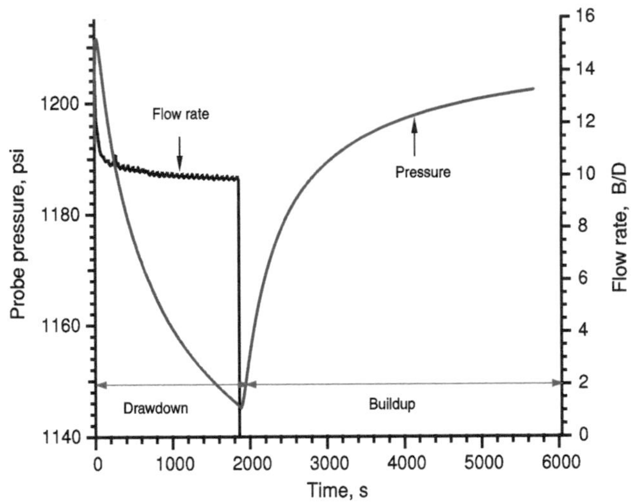

Figure 4.58 Measured probe pressure and flow-rate data for Field Example 1.

points remain in each pressure data set. We then apply p-p deconvolution to the entire test data to the reconstruct $\mathcal{G}$- and $t\mathcal{G}$-functions and allow the algorithm to adjust the initial probe pressure. As they are accurately measured, packer pressure changes (computed using the initial pressure measured at the packer interval and then approximated as stepwise constants) are fixed in deconvolution (Figure 4.59). In deconvolution, we ignored probe pressure data for times less than 23 seconds (0.0064 hr) by assigning zero weights to such data because they were affected by tool operations.

The initial probe pressure estimated by the algorithm is 1211.54 psi, and the deconvolved $\mathcal{G}$- and $t\mathcal{G}$-functions are shown in Figures 4.60 and 4.61. Figure 4.59 also compares the computed and measured probe pressures. The RMS error for the probe pressure match is 0.035 psi. As can be seen in Figures 4.60 and 4.61, the radial flow regime can clearly be identifiable at the late time from the reconstructed $\mathcal{G}$ and $t\mathcal{G}$, but not the spherical (or hemispherical) flow regime.

For comparison purposes, we also present the conventional buildup derivatives based on the superposition time function using measured rates of both packer and probe, as shown in Figure 4.61. As can be seen from this figure, the buildup derivatives for both packer and probe data converge to the same radial flow regime established at about 0.15 hr due to the no-flow top and bottom boundaries of the formation. The radial flow regime is also confirmed by the $\mathcal{G}$- and $t\mathcal{G}$-functions reconstructed from p-p convolution without flow-rate data. On the other hand, the buildup derivative for the packer exhibits a hemispherical flow regime ($m = -1/2$ slope) from 0.02 to 0.1 hr. This is because the packer open interval is very close to the bottom of

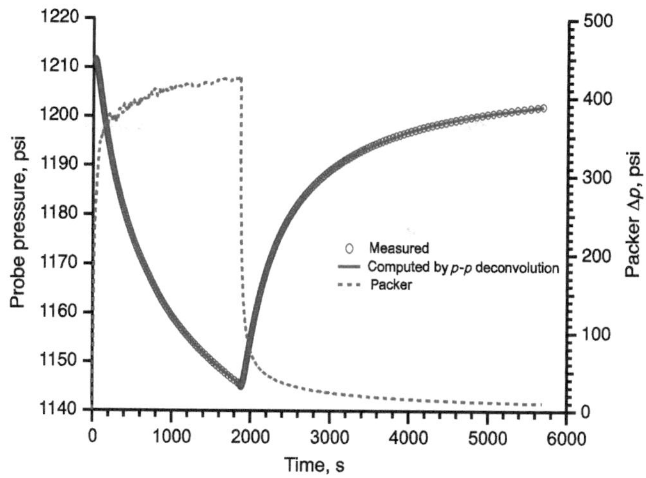

Figure 4.59 Comparison of measured and computed probe pressures (left), and the packer pressure change data (right), which are used in p–p deconvolution.

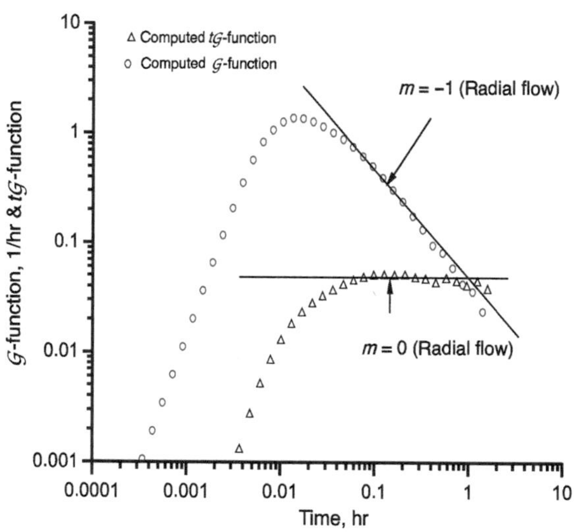

Figure 4.60 Reconstructed $\mathcal{G}$- and $t\mathcal{G}$-functions from p–p deconvolution.

the zone, therefore the response goes almost immediately from a spherical to a hemispherical flow regime without any discernible transition. It should be pointed out that neither the conventional buildup derivative for the probe nor the reconstructed $\mathcal{G}$- or $t\mathcal{G}$-function from p–p deconvolution exhibits a hemispherical flow regime ($m = -1/2$ slope). This is not surprising because $\mathcal{G}$ and $t\mathcal{G}$, by definition, is a ratio of unit-impulse responses at the packer and probe. As can be seen in Figures 4.60 and 4.61, the $t\mathcal{G}$-function behavior is

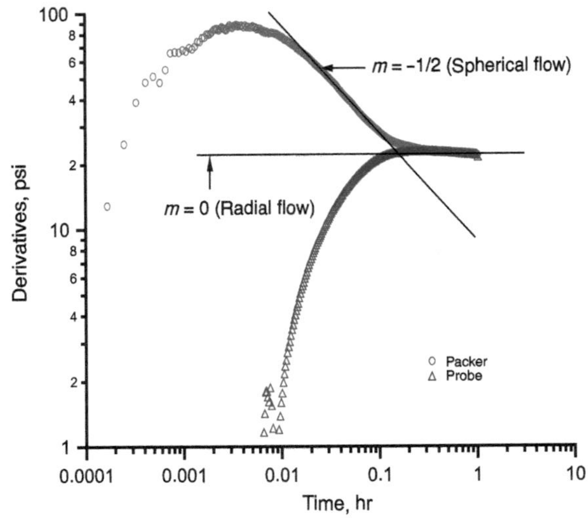

Figure 4.61 Conventional buildup derivatives for the packer and probe.

very similar to that of the probe derivative. Perhaps, this may be considered as one of weaknesses of the $\mathcal{G}$- and $t\mathcal{G}$-functions because it mixes different flow regime information from the unit-impulse responses at two different locations in the same time interval. Nevertheless, the $t\mathcal{G}$ and conventional buildup derivative signatures are typical of a single-layer formation with no-flow top and bottom boundaries where the limited interval is very close to the no-flow bottom boundary of the formation; this is consistent with geological and log data, and verifies the radial flow regime.

4.9.3. Vertical well IPTT field Example 2

The field example is a packer-probe IPTT conducted in a vertical well in a three-layer system (Figure 4.62). The layer definitions shown in Figure 4.62 were determined from openhole logs and local geological information.

Figures 4.63 and 4.64 present pressures at the packer and probe, which were recorded with high-resolution quartz gauges, and flow-rate measurements from the pump displacement. For this test, the flow-rate data are reasonably accurate to use the p-r deconvolution, however we will also apply p-p deconvolution to this IPTT for verification purposes. The initial pressure measured at the probe and packer locations are 2496 and 2500 psi, respectively. There are 553 data points in each pressure data set after filtering.

The p-p deconvolution algorithm applied the probe pressure data using the packer pressure change data, for which the initial pressures are fixed, but the initial probe pressure is estimated. The estimated initial probe pressure is obtained as 2495.88 psi, which is close to the initial guess of 2496.0 psi. The reconstructed $\mathcal{G}$ and $t\mathcal{G}$ are shown in Figure 4.65. For

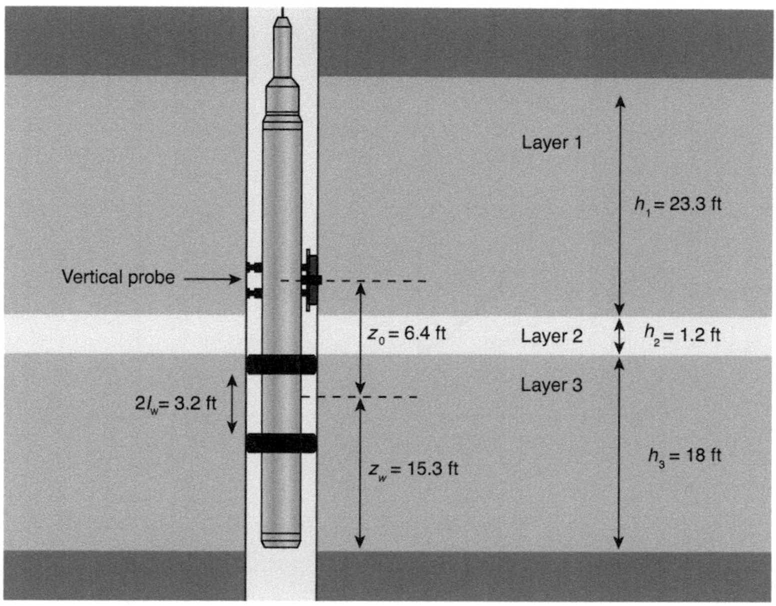

Figure 4.62 Schematic of the packer and probe set up in a three-layer system for IPTT Field Example 2.

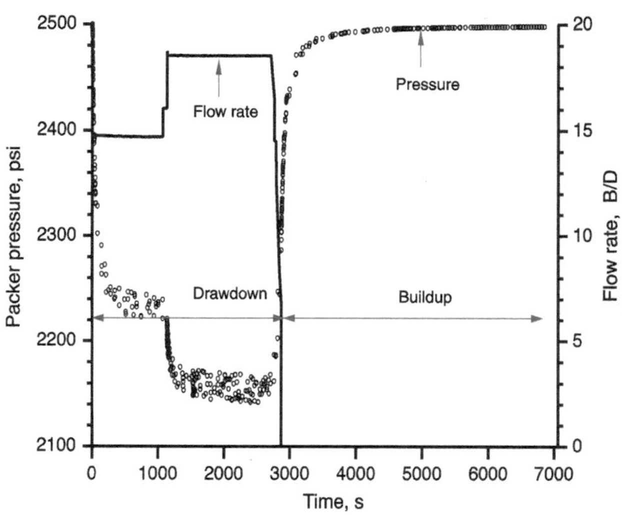

Figure 4.63 Packer pressure and flow-rate data for IPTT Field Example 2.

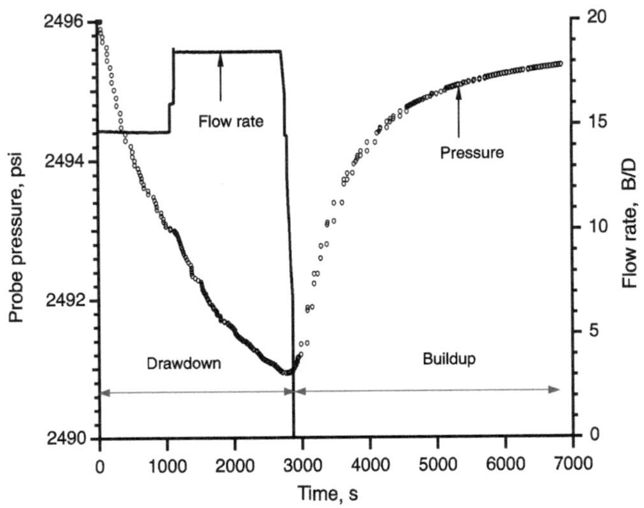

Figure 4.64 Probe pressure and flow-rate data for IPTT Field Example 2.

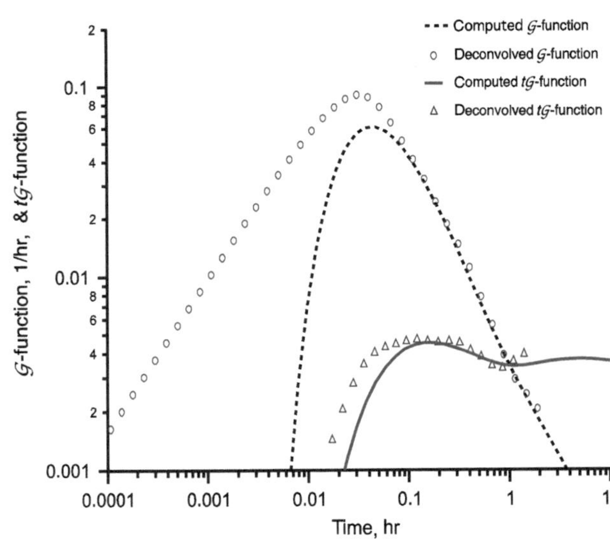

Figure 4.65 Reconstructed $\mathcal{G}$- and $t\mathcal{G}$-functions from p-p deconvolution.

comparison purposes, we also performed pressure-rate deconvolutions for packer and probe pressures, where flow-rate history and initial pressures were fixed during the deconvolution computation. The estimated unit-impulse responses $tg(t)$ for packer and probe are shown in Figure 4.66. As can be seen from Figures 4.65 and 4.66, the estimated responses from p-p and p-r deconvolution are consistent with each other and seem not to exhibit any

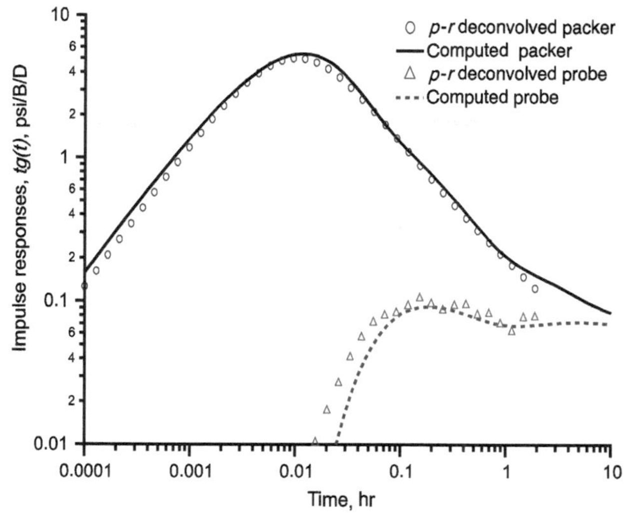

Figure 4.66 Reconstructed unit-impulse ($tg(t)$) responses for the packer and probe from p-r deconvolution.

discernible flow regimes. However, the reconstructed responses, particularly the unit-impulse responses $tg(t)$s for packer and probe give the signature of a layered system where the probe is located in a high permeable layer, and the pseudo-radial flow regime due to no-flow top and bottom boundaries of the whole system has not been established because the $tg(t)$ responses clearly do not merge at late times.

Based on the layer properties estimated for the three layer model given by Figure 4.62 from the MLE nonlinear regression (given in Chapter 5), the model unit-impulse responses $[tg(t)]$ for packer and probe as well as the model $\mathcal{G}$- and $t\mathcal{G}$-functions are also shown in Figures 4.65 and 4.66. These model responses agree quite well, particularly at times greater than 0.03 hr, with those reconstructed directly from pressure and rate data by p-r and p-p deconvolution.

CHAPTER 5

Nonlinear Parameter Estimation

Contents

Developments in Petroleum Science, Volume 57
ISSN 0376-7361, DOI: 10.1016/S0376-7361(10)05711-0

5.1. Introduction

In the last two decades, computer-aided nonlinear parameter estimation coupled with statistical methods (or simply referred to as nonlinear regression analysis) has become a standard procedure for interpreting pressure-transient data. The reasons are mainly due to: (1) Enhancements in computer hardware, software and the instrumental resolution of pressure and flow-rate measurements (e.g., quartz pressure gauges, flow-meters and spinners), and (2) Availability of sophisticated well/reservoir models. As we said earlier, for well testing, reservoir models normally include a few wells but not the entire field, unless it is very small. For wireline formation testing, reservoir or formation models (both will be used interchangeably) typically include a one hundred-ft formation around the wellbore with a high degree of vertical resolution.

The most important feature of nonlinear regression analysis over the conventional graphical methods, such as manual type-curve, semi-log, and log analyses [for example see Bourdet (2002) and Earlougher (1977)], is that it allows one to quantify the uncertainty in the final estimated formation parameters and model uniqueness in the presence of noisy (or inexact) data and uncertainty about the true, but unknown, reservoir model.

The purpose of this chapter is to provide a detailed review of fundamental concepts, theory, definitions, and methods that are used in nonlinear regression analysis of pressure-rate data as well as to present the applications of nonlinear regression analysis specific to synthetic and field data sets of pressure transient well testing and Interval Pressure Transient Testing (IPTT). However, it is not our intention to attempt a thorough treatment of the subject. For a more complete presentation on all aspects of nonlinear regression, we refer the interested reader to the textbooks by (Bard, 1974; Bates & Watts, 1988; Draper & Smith, 1966; Seber & Wild, 1989). In addition, it is assumed that readers have a modest background on some elementary vector/matrix algebra and multivariable differential calculus as well as on the ideas of statistics and probability; such as parameters, estimates, distributions, mean and variance of a random variable, covariance between two variables, and covariance matrices for multivariate random variables.

First, we begin with an overview of the parameter estimation problem for pressure-transient data interpretation. Next, the parameter estimation method based on the maximum likelihood estimation (MLE) is introduced and compared with the least-squares estimation (LSE), which is the most widely used and known estimation method in petroleum engineering. Although MLE is not widely used in petroleum engineering, it is, by far, the most commonly used method of parameter estimation in statistics literature

(Bard, 1974; Barlow, 1989; Bates & Watts, 1988; Carroll & Ruppert, 1988; Draper & Smith, 1966). In fact, it can be shown that LSE is a special case of MLE if the errors in observed pressure and rate data follow a normal (or Gaussian) distribution. Next, we discuss minimization techniques and algorithms (while constraining the parameters within a feasible region) that can be used for minimizing the objective functions arising from MLE and LSE formulations. Computation of various statistics [e.g., 95% confidence intervals for parameters, correlation coefficients for parameters, standard deviation of residuals, root-mean-square (RMS) errors] in assessing the uncertainty in estimated parameters and goodness of fit is provided. The nonlinear parameter estimation problem for the case in which the number of observed pressure data is less than the number of unknown model parameters is also given. Such applications arise in the case where the model takes into account the spatial heterogeneity in rock property fields (e.g., permeability and porosity in each simulator grid block) based on geostatistical models (Journel & Huijbregts, 1978; Oliver, Reynolds, & Liu, 2008). Throughout, we consider models based on a single-phase flow of a slightly compressible fluid of constant viscosity, but the methodology presented here is general in that it can also be used for multiphase flow in the reservoir.

At the end of the chapter, a number of examples are presented using both real field and synthetic tests to illustrate the use of nonlinear regression analysis based on the maximum likelihood and least-squares estimations.

5.2. Parameter Estimation Problem for Pressure-Transient Test Interpretation

The objective of pressure-transient data interpretation is to identify the "appropriate" interpretation model(s) and then obtain "reasonable" estimates of the formation (or reservoir) parameters of interest from pressure and flow-rate measurements. These estimates are defined in terms of a reservoir model that is derived using simplifying assumptions, but still from physical principles (conservation laws for mass, Darcy's law, etc.) governing the behavior of the system under observation. For example, as discussed in Chapter 2, the single-phase pressure transient in porous media is governed by the pressure diffusion equation with appropriate initial and boundary conditions that can be solved either analytically or numerically, depending on the complexity of fluid flow geometry and reservoir geology.

When a mathematical reservoir model is specified (or identified) with its parameters, and pressure and rate data have been collected from the system, one is in a position to evaluate the model's goodness of fit, i.e., how well the model fits the observed pressure and rate data. Goodness of fit is assessed by finding parameter values of a model that "best" fits the data—a procedure called parameter estimation, or also referred to as optimization. In general, the model functions of pressure and rate data used for pressure-transient data interpretation are nonlinear with respect to the model parameters—a model function is called nonlinear if any of the partial derivatives of a model response function (e.g.pressure) is dependent on any of the model parameters. Otherwise, it is linear. For example, let $y_c(t)$ be a model response function of an independent variable t with two model parameters a and b, then $y_c = a + bt$ is a linear model, but $y_c = at^b$ is not a linear model. Nonlinear parameter estimation therefore must be applied for pressure-transient test data interpretation.

Matching nonlinear models to observed pressure and rate data that are contaminated by any level of noise, and in the presence of uncertainty in the model itself, is a notoriously ill-posed inverse problem. In other words, one often finds a number of different models as well as different sets of unknown model parameters, even if the model chosen for history matching is the correct one, that may appear to match the observed data equally well. Therefore, as stated well by Bard (1974), it is not enough to compute the unknown model parameters based on minimizing an objective function and then state that these are the estimated values of the unknown parameters. One must also investigate the reliability and precision of estimates based on statistical inferences coupled with the estimation procedure.

The applications of computer-aided (automated) type-curve matching and nonlinear regression analysis for interpretation of the pressure-transient data have been subject to many studies in petroleum engineering literature. Early application dates go back to the 1970s by Dixon, Seinfeld, Startzman, and Chen (1973), Dogru, Dixon, and Edgar (1977), Earlougher and Kersch (1972), and Padmanabhan and Woo (1976). Later, development of new more general solutions to a wide variety of well/reservoir models (e.g., layered systems, limited-entry along the wellbore, slanted/horizontal wells, vertically stratified formations, fractures, composite systems, etc.) and the reliability of the Stehfest (1970) algorithm for inversion of Laplace transforms have lead to many studies considering computer-aided type-curve matching and nonlinear regression analyses in the pressure-transient testing literature (Abbaszadeh and Hegeman, 1990; Abbaszadeh & Kamal, 1988; Anraku & Horne, 1995; Barua, Horne, Greenstadt, & Lopez, 1988; Barua, Kucuk, & Gomez-Angulo, 1985; Carvalho, Redner, Thompson, & Reynolds, 1992; Hanson, 1986; Horne, 1994; Kuchuk & Onur, 2003; Nanba & Horne, 1992; Onur & Kuchuk, 2000; Onur & Reynolds, 2002; Rosa & Horne, 1983; 1995).

Over the last two decades, applications of nonlinear regression analysis within the maximum likelihood and Bayesian frame work to more complicated models involving geostatistics-based heterogeneous reservoir models using numerical simulators have also been subject to many studies in the pressure-transient testing literature (Chu, Reynolds, & Oliver, 1995; Gok, Onur, & Kuchuk, 2005; He, Oliver, & Reynolds, 2000; Landa, Horne, Kamal, & Jenkins, 2000; Oliver, 1994; Oliver, He, & Reynolds, 1996; Oliver et al., 2008).

Most of the early works cited above are based on the use of standard unweighted least-squares (UWLS) or weighted least-squares (WLS) estimation methods where the weights[1] for observed pressure data sets are assumed to be known *a priori* or assigned arbitrarily. In this book UWLS, WLS, and ML estimations are often called UWLSE, WLSE, and MLE, respectively, or UWLS, WLS, and ML regressions.

However, weights (or error variances) for observed pressure and rate data are often uncertain in practice for pressure-transient testing applications. In any pressure transient testing, there are two main sources of uncertainty: (1) The gauge resolution and (2) Measurement environment. As discussed in Chapter 1, the gauge accuracy and resolutions sometimes are not well characterized as a function of temperature and the magnitude of pressure changes. Furthermore, the associated electronics for data transmission, the digitizer resolution, and specific sampling time affect both gauge accuracy and resolution considerably. Today the high precision quartz gauge could detect pressure changes of less than 0.01 psi. The measurement environment at downhole conditions is normally noisy. For instance, the bursting of a gas bubble in the vicinity of the pressure gauge can create a high frequency noise spike of amplitude greater than 1 psi. This is particularly true in the wellbore during production, while high frequency noise is absent during buildup and at the observation wells and probes during both production and buildup. If the pressure is changing fast and the sampling rate is slow, which is preferred for high resolution, it could be then difficult to separate high frequency noise from measurements.

Error variances (or uncertainties) on flow-rate data are typically larger than those of pressure data because the rate is traditionally and usually not directly measured in well testing applications, though it is now possible to measure downhole or sandface flow rate accurately and to transmit it in real time to the surface. New wireline formation testers (WFT) may have direct flow metering devices, but for older ones flow rate vs. time data must be inferred from other measurements. For example, as discussed in Chapter 1, fluid production at the sink probe or at the packer module can be performed by using a sample chamber module, flow-control module, or pumpout module. In cases where the flow-control and/or pumpout modules are used, reliable flow-rate data can often be determined from direct measurements of the piston displacement and tool characteristics (Kuchuk, Ramakrishnan, & Dave, 1994). The flow rate computed in this way may

[1] As will be shown later, weights actually represent the inverse of variances of errors in data if the parameter estimation problem is viewed within maximum likelihood (ML) estimation.

still be inaccurate or uncertain to some degree because of extraneous factors such as changing fluid compressibility within the tool or friction. Thus, in cases when flow rate is uncertain, one may use a nonlinear regression scheme that allows simultaneous estimation of flow rates and formation parameters. In this optimization scheme, flow rates measured or estimated are used as initial guesses and *a priori* information in the objective function and then simultaneously updated with the unknown formation parameters by history matching measured pressure and rate data (Kuchuk et al., 1994; Onur & Kuchuk, 1999).

It is well known that if the UWLSE (unweighted least-squares estimation) is used for data sets having disparate orders of magnitude, noise or both, then those data sets with large magnitudes and/or noise will dominate those having small magnitudes and/or noise in estimation (Onur & Kuchuk, 2000). Thus, information contained in data sets with small magnitudes and/or noise will be lost. For multiwell interference tests and IPTTs conducted with a dual packer and observation probes, the pressure drop at producing well or the packer could be more than a few hundred psi, while pressure drop at the observation wells and probes can be much less than a few psi depending on the formation characteristics. Also, in the cases where some observations are less reliable than the others, it is required that parameter estimates are less influenced by unreliable observations. To achieve this, the appropriate choice may be to use a weighted least squares (WLS) regression. Although a number of studies have presented algorithms based on WLS regression, all these studies assume that the weights (actually inverse of variances of errors in data) are assumed to be known *a priori*. As mentioned above, working with real field data, it is often difficult to know the error variance and its structure in pressure and rate data and, thus, to determine the proper weights to be used in the WLS regression.

In the following sections, we will consider a more general estimation approach—the maximum likelihood estimation(MLE)—for the interpretation of pressure-transient formation tests. The MLE not only allows one to treat errors in observed data as unknown (or uncertain) in addition to formation parameters but also provides incorporation of prior information into the estimation when it is used within the Bayesian framework. The main advantage of the MLE method over the standard WLS is that it eliminates the trial-and-error procedure required to determine the appropriate weights for transient pressure and rate data sets in history matching. This provides a significant improvement in parameter estimation when working with pressure data sets of disparate orders of magnitude and/or unknown noise levels. Applications of the maximum likelihood estimation for history matching multiple spatial sets of interval pressure transient tests (IPTT) acquired with wireline formation testers have been previously presented by Kuchuk and Onur (2003), Onur and Kuchuk (2000), and Onur, Hegeman, and Kuchuk (2004a; 2004b).

5.3. Parameter Estimation Methods

As discussed above, there are two widely used parameter estimation methods: (1) Least-squares estimation (LSE) and (2) Maximum likelihood estimation (MLE). The former is the most widely used estimation procedure. Its popularity stems from the fact that it can be applied in an *ad hoc* manner directly to the deterministic model (or pure curve fitting models, where the coefficients have no physical significance), without any cognizance being taken of the probability distribution of the observations. On the other hand, MLE treats observations as random variables with certain probability distributions and thus is more suitable for statistical inference about the match. When LSE is viewed as a curve fitting procedure, it has no basis for constructing confidence intervals or testing hypotheses, whereas all these are naturally built into MLE. In addition, MLE is a prerequisite for the Bayesian modeling (to be discussed later). In fact, starting from the principle of maximum likelihood and considering a normal distribution for observed data (or errors in observed data), it can be shown that LSE is a special application of MLE.

From a statistical standpoint, nonlinear regression provides estimates and other inferential results for the unknown parameters $\mathbf{m} = [m_1, m_2, \ldots, m_M]^T$ of the model

$$y_c = f(\mathbf{m}, t), \tag{5.1}$$

where y_c is the model response (or behavior) of the system, but for convenience is also called computed response denoted with the subscript c, and t is an independent variable. Because all measurements contain errors or noises, therefore with an error term ϵ, an equation for measurement y_m can be written in terms of y_c as

$$\begin{aligned} y_m &= f(\mathbf{m}, t) + \epsilon(t) \\ &= y_c + \epsilon(t), \end{aligned} \tag{5.2}$$

where y_m is the observed (measured) response (or behavior) of the system denoted with subscript m and is considered as a random sample from an unknown population, and hence has both a deterministic part y_c and a stochastic part ϵ. For a given set of $\mathbf{t} = \left[t_1, t_2, \ldots, t_{N_d}\right]^T$ and $\mathbf{m}$, Equations (5.1) and (5.2) yield a set of y_c and y_m, and in a vectorial form they can be expressed as $\mathbf{y}_c = \left[(y_c)_1, (y_c)_2, \ldots, (y_c)_{N_d}\right]^T$ and $\mathbf{y}_m = \left[(y_m)_1, (y_m)_2, \ldots, (y_m)_{N_d}\right]^T$, where both are N_d-dimensional vectors. The deterministic part y_c denotes the nonlinear functional relationship between the model parameter vector $\mathbf{m}$ and independent variable t. The stochastic ϵ term in Equation (5.2) takes into account the noise (or disturbance) in measured (observed) data $\mathbf{y}_m$ due to many unknown phenomena. It can also be written for discrete values of the independent variable t as an N_d-dimensional measurement error vector $\boldsymbol{\epsilon} = \left[\epsilon_1, \epsilon_2, \ldots, \epsilon_{N_d}\right]^T$. As will be discussed, $\boldsymbol{\epsilon}$ is normally assumed to be a multivariate Gaussian or normal distribution with zero (vector) mean and the $N_d \times N_d$ covariance matrix $\mathbf{C}_D$. Here, and throughout, the superscript T denotes transpose of a vector or matrix. In addition, lower case letters in bold type refer to vectors, generally column vectors, while capital case letters in bold type refer to matrices.

For the pressure-transient testing applications such as well tests and IPTTs, $\mathbf{y}_m$ in Equation (5.2) may represent a set of observed (measured) pressure values at the time vector $\mathbf{t}$; i.e., $\{(y_m)_i, t_i\}_{i=1}^{N_d}$. Here, $\mathbf{m}$ is an M-dimensional model parameter vector which could represent unknown model parameters such as horizontal and vertical permeability of layers, mechanical skin, wellbore storage coefficient, static pressures, etc.

From a statistical point of view, the observed data vector $\mathbf{y}_m$ is a random sample as an outcome of an experiment from an unknown population. The objective of statistical data analysis is to identify the population that is most likely to have generated the sample $\mathbf{y}_m$. In statistics, each population is identified by a corresponding probability distribution function. Associated with each probability distribution is a unique value of the model parameters. As the model parameter $\mathbf{m}$ changes in value, different probability distributions are generated. Formally, a model is defined as the family of probability distributions indexed by the model parameters. Therefore, we denote the probability density function (**pdf**) by $p(\mathbf{y}_m|\mathbf{m})$ that specifies the probability of observing data $\mathbf{y}_m$ given the model parameter vector $\mathbf{m}$. Throughout we will consider a Gaussian distribution for $\mathbf{y}_m$ because the most commonly used **pdf** is of a Gaussian (also referred to as normal) distribution.

There are several reasons for considering a Gaussian distribution function for errors in data. First, a Gaussian distribution is the one in which the information content is least (specifying mean and variance, or covariance in the multivariate case, determines the entire normal distribution). This is preferable in cases in which no specific justification for the distribution exists. It is also the one frequently observed in nature, and tractable mathematically. Moreover, if the observed value of the random variable is the resultant of many additive, independent, effects, the resulting distribution is likely to be a Gaussian distribution. This is a result of the central limit theorems (Bard, 1974; Barlow, 1989).

Hence, we can assume that $\mathbf{y}_m$ can be described by a Gaussian distribution **pdf** with mean $\mathbf{y}_c$ and covariance matrix $\mathbf{C}_D$. This distribution is designated by $N(\mathbf{y}_c, \mathbf{C}_D)$ denoting a normal distribution with mean $\mathbf{y}_c$

and covariance matrix $\mathbf{C}_D$. Thus the **pdf** for observed data $\mathbf{y}_m$ can be expressed as

$$p(\mathbf{y}_m|\mathbf{m}, \mathbf{C}_D) = \frac{1}{(2\pi)^{N_d/2} (\det \mathbf{C}_D)^{1/2}} \times \exp\left\{-\frac{1}{2}\left[\mathbf{y}_m - \mathbf{y}_c\right]^T \mathbf{C}_D^{-1}\left[\mathbf{y}_m - \mathbf{y}_c\right]\right\}, \tag{5.3}$$

for a given $\mathbf{m}$ and $\mathbf{C}_D$. Here $\mathbf{C}_D$ is viewed as the distribution parameter.

In Equation (5.3), $\mathbf{C}_D$ is an $N_d \times N_d$ data (or error) matrix describing the correlation structure between pairs of errors in observed data, and det($\mathbf{C}_D$) denotes the determinant of $\mathbf{C}_D$. By definition, covariance matrices are symmetric and positive definite so their inverses exist. $\mathbf{C}_D^{-1}$ represents the inverse of the covariance matrix (or also referred to as the precision matrix in statistics). The diagonal elements of $\mathbf{C}_D$ are just the variances. In the case that we treat random errors as distributed independently, then $\mathbf{C}_D$ is a diagonal matrix with diagonal elements σ_i^2, $i = 1, 2, \ldots, N_d$, representing the variance of error at each observed $(y_m)_i$. In addition, if we assume that all errors are distributed identically (or are equally accurate) so that $\sigma_i^2 = \sigma^2$ for all i, then $\mathbf{C}_D = \sigma^2\mathbf{I}$, where $\mathbf{I}$ is an $N_d \times N_d$ identity matrix.

5.4. Likelihood Function and Maximum Likelihood Estimate

Given the observed data $\mathbf{y}_m$ and assumed model $\mathbf{y}_c$, we are interested in finding the probability density function, among all probability distributions the model parameter $\mathbf{m}$ describes, that is most likely to have produced the observed data. This poses an inverse problem to be solved. To solve this inverse problem, we define the likelihood function as the probability density function of the model parameter and distribution parameter given the observed data. The likelihood function of observed data denoted by L is then defined as a quantity proportional to the joint probability (or probability density) of the observed data vector $\mathbf{y}_m$ (or the error vector $\boldsymbol{\epsilon}$) as a function of the model parameter $\mathbf{m}$ and the distribution parameter $\boldsymbol{\Psi}$ (e.g., it can be the covariance matrix $\mathbf{C}_D$ in Equation (5.3)) (Bard, 1974). It can be expressed as

$$L(\mathbf{m}, \boldsymbol{\Psi}) = p(\mathbf{y}_m|\mathbf{m}, \boldsymbol{\Psi}) \tag{5.4}$$

and viewed as the conditional probability of $\mathbf{m}$ and the parameters of the specified distribution to given observed data $\mathbf{y}_m$. As can be seen from Equation (5.4), the likelihood function has the same form as the probability density function, except that the arguments of the likelihood function are reversed. The likelihood function expresses the values of the model parameter vector $\mathbf{m}$ and the distribution parameter $\boldsymbol{\Psi}$ in terms of known, fixed values of the observed data $\mathbf{y}_m$.

The likelihood function requires a specific distribution for observed data. Assuming that $\mathbf{y}_m$ follows a Gaussian distribution with mean $\mathbf{y}_c$ and the covariance matrix $\mathbf{C}_D$ or, equivalently, the errors (ϵ in Equation (5.2)) on observed data are distributed as $N(\mathbf{0}, \mathbf{C}_D)$, then it follows from Equations (5.3) and (5.4) that the likelihood function for this case can be expressed as

$$L(\mathbf{m}, \mathbf{C}_D) = \frac{1}{(2\pi)^{N_d/2}\sqrt{\det \mathbf{C}_D}} \exp\left\{-\frac{1}{2}\left[\mathbf{y}_m - \mathbf{y}_c\right]^T \mathbf{C}_D^{-1}\left[\mathbf{y}_m - \mathbf{y}_c\right]\right\}. \tag{5.5}$$

Once data have been measured (or observed) and the likelihood function of a model given the data is determined, we are in a position to make statistical inferences about the population, that is, the probability distribution underlying the observed data $\mathbf{y}_m$. Given that different values of $\mathbf{m}$ and $\mathbf{C}_D$ index different probability distributions to be generated by using Equation (5.5), we are interested in finding the values of $\mathbf{m}$ and $\mathbf{C}_D$ that correspond to the desired probability distribution function.

The principle of maximum likelihood estimation, originally developed by R.A Fisher in the 1920s (Fisher, 1950), states that the desired probability distribution is the one that makes the observed data $\mathbf{y}_m$ most likely. This distribution is obtained by seeking the value of the parameter vector ($\mathbf{m}$ and $\mathbf{C}_D$ in the case of Equation (5.5)) that maximizes the likelihood function $L(\mathbf{m}, \mathbf{C}_D)$. The resulting estimate of the parameter vector, which is sought by searching the multidimensional parameter space, is called the maximum likelihood estimate. It should be noted that one can keep the distribution parameter $\mathbf{C}_D$ as known and treat the model parameter $\mathbf{m}$ as the only unknown in the likelihood function.

In the following two sections, we present two simple examples to explain the concept of maximum likelihood. First, we consider a single-parameter linear model, followed by an example of a single-parameter nonlinear model.

5.4.1. Single-parameter linear model

Suppose that the observed dependent variable y_m is defined by a one-parameter linear model (also known a straight-line model with a zero intercept) plus the stochastic part as

$$(y_m)_i = (y_c)_i + \epsilon_i = mt_i + \epsilon_i, \tag{5.6}$$

where m is the slope (a model parameter) of a straight line. Here, the random

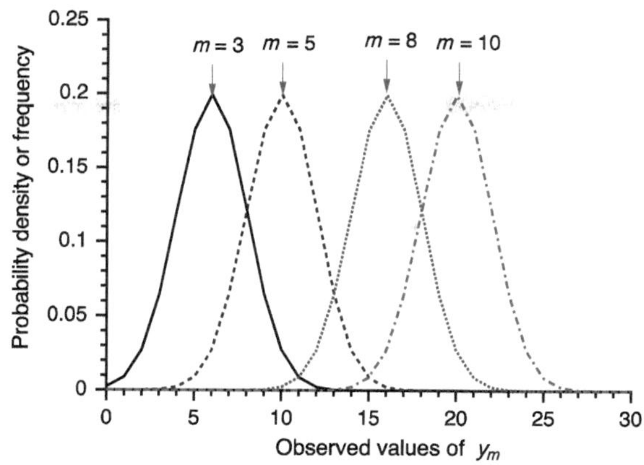

Figure 5.1 Probability density function $p(y_m|m, \sigma^2 = 4)$ versus y_m for the linear model of Equation (5.6) for $m = 3, 5, 8$, and 10.

error ϵ_i, for $i = 1, 2, \ldots, N_d$, is normal with zero mean and known variance, σ^2 (i.e., each error is uncorrelated, and has identical mean and variance). In other words, $E[\epsilon_i] = 0$, $\text{var}(\epsilon_i) = \sigma^2$ for all i, where E and var denote the expectation and variance operators, respectively. m is the slope (or the model parameter) of the straight line. Then, for a given m and σ^2, the probability density function of each y_m can be written as

$$p\left(y_m|m, \sigma^2\right) = \frac{1}{\sqrt{2\pi\sigma^2}} \exp\left[-\frac{1}{2\sigma^2}(y_m - mt)^2\right]. \quad (5.7)$$

Furthermore, suppose that t is known exactly and equal to 2, and the variances of all y_ms are also known and equal to $\sigma^2 = 4$. Figure 5.1 shows a plot of $p\left(y_m|m, \sigma^2\right)$ versus y_m (for given 30 values of y_m in the interval from 0 to 30) for four different values of $m = 3, 5, 8, 10$. As can be seen from Figure 5.1, we have four probability density curves corresponding to each value of m. If we varied variance σ^2 and the independent variable t in addition to the model parameter m, then we would consider a number of different probability density curves, each changing with the values of m, σ^2 and t.

Now, suppose that we have three measurements of y_m at three different values of $t = 1, 2$, and 3, which are equal to 3.012, 10.744, and 14.072, respectively. Let us use the principle of likelihood to find the maximum likelihood estimate (or the most likely value) of m (assuming that the variance $\sigma^2 = 4$ is known) that generates the above three y_m values. For this case, we construct the likelihood function L as m being the random variable given (or conditional to) y_m by using Equation (5.5), and L can be written as the

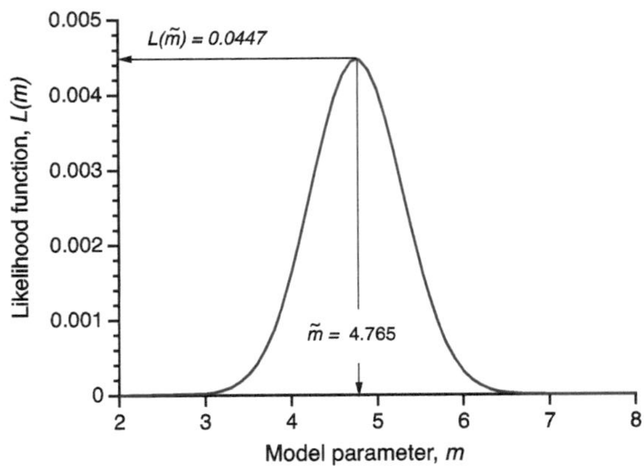

Figure 5.2 Likelihood function $L(m)$ versus the model parameter m for the linear model of Equation (5.6) for three observed values of $(y_m) = \{3.012, 10.744, 14.072\}$ with variance of $\sigma^2 = 4$.

joint probability density of three measurements of y_m:

$$L(m) = \frac{1}{\sqrt{\left(2\pi\sigma^2\right)^3}} \exp\left[-\frac{1}{2\sigma^2}\sum_{i=1}^{3}[(y_m)_i - mt_i]^2\right]. \tag{5.8}$$

As we assume that σ^2 is known and interpret L as the conditional probability to given three measured values of y_m, only m is written as the argument of L in Equation (5.8). Hence, as expected, the plot of $L(m)$ versus m displays a normal distribution as shown in Figure 5.2. The same figure shows that the most likely value of m (called the mode of L and denoted by $\tilde{m}$) that have could produced the values of $(y_m) = \{3.012, 10.744, 14.072\}$ is $\tilde{m} \approx 4.765$.

As mentioned previously, the maximum likelihood allows one to estimate the most likely value of the variance σ^2 in observed y_m in addition to the most likely value of the model parameter m. We could do that by treating the likelihood function L as a function of both σ^2 and m (see Equation (5.9)). For the simple linear example considered above, we can write the L (conditional to the three observed values of y_m) by using Equation (5.5) as

$$L\left(m, \sigma^2\right) = \frac{1}{\sqrt{\left(2\pi\sigma^2\right)^{N_d}}} \exp\left[-\frac{1}{2\sigma^2}\sum_{i=1}^{N_d}[(y_m)_i - mt_i]^2\right]. \tag{5.9}$$

where $N_d = 3$.

A three-dimensional plot of the above L along with its contour plots as a function of σ^2 and m are shown in Figures 5.3 and 5.4. Note that the

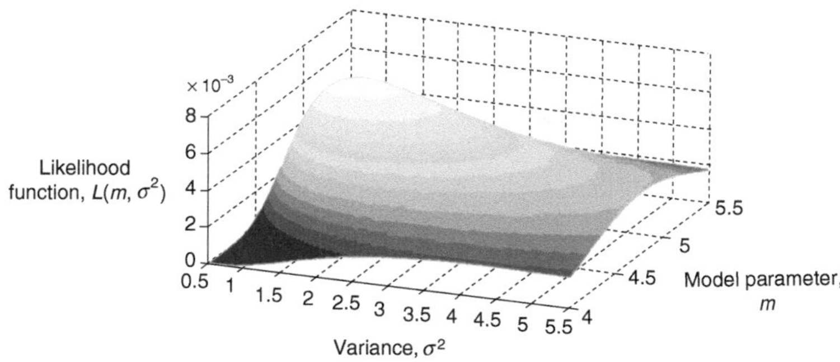

Figure 5.3 Likelihood function $L(m, \sigma^2)$ in a three-dimension space for the linear model given by Equation (5.6), conditional to the observed values of $(y_m) = \{3.012, 10.744, 14.072\}$.

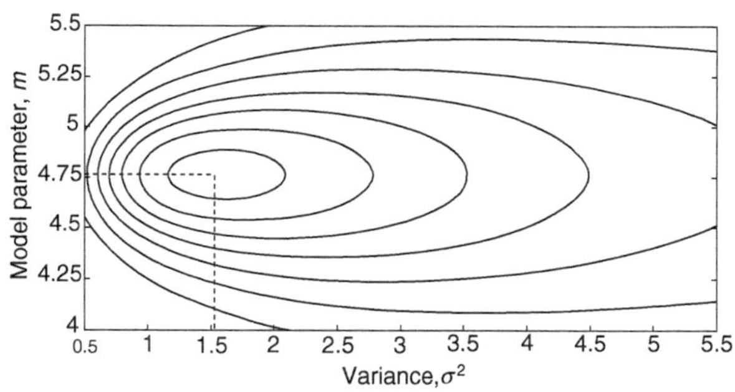

Figure 5.4 Likelihood function $L(m, \sigma^2)$ as a contour plot for the linear model given by Equation (5.6), conditional to the observed values of $(y_m) = \{3.012, 10.744, 14.072\}$.

likelihood function L for this example is *not* a bell-shaped surface in a three-dimensional space. The most likely estimates of σ^2 and m (denoted by $\tilde{\sigma}^2$ and $\tilde{m}$, respectively) are $\tilde{\sigma}^2 = 1.532$ and $\tilde{m} \approx 4.765$. The true values of σ^2 and m (denoted by $\breve{\sigma}^2$ and $\breve{m}$) used to generate three measured values of (y_m) from Equation (5.6) are $\breve{\sigma}^2 = 4$ and $\breve{m} = 5$, respectively. It is not surprising that the maximum likelihood estimates of σ^2 and m are slightly different from the corresponding true values because we have a small number of measurements of (y_m).

For this simple linear model example, one can derive explicit formulas for direct computation of the maximum likelihood (ML) estimates of σ^2 and m. These estimates are derived by first taking the natural logarithm of Equation (5.9) and then differentiating the resulting equation with respect to m and σ^2. Next, we set these equations to zero and then solve them simultaneously for $\tilde{\sigma}^2$ and $\tilde{m}$. The log-likelihood function $\ln(L)$ and the ML estimates of

$\tilde{\sigma}^2$ and $\tilde{m}$ for this example are given by

$$\ln\left[L\left(m,\sigma^2\right)\right] = -\frac{1}{2\sigma^2}\sum_{i=1}^{N_d}[(y_m)_i - mt_i]^2 - \frac{N_d}{2}\ln\left(2\pi\sigma^2\right), \quad (5.10)$$

$$\tilde{m} = \frac{\sum_{i=1}^{N_d} t_i\,(y_m)_i}{\sum_{i=1}^{N_d} t_i^2}, \quad (5.11)$$

and

$$\tilde{\sigma}^2 = \frac{1}{N_d}\sum_{i=1}^{N_d}\left[(y_m)_i - \tilde{m}t_i\right]^2. \quad (5.12)$$

The ML estimate $\tilde{m}$ (Equation (5.11)) is both unbiased and consistent. On the other hand, the ML estimate $\tilde{\sigma}^2$ defined by Equation (5.12) is biased, but consistent. Normally an estimate is said to be unbiased if its expected value is equal to the true value; otherwise, it is called biased. An estimate is said to be consistent if the expected value of the estimate converges to the true value as the number of samples (N_d) in the limit goes to infinity. Otherwise, it is inconsistent. For this specific example, the unbiased ML estimate [denoted by $\hat{\sigma}^2$] can be computed by multiplying the right-hand side of Equation (5.12) with a factor of $N_d/(N_d - 1)$, that is,

$$\hat{\sigma}^2 = \frac{N_d}{(N_d - 1)}\tilde{\sigma}^2 = \frac{1}{(N_d - 1)}\sum_{i=1}^{N_d}\left[(y_m)_i - \tilde{m}t_i\right]^2. \quad (5.13)$$

Thus by using Equation (5.13), an unbiased and consistent ML estimate of σ^2 for the above linear model is computed to be $\hat{\sigma}^2 = 2.298$, compared to the biased estimate of $\tilde{\sigma}^2 = 1.532$. The unbiased ML estimate of the variance is still different from the true value of 4, though it is more close to the true value compared to the biased one. Nevertheless, this is expected because we have used only a small number of samples of y_m to determine this value. As the number of samples increases, we would expect the ML estimate of variance to be closer to the true, unknown variance. The readers may easily verify this by repeating the above example for N_d greater than three.

5.4.2. Single-parameter nonlinear model

Now, we demonstrate the likelihood principle for a simple single-parameter nonlinear model given as

$$(y_m)_i = (y_c)_i + \epsilon_i = t_i + 4\,(m-2)^2 + \epsilon_i. \quad (5.14)$$

As in the case of the linear one-parameter model considered, we assume that the random error ϵ_i is normal with zero mean and has a known variance σ^2. In other words, each error is uncorrelated, and has identical mean and variance, i.e., $E[\epsilon_i] = 0$, $\mathrm{var}(\epsilon_i) = \sigma^2$ for all i. Then, given m and σ^2, the probability density function (**pdf**) of observing each (y_m) for the nonlinear model of Equation (5.14) is given as

$$p\left(y_m|m, \sigma^2\right) = \frac{1}{\sqrt{(2\pi\sigma^2)}} \exp\left(-\frac{1}{2\sigma^2}\left\{y_m - \left[t + 4(m-2)^2\right]\right\}^2\right). \tag{5.15}$$

The **pdf** given by Equation (5.15) (its graph not shown here) will still be normal as a function of y_m, given m and σ^2, as in the case of the linear one-parameter model shown in Figure 5.2. However, unlike the likelihood function for the linear model (Equation (5.6)), the likelihood function L given by Equation (5.15) for the nonlinear model (Equation (5.14)) will not be normal, but bimodal, as a function of the model parameter m, which will be shown next.

Now, suppose that we have three measurements of y_m at three different values of $t = 0.2, 0.5$, and 1.0, which are equal to 1.096, 1.649, and 1.761, respectively. Further, we assume that the variance σ^2 is known and equal to 0.05 for all these three measurements. For this case, we construct the likelihood function L by keeping m as a random variable, which corresponds to three observed values of y_m as

$$L(m) = \frac{1}{\sqrt{(2\pi\sigma^2)^{N_d}}} \exp\left\{-\frac{1}{2\sigma^2}\sum_{i=1}^{N_d}\left[(y_m)_i - \left(t_i + 4(m-2)^2\right)\right]^2\right\}, \tag{5.16}$$

where $N_d = 3$ and $\sigma^2 = 0.05$. Figure 5.5 presents the plot of $L(\mathbf{m})$ versus m and shows that the likelihood function for this nonlinear example is bimodal (i.e., two peaks or with two most likely estimates) and hence is not normal. The most likely values for this nonlinear example (given the variance of $\sigma^2 = 0.05$ and three observed values of y_m) are $\tilde{m} = 1.516$ and 2.484. The likelihood function (Figure 5.5) indicates that these two ML estimates of $\tilde{m}$ will produce three values of the observed data equally likely. Hence, we have two solutions for the model parameter m.

Next, we treat the variance σ^2 as unknown in addition to the model parameter m for the nonlinear model of Equation (5.14), given the three measured values of y_m. Figures 5.6 and 5.7 present the plot of the likelihood function. As shown in these figures, it is clear that we have two peaks, but not bell-shaped surfaces, centered around the two pairs of most likely values of σ^2 and m that are $\tilde{\sigma}^2 = 0.026$, $\tilde{m} = 1.516$ and $\tilde{\sigma}^2 = 0.026$, $\tilde{m} = 2.484$.

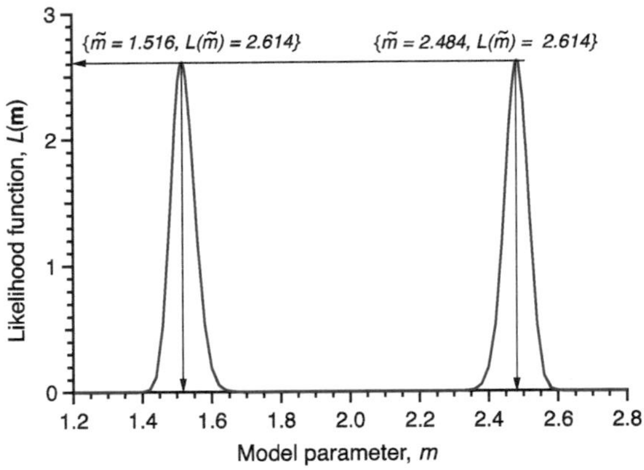

Figure 5.5 Likelihood function $L(\mathbf{m})$ versus model parameter m for the nonlinear model of Equation (5.14) for three observed values of $(y_m) = \{1.096, 1.649, 1.761\}$ with variance of $\sigma^2 = 0.05$.

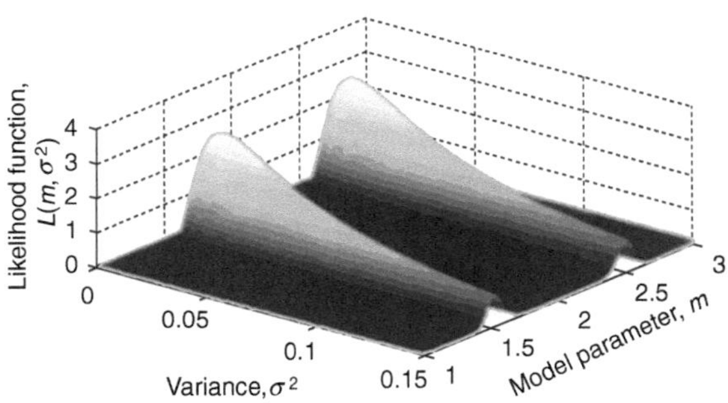

Figure 5.6 Likelihood function $L(m, \sigma^2)$ in a three-dimension space for the nonlinear model of Equation (5.14), conditional to the three observed values of $(y_m) = \{1.096, 1.649, 1.761\}$.

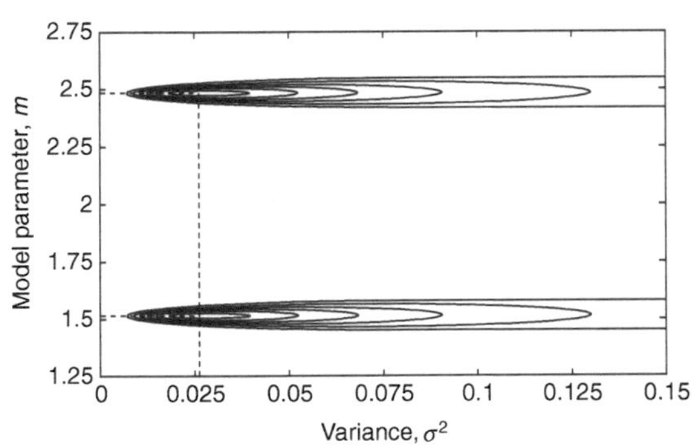

Figure 5.7 Likelihood function $L(m, \sigma^2)$ as a contour plot for the nonlinear model of Equation (5.14), conditional to the three observed values of $(y_m) = \{1.096, 1.649, 1.761\}$.

The true values of σ^2 and m used to generate three measured values of y_m from Equation (5.14) are $\hat{\sigma}^2 = 0.05$ and $\hat{m} = 1.5$, respectively.

It should be noted that for this nonlinear model example, one cannot derive explicit formulas for direct computation of the ML estimates of σ^2 and m. We have used the Levenberg-Marquardt (Levenberg, 1944; Marquardt, 1963) (L-M) optimization method (to be discussed later) to determine the ML estimates of $\tilde{m}$ and $\tilde{\sigma}^2$, which maximize the log-likelihood function $\ln\left[L\left(m, \sigma^2\right)\right]$ given by Equation (5.16). The ML estimates computed from the L-M method are two pairs of $\tilde{m}$ and $\tilde{\sigma}^2$; $\left(\tilde{\sigma}^2 = 0.02586, \tilde{m} = 1.51643\right)$ and $\left(\tilde{\sigma}^2 = 0.02586, \tilde{m} = 2.48356\right)$, depending on the initial guesses used for m and σ^2 in the L-M method. Note that we have two different estimates for m, but a unique estimate for σ^2 because we have a quadratic nonlinear model in m, but the variance is identical for all three measurements (see Equation (5.16)). The ML estimates computed from the L-M method are essentially identical to those determined from the three-dimensional plots of $L\left(m, \sigma^2\right)$ shown in Figures 5.6 and 5.7.

As in the linear model considered previously, the ML estimate $\tilde{m}$ is both unbiased and consistent, while the ML estimate $\tilde{\sigma}^2$ is consistent, but biased. This unbiased ML estimate of the variance can be computed by using the value of $\tilde{\sigma}^2$ in Equation (5.13) and is found to be $\left(\hat{\sigma}\right)^2 = 0.03880$, as compared to the "true" value $\breve{\sigma}^2 = 0.05$. It is much closer to the true value of the variance than the biased estimate $\tilde{\sigma}^2$.

5.5. Extension of Likelihood Function to Multiple Sets of Observed Data

There are often cases where we need to simultaneously history match multiple sets of observed data with different noise levels. These data can also have different physical dimensions and/or are measured at different scales. For instance, for pressure-transient testing, we may have different sets of spatial pressure measurements obtained from multiwell interference tests or local interval transient tests (IPTT) along the same wellbore.

The maximum likelihood estimation can be extended to the more general case of N_s different sets of observed data, $(\mathbf{y}_m)_j$, for $j = 1, 2, \ldots, N_s$. Each data set contains N_{dj}, $j = 1, 2, \ldots, N_s$, observations. So, each $(\mathbf{y}_m)_j$ is an N_{dj}-dimensional vector and can be expressed as

$$(\mathbf{y}_m)_j = \left[(y_m)_{1,j}, (y_m)_{2,j}, \ldots, (y_m)_{N_{dj},j}\right]^T. \qquad (5.17)$$

For instance, for IPTTs, $(\mathbf{y}_m)_j$ can represent the set of dual-packer or observation probe pressure measurements. The observed data set $(\mathbf{y}_m)_j$ is given in Equation (5.2) and related to the model function $(\mathbf{y}_c)_j$ given by Equation (5.1). The probability density function of each $(\mathbf{y}_m)_j$ can be written as

$$p_j((\mathbf{y}_m)_j|\mathbf{m}, \mathbf{C}_{D,j}) = \frac{1}{(2\pi)^{N_{dj}/2}\left(\det \mathbf{C}_{D,j}\right)^{1/2}} \exp \times \left[-\frac{1}{2}\left[(\mathbf{y}_m)_j - (\mathbf{y}_c)_j\right]^T \mathbf{C}_{D,j}^{-1}\left[(\mathbf{y}_m)_j - (\mathbf{y}_c)_j\right]\right] \tag{5.18}$$

for $j = 1, 2, \ldots, N_s$, where the error term $\boldsymbol{\epsilon}_j = (\mathbf{y}_m)_j - (\mathbf{y}_c)_j$ for each $(\mathbf{y}_m)_j$ is statistically independent, and $\boldsymbol{\epsilon}_j$ is a $N_{d,j}$-dimensional vector of a Gaussian (normal) errors with zero mean (zero vector) and covariance matrix $\mathbf{C}_{D,j}$ of size $N_{d,j} \times N_{d,j}$. Then, the likelihood function for the whole observed data sets $(\mathbf{y}_m)_j$ for $j = 1, 2, \ldots, N_s$ is defined as the joint **pdf** of the individual **pdf**s of the observed data sets:

$$L(\mathbf{m}, \mathbf{C}_{D,1}, \ldots, \mathbf{C}_{D,N_s}) = \prod_{j=1}^{N_s} p_j((\mathbf{y}_m)_j|\mathbf{m}, \mathbf{C}_{D,j}) \tag{5.19}$$

and then substituting Equation (5.18) in Equation (5.19) gives

$$L(\mathbf{m}, \mathbf{C}_{D,1}, \ldots, \mathbf{C}_{D,N_s}) = \frac{\prod_{j=1}^{N_s}\left(\det \mathbf{C}_{D,j}\right)^{-1/2}}{(2\pi)^{\sum_{j=1}^{N_s} N_{dj}/2}} \times \exp\left[-\frac{1}{2}\sum_{j=1}^{N_s}\left[(\mathbf{y}_m)_j - (\mathbf{y}_c)_j\right]^T \mathbf{C}_{D,j}^{-1}\left[(\mathbf{y}_m)_j - (\mathbf{y}_c)_j\right]\right]. \tag{5.20}$$

5.6. Least-Squares Estimation Methods

The least-squares estimation methods can be derived from the principle of maximum likelihood if the observed data are sampled from a Gaussian distribution. First, we consider a single observed data set case in Equation (5.5), and then the more general case of N_s observed data sets (Equation (5.20)).

In practice, it is easier to maximize the logarithm of L. Considering a single observed data set of the same quantities and thus taking the natural logarithm of both sides of Equation (5.5), we obtain

$$\ln[L(\mathbf{m}, \mathbf{C}_D)] = -\frac{N_d}{2}\ln(2\pi) - \frac{1}{2}\ln(\det \mathbf{C}_D) - \frac{1}{2}\left[\mathbf{y}_m - \mathbf{y}_c\right]^T \mathbf{C}_D^{-1}\left[\mathbf{y}_m - \mathbf{y}_c\right]. \quad (5.21)$$

If we assume that $\mathbf{C}_D$ is known, then the likelihood function can be expressed as a function of the model parameter vector $\mathbf{m}$ only, that is,

$$\ln[L(\mathbf{m})] = -\frac{N_d}{2}\ln(2\pi) - \frac{1}{2}\ln(\det \mathbf{C}_D) - \frac{1}{2}\left[\mathbf{y}_m - \mathbf{y}_c\right]^T \mathbf{C}_D^{-1}\left[\mathbf{y}_m - \mathbf{y}_c\right]. \quad (5.22)$$

Thus, with a known $\mathbf{C}_D$, obtaining the values of $\mathbf{m}$ that maximize Equation (5.22) is equivalent to minimizing the following objective function

$$O(\mathbf{m}) = \frac{1}{2}\left[\mathbf{y}_m - \mathbf{y}_c\right]^T \mathbf{C}_D^{-1}\left[\mathbf{y}_m - \mathbf{y}_c\right], \quad (5.23)$$

which defines the objective function for the general weighted least-squares (WLS) that considers correlation in observed data errors. Thus, for a normal distribution with known covariance matrix for errors in $\mathbf{y}_m$, the MLE reduces to weighted least-squares, with weights given by the elements of the inverse of the covariance matrix $\mathbf{C}_D$.

To derive more familiar and simplified forms of least-squares methods, we can assume that the known covariance matrix $\mathbf{C}_D$ is diagonal with diagonal entries equal to σ_i^2, $i = 1, 2, \ldots, N_d$, representing the variance of error at each observed $(y_m)_i$, then Equation (5.23) can be written as

$$O(\mathbf{m}) = \frac{1}{2}\sum_{i=1}^{N_d}\left[\frac{(y_m)_i - (y_c)_i}{\sigma_i}\right]^2, \quad (5.24)$$

which is the objective function for the well-known ordinary WLSE. This clearly shows that for a normal distribution with a known diagonal covariance, the MLE reduces to the ordinary WLSE, with weights given by the reciprocal of the variances σ_i^2.

For completeness, it should be pointed out that the ML estimate of $\mathbf{m}$ is identical to the unweighted least-squares (UWLS) estimate of $\mathbf{m}$ for the case in which measurements are statistically distributed independently with identical variances; i.e., $\sigma_i^2 = \sigma^2$ for all i in Equation (5.24), regardless of whether σ is known or unknown. In this case, it can be shown that Equation (5.22) can be written as

$$\ln[L(\mathbf{m})] = -\frac{N_d}{2}\ln(2\pi) - N_d \ln(\sigma) - \frac{1}{2\sigma^2}\sum_{i=1}^{N_d}[(y_m)_i - (y_c)_i]^2, \quad (5.25)$$

which indicates that whether σ is known or unknown, maximizing $\ln(L)$ is equivalent to minimizing the objective function given by

$$O(\mathbf{m}) = \frac{1}{2}\sum_{i=1}^{N_d}[(y_m)_i - (y_c)_i]^2 . \quad (5.26)$$

This is the objective function for the well-known standard (ordinary) UWLSE. Here MLE is equivalent to UWLSE. The positive constant "1/2" in the right-hand side of Equation (5.26) could be dropped because multiplying a function with a positive constant does not change its minimum with respect to the model parameters.

Now, we treat the more general case where one may wish to simultaneously match different and independent observed data sets. In this general case, the likelihood function is given by Equation (5.20). Taking the natural logarithm of Equation (5.20) yields the following log-likelihood function for N_s data sets as

$$\ln(L) = -\frac{1}{2}\left(\sum_{j=1}^{N_s} N_{dj}\right)\ln(2\pi) - \frac{1}{2}\sum_{j=1}^{N_s}\ln\left(\det \mathbf{C}_{D,j}\right)$$
$$-\frac{1}{2}\sum_{j=1}^{N_s}\left[(\mathbf{y}_m)_j - (\mathbf{y}_c)_j\right]^T \mathbf{C}_{D,j}^{-1}\left[(\mathbf{y}_m)_j - (\mathbf{y}_c)_j\right]. \quad (5.27)$$

If all $\mathbf{C}_{D,j}$, $j = 1, 2, \ldots, N_s$, are known, finding the value of $\mathbf{m}$ that maximizes Equation (5.27) is equivalent to minimizing the following WLS objective function:

$$O(\mathbf{m}) = \frac{1}{2}\sum_{j=1}^{N_s}\left[(\mathbf{y}_m)_j - (\mathbf{y}_c)_j\right]^T \mathbf{C}_{D,j}^{-1}\left[(\mathbf{y}_m)_j - (\mathbf{y}_c)_j\right]. \quad (5.28)$$

In summary, starting from the principle of maximum likelihood, we showed that the covariance matrix provides the correct relative weighting of the data terms in the weighted least-squares objective functions. In other words, to estimate statistically optimal values of unknown model parameters from the weighted least-squares method one should consider weights as the elements of the inverse of the covariance matrix for observed data errors. In addition, we showed that the least-squares estimation assumes that these weights (variances or covariance matrices for in the more general case) are known *a priori*, except in the case where measurements are statistically distributed independently with identical variances.

Although a few papers have been published (Aanonsen et al., 2003; Onur & Reynolds, 2002; Zao, Li, & Reynolds, 2007) on the non-diagonal data error covariance matrix for the least-squares estimation, most of the work in the petroleum engineering literature assumes a diagonal covariance matrix for the observed data set, where each data point is represented by the same single error variance. The main reason for these assumptions is that we often do not know the proper weights or the statistical structure of the noise *a priori* so that one can consider unequal variances or non-diagonal covariance matrices. One way to obtain a direct estimate of the error covariance matrix (assuming that errors in Equation (5.2) are only due to measurement errors, not also due to the model itself) is to replicate the measurements. This is usually not achievable for pressure-transient tests, i.e., we have to work with only one sample collected. In addition, inclusion of a non-diagonal error data covariance matrix in Equation (5.23) or Equation (5.28) requires special algorithms for its storage and inverse in computations, particularly in the cases for which the number of observed data points is quite large, in which case $\mathbf{C}_D$ is quite large.

There are a few new techniques based on local polynomial regression, smoothing splines, and wavelet transforms (Aanonsen et al., 2003; Athichanagorn, Horne, & Kikani, 2002; Horne, 2007; Zao et al., 2007) that can be used to estimate weights (or variances for each data point) or in general the covariance matrix of noise from the observed data itself. Although these techniques seem promising, they require sophisticated smoothing algorithms to be applied before the history matching process. Another alternative is to use the generalized least-squares (GLS) estimation (Carroll & Ruppert, 1988) requiring iterative estimation of the weights and the unknown model parameters, simultaneously, but the use of GLS may become quite tedious, especially in the cases when we have to simultaneously match different sets of observed pressure data sets, e.g., spatial pressure-transient data sets recorded at different spatial locations: either along the wellbore of a single well or in among wells in the system.

Another simple alternative is to use the maximum likelihood function (Equations (5.5) and (5.20)) which, under some simplifying assumptions to be discussed next, enables one to estimate the weights (or inverse of the error variances) along with the unknown model parameters. Due to the difficulties with the treatment of non-diagonal covariance matrices in the cases where we consider different sets of pressure data sets, and our desire to make the estimation procedure feasible for the unknown model parameters as well as unknown weights (or variances), we will consider only applications of the maximum likelihood estimation for the cases regarding diagonal covariance matrices where each data set can be characterized by a statistically independent diagonal error data covariance matrix. Treatment of the general non-diagonal covariance cases, based on the assumption that each data set has the same covariance matrix, is given in Bard (1974).

5.7. Maximum Likelihood Estimation for Unknown Diagonal Covariance

As discussed in Section 5.4, the concept of maximum likelihood allows one to view the likelihood function as a function of the vector of unknown model parameters, **m**, and the distribution parameter such as the covariance matrix, $\mathbf{C}_D$, and hence can be used to treat the covariance matrix as unknown in estimating the model parameter. Therefore, the maximum likelihood offers an alternative to the WLS in cases where the covariance matrix (or variance in diagonal covariance case) is unknown. Here, we will deal only with the unknown diagonal covariance case.

Let us consider the likelihood function given by Equation (5.20), which applies for the more general case of N_s different, statistically independent sets of observed data, $(\mathbf{y}_m)_j$, for $j = 1, 2, \ldots, N_s$. In addition, let us consider that the total number of data points in each data set (N_{dj}) is greater than the number of unknown model parameters, M. In other words, we consider an over-determined problem such that $N_{dj} > M$ for all $j = 1, 2, \ldots, N_s$. (The readers should refer to the book of Oliver et al. (2008) for detailed and rigorous mathematical definitions of the overdetermined and underdetermined problems in history matching applications.) This corresponds to the cases where we use the models that can be parameterized by a few unknown parameters, such as the conventional analytical models used for pressure transient applications.

Taking the logarithm of both sides of Equation (5.20) gives

$$\ln(L) = -\frac{1}{2}\left(\sum_{j=1}^{N_s} N_{dj}\right)\ln(2\pi) - \frac{1}{2}\sum_{j=1}^{N_s}\ln\left(\det \mathbf{C}_{D,j}\right) - \frac{1}{2}\sum_{j=1}^{N_s}\left[(\mathbf{y}_m)_j - (\mathbf{y}_c)_j\right]^T \mathbf{C}_{D,j}^{-1}\left[(\mathbf{y}_m)_j - (\mathbf{y}_c)_j\right]. \quad (5.29)$$

It is normally difficult to treat each covariance matrix in Equation (5.29) as an unknown. To simplify the estimation problem, we will assume that all covariance matrices are diagonal with diagonal entries equal to $\sigma_{i,j}^2$, i.e., off diagonal terms of each $\mathbf{C}_{D,j}$ are zero. Then, Equation (5.29) can be written as

$$\ln(L) = -\frac{1}{2}\left(\sum_{j=1}^{N_s} N_{dj}\right)\ln(2\pi) - \frac{1}{2}\sum_{j=1}^{N_s}\sum_{i=1}^{N_{dj}}\ln(\sigma_{i,j}^2)$$
$$-\frac{1}{2}\sum_{j=1}^{N_s}\sum_{i=1}^{N_{dj}}\frac{\left[(y_m)_{i,j} - f(\mathbf{m}, t_{i,j})\right]^2}{\sigma_{i,j}^2}, \tag{5.30}$$

where $\sigma_{i,j}^2$ denotes the error variance for the observed $(y_m)_{i,j}$ at $t_{i,j}$.

We will maximize $\ln(L)$ given by Equation (5.30) with respect to unknown error variances $\sigma_{i,j}^2$ and the model parameter $\mathbf{m}$ because error variances will be treated as unknown in regression. However, in order for the estimation of the variances to be feasible, we must additionally assume that error variances for all observations in a given data set $(\mathbf{y}_m)_j$ are all equal, i.e., $\sigma_{i,j}^2 = \sigma_j^2$ for $i = 1, \ldots, N_{dj}$ for a given j, then $\ln(L)$ is simplified to

$$\ln(L) = -\frac{1}{2}\left(\sum_{j=1}^{N_s} N_{dj}\right)\ln(2\pi) - \frac{1}{2}\sum_{j=1}^{N_s} N_{dj}\ln\sigma_j^2$$
$$-\frac{1}{2}\sum_{j=1}^{N_s}\frac{1}{\sigma_j^2}\sum_{i=1}^{N_{dj}}\left[(y_m)_{i,j} - f(\mathbf{m}, t_{i,j})\right]^2. \tag{5.31}$$

To maximize Equation (5.31), we can proceed by the stage-wise maximization method (Bard, 1974). This method involves finding the values of σ_j^2 that maximize $\ln(L)$ for any value of $\mathbf{m}$. These σ_j^2 values will be some functions of $\mathbf{m}$, say $\tilde{\sigma}_j^2(\mathbf{m})$. Substitution of $\tilde{\sigma}_j^2(\mathbf{m})$ for σ_j^2 in Equation (5.31) reduces $\ln(L)$ to a function $\tilde{L}$ of $\tilde{\mathbf{m}}$ alone, and we seek $\tilde{\mathbf{m}}$ so as to maximize $\tilde{L}(\mathbf{m})$.

The first step in this maximization procedure, then, is to differentiate Equation (5.31) with respect to σ_j^2 while holding other variables fixed, and then to equate the resulting derivatives to zero to obtain

$$\tilde{\sigma}_j^2(m) = \frac{1}{N_{dj}}\sum_{i=1}^{N_{dj}}\left[(y_m)_{i,j} - f(\mathbf{m}, t_{i,j})\right]^2 \tag{5.32}$$

for $j = 1, \ldots, N_s$.

Using Equation (5.32) for σ_j^2 in Equation (5.31) and simplifying the resulting equation, we obtain the so-called *concentrated likelihood* function as

$$\tilde{L}(\mathbf{m}) = \frac{1}{2}\left\{\sum_{j=1}^{N_{dj}} N_{dj}\left[\ln\left(\frac{N_{dj}}{2\pi}\right) - 1\right]\right\} - \frac{1}{2}\sum_{j=1}^{N_s} N_{dj} \ln\left\{\sum_{i=1}^{N_{dj}} \left[(y_m)_{i,j} - f(\mathbf{m}, t_{i,j})\right]^2\right\}. \quad (5.33)$$

Note that maximizing Equation (5.33) is equivalent to minimizing the following objective function as

$$\tilde{O}(\mathbf{m}) = \frac{1}{2}\sum_{j=1}^{N_s} N_{dj} \ln\left\{\sum_{i=1}^{N_{dj}} \left[(y_m)_{i,j} - f(\mathbf{m}, t_{i,j})\right]^2\right\}. \quad (5.34)$$

In summary, in order to solve the estimation problem given by Equation (5.31), we proceed as follows: (1) Find $\tilde{\mathbf{m}}$ that maximizes $\tilde{L}(\mathbf{m})$ (Equation (5.33)) or minimizes $\tilde{O}(\mathbf{m})$ (Equation (5.34)) and (2) Estimate error variances from Equation (5.32) using $\mathbf{m} = \tilde{\mathbf{m}}$. It is important to note that variances obtained from this procedure are biased, but consistent. The bias can be eliminated approximately by using

$$\hat{\sigma}_j^2 = \frac{N_{dj}}{N_{dj} - \frac{M}{N_s}}\tilde{\sigma}_j^2, \quad (5.35)$$

where $\tilde{\sigma}_j^2$ is computed from Equation (5.32) with $\mathbf{m} = \tilde{\mathbf{m}}$ (Bard, 1974). As a special case, we consider that each observed data set contains the same number of observations, i.e., $N_{dj} = N_d$, for $j = 1, 2, \ldots, N_s$. Then, for this case, Equation (5.33) can be written as

$$\tilde{L}(\mathbf{m}) = \frac{N_s N_d}{2}\left[\ln\left(\frac{N_d}{2\pi}\right) - 1\right] - \frac{N_d}{2}\sum_{j=1}^{N_s} \ln\left\{\sum_{i=1}^{N_d} \left[(y_m)_{i,j} - f(\mathbf{m}, t_{i,j})\right]^2\right\}. \quad (5.36)$$

Maximizing Equation (5.36) is equivalent to minimizing the following objective function:

$$\tilde{O}(\mathbf{m}) = \frac{N_d}{2}\sum_{j=1}^{N_s} \ln\left\{\sum_{i=1}^{N_d} \left[(y_m)_{i,j} - f(\mathbf{m}, t_{i,j})\right]^2\right\} \quad (5.37)$$

and an unbiased estimate of σ_j^2, using the minimizer $\tilde{\mathbf{m}}$ of Equation (5.37), can be computed from

$$\hat{\sigma}_j^2 = \frac{N_d}{N_d - \frac{M}{N_s}} \tilde{\sigma}_j^2, \tag{5.38}$$

where $\tilde{\sigma}_j^2$ is computed from Equation (5.32) with N_{dj} replaced by N_d.

5.7.1. Single-parameter linear model case

In this section, we provide a comparison of the estimates to be obtained by the application of the MLE, WLSE, and UWLSE methods for a simple single-parameter linear model, $f(m) = m$. We have chosen this model because it is simple enough so that the readers could easily captures the essential features of these methods.

Suppose that we have two sets of measurements $(\mathbf{y}_m)_j$, where the data set 1 ($j = 1$) contains N_{d1} measurements with noise having zero mean and variance σ_1^2, and the data set 2 ($j = 2$) contains N_{d2} measurements with noise having zero mean and variance σ_2^2. Then, the log-likelihood function for this problem (Equation (5.31) with $N_s = 2$) can be written as

$$\begin{aligned} \ln(L) = &-\frac{1}{2}\left(\sum_{j=1}^{2} N_{dj}\right)\ln(2\pi) - \frac{1}{2}\sum_{j=1}^{2} N_{dj} \ln \sigma_j^2 \\ &-\frac{1}{2}\sum_{j=1}^{2}\frac{1}{\sigma_j^2}\sum_{i=1}^{N_{dj}}\left[(y_m)_{i,j} - m\right]^2. \end{aligned} \tag{5.39}$$

If we assume that the variances in Equation (5.39) are known, then, as shown previously, maximizing $\ln(L)$ is equivalent to minimizing the following WLS objective function as

$$O(m) = \frac{1}{2}\sum_{j=1}^{2}\frac{1}{\sigma_j^2}\sum_{i=1}^{N_{dj}}\left[(y_m)_{i,j} - m\right]^2. \tag{5.40}$$

It can be shown that the estimate of m that minimizes Equation (5.40) is

$$\tilde{m} = \frac{\frac{1}{\sigma_1^2}\sum_{i=1}^{N_{d1}}(y_m)_{i,1} + \frac{1}{\sigma_2^2}\sum_{i=1}^{N_{d2}}(y_m)_{i,2}}{\frac{N_{d1}}{\sigma_1^2} + \frac{N_{d2}}{\sigma_2^2}} = \frac{\frac{N_{d1}}{\sigma_1^2}(\bar{y}_m)_1 + \frac{N_{d2}}{\sigma_2^2}(\bar{y}_m)_2}{\frac{N_{d1}}{\sigma_1^2} + \frac{N_{d2}}{\sigma_2^2}}, \tag{5.41}$$

where $(\bar{y}_m)_j = \left(\sum_{i=1}^{N_{dj}}(y_m)_{i,j}\right)/N_{dj}$ for $j = 1, 2$, and represents an arithmetic average of the observed data points in the data set j, and is also referred to as the sample evidence for the data set j (Guttman, Wilks, & Hunter, 1982).

Equation (5.41) shows that the WLS estimate of the model parameter m is a weighted combination of the two sample (data sets') evidences, namely $\bar{y}_{m,j}$ values for $j = 1, 2$; the weights being the precision[2] of $(\bar{y}_m)_j$ values for $j = 1, 2$; that is N_{dj}/σ_j^2 for $j = 1$ and 2, respectively.

The variance of $\tilde{m}$ given by Equation (5.41) can be written as

$$\sigma_{\tilde{m}}^2 = \frac{1}{\frac{N_{d1}}{\sigma_1^2} + \frac{N_{d2}}{\sigma_2^2}}, \tag{5.42}$$

which indicates that the variance of $\tilde{m}$ (or uncertainty in $\tilde{m}$) is equal to the reciprocal of the sum of the precisions of observed sample evidence. Note that if all data sets have the same variance (i.e., $\sigma_1^2 = \sigma_2^2 = \sigma^2$), then the variance of $\tilde{m}$ will be equal to the arithmetic average of the data variances, that is, $\sigma_{\tilde{m}}^2 = \sigma^2/N_{dt}$ where $N_{dt} = N_{d1}+N_{d2}$. This indicates that increasing the number of data points decreases the variance of $\tilde{m}$ by a factor of $1/N_{dt}$.

The UWLSE finds the estimate of the model parameter m by minimizing the following objective function for the simple example case as

$$O(m) = \frac{1}{2}\sum_{j=1}^{2}\sum_{i=1}^{N_{dj}}\left[(y_m)_{i,j} - m\right]^2, \tag{5.43}$$

which is obtained by assuming that all data have the same variance, i.e., $\sigma_j^2 = \sigma^2$ for $j = 1, 2$, in Equation (5.39) or Equation (5.40). The value of σ^2 is immaterial for the UWLSE, and hence the UWLSE considers that each data point has the same quality, though obviously not true for the simple case considered here. The UWLS estimate of the model parameter m that minimizes Equation (5.43) is given as

$$\tilde{m} = \frac{\sum_{i=1}^{N_{d1}}(y_m)_{i,1} + \sum_{i=1}^{N_{d2}}(y_m)_{i,2}}{N_{d1} + N_{d2}} = \frac{N_{d1}(\bar{y}_m)_1 + N_{d2}(\bar{y}_m)_2}{N_{d1} + N_{d2}}, \tag{5.44}$$

which shows that the UWLS estimate of the model parameter m is a weighted combination of the two sample evidences, namely $(\bar{y}_m)_j$s for $j = 1, 2$, and 3, but in this case, the weights being the number of data points

[2] In statistics, precision is defined as the reciprocal of the variance.

for each data set, that is, N_{d1} and N_{d2}, respectively. So, unlike WLSE, the UWLSE ignores the precision of the data sets, and hence one cannot expect that the WLS and UWLS estimates should be the same unless the precision of each data point is identical (see Equations (5.41) and (5.44)).

For the example case, the UWLSE provides an unbiased estimate of the overall variance in data by using the following equation

$$\hat{\sigma}^2 = \frac{\sum_{i=1}^{N_{d1}} \left[(y_m)_{i,1} - \tilde{m}\right]^2 + \sum_{i=1}^{N_{d2}} \left[(y_m)_{i,2} - \tilde{m}\right]^2}{N_{d1} + N_{d2} - 1}. \tag{5.45}$$

Thus, the variance (uncertainty) of $\tilde{m}$ is computed from

$$\sigma_{\tilde{m}}^2 = \frac{\hat{\sigma}^2}{N_{d1} + N_{d2}}. \tag{5.46}$$

The MLE methods (either based on direct maximization of $\ln(L)$ or based on the stage-wise maximization of $\ln L$) treat both the model parameter m and the variances of each data set; namely, σ_1^2 and σ_2^2 as unknown. So, the MLE maximizes Equation (5.39) with respect to three parameters; m, σ_1^2, and σ_2^2. The MLE estimates of m, σ_1^2, and σ_2^2 are then to be computed by solving the following three nonlinear equations simultaneously:

$$m = \frac{\frac{1}{\sigma_1^2(m)} \sum_{i=1}^{N_{d1}} (y_m)_{i,1} + \frac{1}{\sigma_2^2(m)} \sum_{i=1}^{N_{d2}} (y_m)_{i,2}}{\frac{N_{d1}}{\sigma_1^2(m)} + \frac{N_{d2}}{\sigma_2^2(m)}}, \tag{5.47}$$

where

$$\sigma_j^2(m) = \frac{1}{N_{dj}} \sum_{i=1}^{N_{dj}} \left[(y_m)_{i,j} - m\right]^2 \quad \text{for } j = 1 \text{ and } 2. \tag{5.48}$$

As discussed previously, the variances to be estimated from Equation (5.48) will be biased, but could be adjusted to obtain unbiased values by using Equation (5.35). Of course, if N_{dj} is large enough such that $N_{dj} \gg M/N_s$, then Equation (5.48) provides a very accurate estimate of the true, unknown variances.

The variance $\sigma_{\tilde{m}}^2$ is computed from Equation (5.42) with σ_j^2, $j = 1$ and 2, replaced by their unbiased estimates $\hat{\sigma}_j^2$, $j = 1$ and 2, computed from Equation (5.35), after solving nonlinear Equations (5.47) and (5.48) simultaneously for $\tilde{m}$, $\tilde{\sigma}_1^2$, and $\tilde{\sigma}_2^2$.

5.7.2. An example application

In this section, we present a numerical example to demonstrate the results obtained by minimization of the objective function of Equation (5.34) (or equivalently, the stage-wise maximization method for Equation (5.31)) and compare the estimates of variances and the model parameters obtained from this method with those obtained by maximizing Equation (5.31) directly by simultaneously treating all variances and the model parameter vector as unknown. We will further compare the ML estimates with those to be obtained from the WLSE and UWLSE methods.

For this example, we consider three sets of observed data with different error variances, sampled from the same two-parameter linear model of Equation (5.51). The log-likelihood function (Equation (5.31) with $N_s = 3$) for this example is given as

$$\ln(L) = -\frac{1}{2}\left(\sum_{j=1}^{3} N_{dj}\right)\ln(2\pi) - \frac{1}{2}\sum_{j=1}^{3} N_{dj}\ln(\sigma_j^2) - \frac{1}{2}\sum_{j=1}^{3}\frac{1}{\sigma_j^2}\sum_{i=1}^{N_{dj}}\left[(y_m)_{i,j} - f(\mathbf{m}, t_{i,j})\right]^2 \tag{5.49}$$

and the objective function to be minimized with respect to the model parameters only for the stage-wise maximization is given as

$$O(\mathbf{m}) = \frac{1}{2}\sum_{j=1}^{3} N_{dj}\ln\left\{\sum_{i=1}^{N_{dj}}\left[(y_m)_{i,j} - f(\mathbf{m}, t_{i,j})\right]^2\right\}, \tag{5.50}$$

where the model functions $f(\mathbf{m}, t_{i,1})$, $f(\mathbf{m}, t_{i,2})$, and $f(\mathbf{m}, t_{i,3})$ are

$$f\left(\mathbf{m}, t_{i,j}\right) = m_1 + m_2 \ln t_{i,j}, \quad \text{for } j = 1, 2, 3. \tag{5.51}$$

Note that the model function given by Equation (5.51) represents a two-parameter semilog-straight-line model with the intercept m_1 and slope m_2 and is linear with respect to the model parameters m_1 and m_2. We assume that the true values of the model parameters m_1 and m_2 are $\breve{m}_1 = 10$ and $\breve{m}_2 = 5$.

Suppose that we have 15 observed values of $(y_m)_{i,1}$, $i = 1, 2, \ldots, 15$ (i.e., $N_{d1} = 15$ in Equation (5.49)), with random errors having normal

Table 5.1 Observed data sets for the synthetic example

Data set 1		Data set 2		Data set 3	
$t_{i,1}$	$(y_m)_{i,1}$	$t_{i,2}$	$(y_m)_{i,2}$	$t_{i,3}$	$(y_m)_{i,3}$
0.2	2.84903	3.2	15.26713	5.2	18.24777
0.4	4.84396	3.4	16.23261	5.4	18.42912
0.6	8.20325	3.6	15.86329	5.6	18.61761
0.8	8.14101	3.8	16.64221	5.8	18.78557
1.0	8.12388	4.0	17.04152	6	18.94941
1.2	8.94669	4.2	16.95869	6.2	19.11292
1.4	11.91136	4.4	18.02271	6.4	19.28263
1.6	13.95758	4.6	17.67031	6.6	19.44338
1.8	9.93653	4.8	17.7389	6.8	19.56960
2.0	11.47588	5.0	19.18386	7.0	19.71960
2.2	15.30103			7.2	19.87719
2.4	17.17171			7.4	20.02137
2.6	14.38451			7.6	20.13877
2.8	17.73598			7.8	20.28355
3.0	19.45149			8.0	20.41699

distribution with zero mean and variance $\sigma_1^2 = 4.0$. Similarly, we have 10 observed values of $(y_m)_{i,2}$, $i = 1, 2, \ldots, 10$ (i.e., $N_{d2} = 10$) for the second set with random errors having normal distribution with zero mean and variance $\sigma_2^2 = 0.25$, and 15 observed values of $(y_m)_{i,3}$, $i = 1, 2, \ldots, 15$ (i.e., $N_{d3} = 15$) for the third set with random errors having normal distribution with zero mean and variance $\sigma_3^2 = 10^{-4}$. The observed values of $(y_m)_{i,1}$, $(y_m)_{i,2}$, and $(y_m)_{i,3}$ are tabulated in Table 5.1 for the corresponding values of independent variables $t_{i,1}$, $t_{i,2}$, and $t_{i,3}$.

Table 5.2 presents the values of the model parameters and variances simultaneously estimated by maximizing the log-likelihood function given by Equation (5.49) (see 3-rd row of Table 5.2) and by minimizing the objective function given by Equation (5.50) (see 4-th row of Table 5.2). Minimizing Equation (5.50) is equivalent to maximizing the concentrated log-likelihood function (Equation (5.33)) in the stage-wise maximization procedure discussed previously. Both maximization of Equation (5.49) and minimization of Equation (5.50) require nonlinear optimization, though the model functions (Equation (5.51)) are linear with respect to the model parameters m_1 and m_2 because variances are treated as unknown in both methods.

As can be seen from Table 5.2, the stage-wise maximization method (or equivalently the minimization of the objective function given by

Table 5.2 Summary of results for the synthetic example from the maximum likelihood (ML), weighted (WLS) and unweighted (UWLS) least-squares estimations

	m_1	m_2	$\tilde{\sigma}_1^2$	$\tilde{\sigma}_2^2$	$\tilde{\sigma}_3^2$	$\hat{\sigma}_1^2$	$\hat{\sigma}_2^2$	$\hat{\sigma}_3^2$
True values	10	5	–	–	–	4	0.25	1×10^{-4}
MLE (Equation (5.49))	9.9453	5.0298	3.8325[a]	0.2340[a]	7.863×10^{-5}[a]	4.0108[b]	0.2507[b]	8.228×10^{-5}[b]
MLE (Equation (5.50))	9.9453	5.0298	3.8325[d]	0.2340[d]	7.863×10^{-5}[d]	4.0108[b]	0.2507[b]	8.228×10^{-5}[b]
WLSE (Equation (5.52))	9.9453	5.0298	–	–	–	4.0[c]	0.25[c]	1×10^{-4}[c]
UWLSE (Equation (5.53))	9.9989	5.0931	–	–	–	3.9302[b]	0.2452[b]	3.133×10^{-2}[b]

[a] Obtained by maximization of Equation (5.49).
[b] Computed from Equation (5.35).
[c] Assumed as known.
[d] Computed from Equation (5.32).

Equation (5.50)) and the direct maximization of the log-likelihood function given by Equation (5.49) yield the same values of the model parameters as well as the biased and unbiased estimates of the variances of the noise in data sets, obtained by maximization of the log-likelihood function by Equation (5.49) for the specific example. This result proves that the stage-wise maximization method (which is based on minimization of the objective function given by Equation (5.50) with respect to only the model parameters) is equivalent to simultaneously estimating the model parameters and biased data error variances by direct maximization of the general log-likelihood function (Equation (5.49)). The biased estimates of data error variances can be further adjusted by the use of Equation (5.35) to obtain unbiased estimates of data error variances. Note that the ML estimates of the model parameters and variances (unbiased ones) are very close to the true values in Table 5.2.

The results in Table 5.2 (see 5-th row) also prove that if the variances of all data sets are known and used in the WLS objective function given by

$$O(\mathbf{m}) = \frac{1}{2}\sum_{j=1}^{N_s}\frac{1}{\sigma_j^2}\sum_{i=1}^{N_{dj}}\left[(y_m)_{i,j} - f\left(\mathbf{m}, t_{i,j}\right)\right]^2 \tag{5.52}$$

then the WLSE method based on the minimization of Equation (5.52) yields the same estimates of the model parameters as those obtained from the MLE method based on the minimization of Equation (5.50).

For the purpose of comparison, the sixth row of Table 5.2 presents the results for the specific example obtained from the minimization of the UWLS objective function that can be expressed as

$$O(\mathbf{m}) = \frac{1}{2}\sum_{j=1}^{N_s}\sum_{i=1}^{N_{dj}}\left[(y_m)_{i,j} - f\left(\mathbf{m}, t_{i,j}\right)\right]^2. \tag{5.53}$$

The estimated model parameters and variances given for the UWLSE case in Table 5.2 are computed from Equation (5.35) after the model parameters estimated by minimizing the UWLSE objective function given by Equation (5.53). The estimated model parameters and variances for the data sets 1 and 2 agree reasonably well with the corresponding true values, but the estimated variance for the data set 3, which is equal to 3.1×10^{-2}, is quite far from the true value of 1×10^{-4}. The overall (average) data variance estimated from the UWLSE is equal to 1.554, which is very close to the number of data points-averaged variance of the three data sets, that is,

$$\bar{\sigma}^2 = \frac{N_{d1}\sigma_1^2 + N_{d2}\sigma_2^2 + N_{d3}\sigma_3^2}{N_{d1} + N_{d2} + N_{d3}}. \tag{5.54}$$

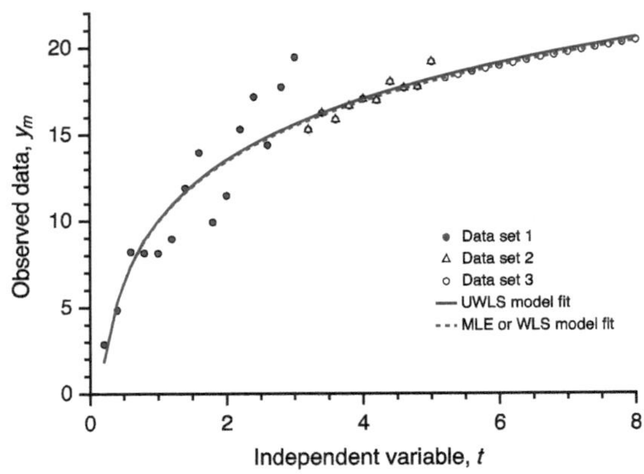

Figure 5.8 Model fits by UWLSE, WLSE, and MLE of the data sets of the synthetic example.

This is an expected result because the UWLS estimation yields the estimates of the model parameters influenced more by the data sets with larger variances than those data sets with small variances. This is further graphically illustrated in Figure 5.8.

As seen from Figure 5.8, the curve fitted by the UWLSE method does not match well the good quality data of the set 3, while the MLE and WLSE (with the appropriate variances) methods honor the good quality data of the set 3. Of course, if the UWLSE is viewed purely as a curve fitting method for estimating the model parameters, we can state that it is quite satisfactory, yielding the model parameter estimates which are very close to the true values for this specific synthetic example. However, confidence intervals for estimation, simulation and prediction based on the use of the UWLSE method for data sets having different variances will be incorrect and misleading, as will discussed later.

5.8. Use of Prior Information in ML Estimation: Bayesian Framework

In many cases, we have *a priori* knowledge or justification for the values of some parameters even before any data are acquired. For instance, we know that permeability and porosity must be positive. Thus, an estimation procedure that came up with negative values for such parameters should be entirely unacceptable. In addition, *a priori* knowledge may lead to rejection of some other values as being entirely implausible, even though they are strictly speaking not impossible. Even among admissible values, we may regard some values as more plausible than others. For instance, *a priori* knowledge may lead us to think that the permeability of a layer has to be high (e.g., 100 ± 10 md).

A priori knowledge or justification can be described by a prior probability distribution of parameters. The prior distribution may be characterized by means of the prior density function $p_0(\mathbf{m})$. For instance, we may consider that the prior density function of the vector of unknown model parameters is a multi-Gaussian with mean $\mathbf{m}_{\text{prior}}$ and covariance matrix $\mathbf{C}_M$, that is,

$$p_0(\mathbf{m}) = \frac{1}{(2\pi)^{M/2} (\det \mathbf{C}_M)^{1/2}} \times \exp\left[-\frac{1}{2} \left(\mathbf{m} - \mathbf{m}_{\text{prior}}\right)^T \mathbf{C}_M^{-1} \left(\mathbf{m} - \mathbf{m}_{\text{prior}}\right) \right]. \tag{5.55}$$

Here, $\mathbf{m}_{\text{prior}}$ denotes the vector of prior means of unknown model parameters or *a priori* knowledge of the mean of the unknown model parameters. In Equation (5.55), $\mathbf{C}_M$ is an $M \times M$ symmetric and positive-definite[3] covariance matrix of model parameters and describes the relation among the individual model parameters:

$$\mathbf{C}_M = \begin{bmatrix} \sigma_{m_1}^2 & \text{cov}(m_1, m_2) & \dots & \text{cov}(m_1, m_M) \\ \text{cov}(m_1, m_2) & \sigma_{m_2}^2 & \dots & \text{cov}(m_2, m_M) \\ \vdots & & \dots & \\ \text{cov}(m_1, m_M) & \text{cov}(m_2, m_M) & \dots & \sigma_{m_M}^2 \end{bmatrix}, \tag{5.56}$$

where $\sigma_{m_j}^2$, $j = 1, 2, \dots, M$, denotes prior variance of the model parameter m_j, while $\text{cov}(m_i, m_j)$, for $i \neq j$, denotes the prior covariance between the model parameter m_i and m_j. In cases where there exists no correlation between the model parameter pairs, then $\mathbf{C}_M$ is a diagonal matrix with diagonal elements equal to $\sigma_{m_j}^2$ for $j = 1, 2, \dots, M$. Here, we assume that a prior model for the vector of unknown parameters is multi-normal with the mean vector $\mathbf{m}_{\text{prior}}$ and covariance matrix $\mathbf{C}_M$, as described by Equation (5.55).

In the applications, where analytical pressure-transient models are used to interpret the observed data, $\mathbf{C}_M$ will usually be small in size because such models normally contain a fewer number of parameters compared to the number of observed pressure data (i.e., an overdetermined problem). Consequently, this should not impose much of a difficulty in terms of computations. However, if one considers geostatistics-based heterogeneous

[3] If all eigenvalues of a symmetric matrix are strictly positive, the matrix is said to be positive definite. If the matrix is positive definite, it is nonsingular and its all diagonal elements are positive.

reservoir models where one wishes to consider heterogeneity in each simulation grid block (Chu et al., 1995; Gok et al., 2005; He et al., 2000; Landa et al., 2000; Oliver, 1994; Oliver et al., 1996) for which the number of observed data can be far less than the number of model parameters (i.e., an underdetermined problem), then $\mathbf{C}_M$ can be quite large, requiring special algorithms for its storage and computations.

Including a prior distribution for the model parameters is advantageous in several ways: First, it provides a regularization (constraining) of model parameters in history matching by improving the convergence properties of minimization methods. Second, as will be described next, the information on the parameters contained in the observed data may be expressed in the form of a so-called posterior distribution. It is proper to use the posterior distribution from the already observed data as the prior distribution for the measurements yet to be completed. As to be discussed in the next section, this is useful for performing sequential matching of more than one pressure-transient test data set performed in the same system.

We have already shown that the information on model parameters contained in the observed data can be represented by means of the likelihood function L and that *a priori* knowledge or information on the model parameters can be represented by means of the prior density $p_0(\mathbf{m})$. Following Bard (1974) and Tarantola (2005), we can combine the two in the so-called posterior density function which is proportional to their product

$$\hat{p}(\mathbf{m}, \boldsymbol{\Psi}|\mathbf{y}_m) = aL(\mathbf{m}, \boldsymbol{\Psi})p_0(\mathbf{m}). \tag{5.57}$$

As in Equation (5.4), the symbol $\boldsymbol{\Psi}$ in Equation (5.57) is used to denote distribution parameters such as variances (or in general covariance matrix) of noise. In Equation (5.57), a denotes a generic constant which may ensure that the posterior **pdf**, $\hat{p}$, integrates to 1 and equal to the integral of the product of likelihood and prior **pdf**s provided that such an integral exists.

It may be also worth noting that in the cases where both $\mathbf{m}$ and $\boldsymbol{\Psi}$ are treated as unknowns, the prior $p_0(\mathbf{m})$ given in Equation (5.57) will be replaced by a joint prior of $\mathbf{m}$ and $\boldsymbol{\Psi}$, that is $p_0(\mathbf{m}, \boldsymbol{\Psi})$. However, if $p_0(\mathbf{m})$ and $p_0(\boldsymbol{\Psi})$ are treated as independent *a priori*, then $p_0(\mathbf{m}, \boldsymbol{\Psi}) = p_0(\mathbf{m})p_0(\boldsymbol{\Psi})$. Furthermore, we may assume a "non-informative" prior for $p_0(\boldsymbol{\Psi})$, such as a constant or "uniform" prior, so that all values of $\boldsymbol{\Psi}$ are equally plausible, and hence we can obtain Equation (5.57). We should also note that the value of a in estimating the model parameter $\mathbf{m}$ and $\boldsymbol{\Psi}$ using Equation (5.57) is immaterial. Throughout this book, we will consider only the use of the posterior given by Equation (5.57) in estimating $\mathbf{m}$ and $\boldsymbol{\Psi}$, whether $\boldsymbol{\Psi}$ is treated as known or unknown. It is beyond the scope of this book to provide a further detailed discussion of informative and non-informative priors as well as proper and improper posterior distributions, but the interested reader is referred to the books by Bard (1974) and Jeffreys (1961).

If $p_0(\mathbf{m})$ is regarded as a probability ascribed to $\mathbf{m}$ before the observations, then $\hat{p}$ is the probability density we must ascribe to $\mathbf{m}$ after the data were obtained. In fact, this is a result of the well-known Bayes' theorem (Bard, 1974; Barlow, 1989). Note that in the absence of prior information (or equivalently it means that there is complete uncertainty, or infinite variance and null correlations, in model parameters), the posterior distribution given by Equation (5.57) is equal to the likelihood because in this case in the limit $\mathbf{C}_M^{-1} \rightarrow \mathbf{0}$, a null matrix, in Equation (5.55) and hence the *a priori* distribution given by Equation (5.55) approaches a constant or "uniform" prior (Tarantola, 2005).

The natural extension of the maximum likelihood method that we summarized in the previous sections to Bayesian estimation problems consists of looking for the mode of the posterior distribution; i.e., we accept the value of $\mathbf{m}$ as our estimate for which $\hat{p}(\mathbf{m})$ is maximum. This method is usually referred to as the maximum posterior distribution (Bard, 1974).

If we assume that the data error covariance matrix $\mathbf{C}_D$ is known, then one can derive the objective functions for the least-squares parameter estimation problem to be used in the Bayesian framework, from Equation (5.57). If we consider a single observed data set and $\mathbf{\Psi} = \mathbf{C}_D$ in Equation (5.57), then we can rewrite Equation (5.57) as

$$\hat{p}(\mathbf{m}, \mathbf{C}_D|\mathbf{y}_m) = aL(\mathbf{m}, \mathbf{C}_D)p_0(\mathbf{m}). \tag{5.58}$$

Using the likelihood function given by Equation (5.5) and the prior density function given by Equation (5.55) in Equation (5.58) yields

$$\hat{p}(\mathbf{m}, \mathbf{C}_D|\mathbf{y}_m) = b\exp\left\{-\frac{1}{2}\left[\left(\mathbf{y}_m - \mathbf{y}_c\right)^T \mathbf{C}_D^{-1}\left(\mathbf{y}_m - \mathbf{y}_c\right) + \left(\mathbf{m} - \mathbf{m}_{\text{prior}}\right)^T \mathbf{C}_M^{-1}\left(\mathbf{m} - \mathbf{m}_{\text{prior}}\right)\right]\right\}, \tag{5.59}$$

where b is a generic constant. Then, maximizing the posterior density is equivalent to minimizing the argument of the exponential function in Equation (5.59):

$$O(\mathbf{m}) = \frac{1}{2}\left(\mathbf{y}_m - \mathbf{y}_c\right)^T \mathbf{C}_D^{-1}\left(\mathbf{y}_m - \mathbf{y}_c\right) + \frac{1}{2}\left(\mathbf{m} - \mathbf{m}_{\text{prior}}\right)^T \mathbf{C}_M^{-1}\left(\mathbf{m} - \mathbf{m}_{\text{prior}}\right), \tag{5.60}$$

which is the LS objective function incorporating prior information in the Bayesian setting, but assuming that the data error covariance matrix $\mathbf{C}_D$ is known *a priori*.

For the general case of N_s independent observed data sets, each with a known data error covariance matrix, $\mathbf{C}_{D,j}$, $j = 1, 2, \ldots, N_s$, similarly, we can obtain the following least-squares objective function:

$$O(\mathbf{m}) = \sum_{j=1}^{N_s} \frac{1}{2} \left[(\mathbf{y}_m)_j - (\mathbf{y}_c)_j \right]^T \mathbf{C}_{D,j}^{-1} \left[(\mathbf{y}_m)_j - (\mathbf{y}_c)_j \right] + \frac{1}{2} \left(\mathbf{m} - \mathbf{m}_{\text{prior}} \right)^T \mathbf{C}_M^{-1} \left(\mathbf{m} - \mathbf{m}_{\text{prior}} \right). \tag{5.61}$$

The applications based on minimization of the objective functions Equations (5.60) and (5.61) have been presented extensively in the literature (Chu et al., 1995; Gok et al., 2005; He et al., 2000; Landa et al., 2000; Oliver, 1994; Oliver et al., 1996). As mentioned previously, for the parameter estimation based on the use of the objective functions of Equations (5.60) and (5.61), it is necessary to characterize the measurement errors *a priori* by the covariance functions because the covariance functions in Equations (5.60) and (5.61) provide the correct relative weighting of different data sets in the objective function (Zao et al., 2007).

Next, we focus on the MLE in the Bayesian setting and treating data error variances as unknown in parameter estimation. It is more convenient to work with the logarithm of the posterior function. Taking the natural logarithm of Equation (5.57) gives

$$\ln \left[\hat{p}(\mathbf{m}, \mathbf{\Psi} | \mathbf{y}_m) \right] = \ln(a) + \ln \left[L(\mathbf{m}, \mathbf{\Psi}) \right] + \ln \left[p_0(\mathbf{m}) \right], \tag{5.62}$$

but maximizing the above log-posterior is equivalent to maximizing

$$\Phi(\mathbf{m}, \mathbf{\Psi}) = \ln \left[L(\mathbf{m}, \mathbf{\Psi}) \right] + \ln \left[p_0(\mathbf{m}) \right], \tag{5.63}$$

where we have used the symbol $\Phi(\mathbf{m}, \mathbf{\Psi})$ to denote $\ln \left[\hat{p}(\mathbf{m}, \mathbf{\Psi} | \mathbf{y}_m) \right]$. Because a is a generic positive constant, it can be deleted in Equation (5.62) to obtain Equation (5.63).

We directly start with the general case of N_s independent observed data sets for which the log-likelihood function is given by Equation (5.31). We assume that the *a priori* model is given by Equation (5.55) and treat the error variances σ_j^2, $j = 1, 2, \ldots, N_s$, for each data set as unknown, then, by using

Equations (5.31), (5.55) and (5.62), we can obtain:

$$\Phi(\mathbf{m}, \sigma_1^2, \ldots, \sigma_{N_s}^2) = -\frac{1}{2}\sum_{j=1}^{N_s} N_{dj} \ln \sigma_j^2 - \frac{1}{2}\sum_{j=1}^{N_s} \frac{1}{\sigma_j^2}\left[(\mathbf{y}_m)_j - (\mathbf{y}_c)_j\right]^T \left[(\mathbf{y}_m)_j - (\mathbf{y}_c)_j\right] - \frac{1}{2}\left(\mathbf{m} - \mathbf{m}_{\text{prior}}\right)^T \mathbf{C}_M^{-1}\left(\mathbf{m} - \mathbf{m}_{\text{prior}}\right). \quad (5.64)$$

To maximize Equation (5.64), we can proceed by the stage-wise maximization method as previously discussed. When this method is applied to Equation (5.64), it consists of finding, for any value of $\mathbf{m}$, the values of the σ_j^2 that maximize Equation (5.64). These σ_j^2 values will be a function of $\mathbf{m}$, say $\tilde{\sigma}_j^2(m)$. Substitution of $\tilde{\sigma}_j^2(\mathbf{m})$ for σ_j^2 in Equation (5.64) gives the concentrated log-likelihood function, which is a function of $\mathbf{m}$ alone, and we seek $\tilde{\mathbf{m}}$ so as to maximize this concentrated log-likelihood function.

The first step, then, is to differentiate Equation (5.64) with respect to σ_j^2 while holding other variables fixed. Next, we set the resulting derivatives to zero to obtain

$$\tilde{\sigma}_j^2(m) = \frac{1}{N_{dj}}\left[(\mathbf{y}_m)_j - (\mathbf{y}_c)_j\right]^T \left[(\mathbf{y}_m)_j - (\mathbf{y}_c)_j\right] = \frac{1}{N_{dj}}\sum_{i=1}^{N_{dj}}\left[(y_m)_{i,j} - f(\mathbf{m}, t_{i,j})\right]^2 \quad (5.65)$$

for $j = 1, \ldots, N_s$.

Using Equation (5.65) for σ_j^2 in Equation (5.64) and simplifying the resulting equation, we can derive the so-called *concentrated likelihood* function $\tilde{L}(\mathbf{m})$ with *a priori* information as

$$\tilde{L}(\mathbf{m}) = \frac{1}{2}\left\{\sum_{j=1}^{N_s}\left[\ln(N_{dj}) - 1\right]\right\} - \frac{1}{2}\sum_{j=1}^{N_s} N_{dj} \ln\left\{\left[(\mathbf{y}_m)_j - (\mathbf{y}_c)_j\right]^T \left[(\mathbf{y}_m)_j - (\mathbf{y}_c)_j\right]\right\} - \frac{1}{2}\left(\mathbf{m} - \mathbf{m}_{\text{prior}}\right)^T \mathbf{C}_M^{-1}\left(\mathbf{m} - \mathbf{m}_{\text{prior}}\right). \quad (5.66)$$

Maximizing Equation (5.66) is equivalent to minimizing the following objective function:

$$\tilde{O}(\mathbf{m}) = \frac{1}{2}\sum_{j=1}^{N_s} N_{dj} \ln\left\{\left[(\mathbf{y}_m)_j - (\mathbf{y}_c)_j\right]^T \left[(\mathbf{y}_m)_j - (\mathbf{y}_c)_j\right]\right\} + \frac{1}{2}\left(\mathbf{m} - \mathbf{m}_{\text{prior}}\right)^T \mathbf{C}_M^{-1}\left(\mathbf{m} - \mathbf{m}_{\text{prior}}\right). \tag{5.67}$$

Equation (5.67) represents the objective function to be minimized when adding prior information on the model parameters. (In the absence of prior information; i.e., $\mathbf{C}_M^{-1} \to \mathbf{0}$, a null matrix, in Equation (5.67), Equation (5.67) reverts to Equation (5.34), the objective function without *a priori* information.)

For the special case where $\mathbf{C}_M$ is a diagonal matrix in Equation (5.67), i.e., model parameters are not correlated in the *a priori* model, then Equation (5.67) becomes

$$\tilde{O}(\mathbf{m}) = \frac{1}{2}\sum_{j=1}^{N_s} N_{dj} \ln\left\{\sum_{i=1}^{N_{dj}}\left[(y_m)_{i,j} - f(\mathbf{m}, t_{i,j})\right]^2\right\} + \frac{1}{2}\sum_{i=1}^{M}\left(\frac{\mathbf{m}_i - \mathbf{m}_{\text{prior}}}{\sigma_{m_i}}\right)^2. \tag{5.68}$$

After finding $\tilde{\mathbf{m}}$ minimizing the $\tilde{O}(\mathbf{m})$ given by Equation (5.67) (or Equation (5.68) for the uncorrelated case), the estimates of the error variances for each data set are computed from Equation (5.35) by using the estimated $\tilde{\mathbf{m}}$, that is,

$$\hat{\sigma}_j^2 = \frac{N_{dj}\tilde{\sigma}_j^2}{(N_{dj} - M/N_s)}, \tag{5.69}$$

where $\tilde{\sigma}_j^2$ is the biased estimates of the data error variances, given by

$$\tilde{\sigma}_j^2 = \frac{1}{N_{dj}}\sum_{i=1}^{N_{dj}}\left[(y_m)_{i,j} - f(\tilde{\mathbf{m}}, t_{i,j})\right]^2, \tag{5.70}$$

for $j = 1, 2, \ldots, N_s$.

It should be noted that introducing the *a priori* distribution biases the results of the estimation process so as to favor parameters values for which

$p_0(\mathbf{m})$ is relatively large. However, this bias diminishes as the number of observations increases. In other words, when the amount of observed data is sufficiently large, the effect of the *a priori* distribution on the parameter estimates is negligible (Bard, 1974).

5.8.1. Single-parameter linear model case

Here, we consider the same simple single-parameter linear model $f(m) = m$ considered previously in Section 5.7.1 to illustrate the use of *a priori* information in parameter estimation. Suppose that *a priori* information for the model parameter m is described by a normal distribution with mean m_{prior} and variance σ_m^2, that is,

$$p_0(m) = \frac{1}{\sqrt{2\pi\sigma_m^2}} \exp\left[-\frac{1}{2}\frac{\left(m - m_{\text{prior}}\right)^2}{\sigma_m^2}\right]. \tag{5.71}$$

Further suppose that we have two independent sets of samples of $(\mathbf{y}_m)_j$, $j = 1, 2$, with two different variances; one set containing N_{d1} independent measurements $(y_m)_{1,1}, (y_m)_{2,1}, \ldots, (y_m)_{N_{d1},1}$ with noise having zero mean and variance σ_1^2, and the other containing N_{d2} independent measurements $(y_m)_{1,2}, (y_m)_{2,2}, \ldots, (y_m)_{N_{d2},2}$ with noise with zero mean and variance σ_2^2. Then, the likelihood function treating m, σ_1^2 and σ_2^2 as unknown is given by

$$L\left(m, \sigma_1^2, \sigma_2^2 | \mathbf{y}_m\right) = \frac{1}{\sqrt{\left(2\pi\sigma_1^2\right)^{N_{d1}}}\sqrt{\left(2\pi\sigma_2^2\right)^{N_{d2}}}}$$
$$\times \exp\left\{-\frac{1}{2\sigma_1^2}\sum_{i=1}^{N_{d1}}\left[(y_m)_{i,1} - m\right]^2 \frac{1}{2\sigma_2^2}\sum_{i=1}^{N_{d2}}\left[(y_m)_{i,2} - m\right]^2\right\}. \tag{5.72}$$

Combining Equations (5.71) and (5.72) by Bayes' theorem expressed by Equation (5.58) gives the following posterior **pdf** such that:

$$\hat{p}(m, \sigma_1^2, \sigma_2^2 | \mathbf{y}_m) = b \exp\left\{-\frac{1}{2\sigma_1^2}\sum_{i=1}^{N_{d1}}\left[(y_m)_{i,1} - m\right]^2\right.$$
$$\left. - \frac{1}{2\sigma_2^2}\sum_{i=1}^{N_{d2}}\left[(y_m)_{i,2} - m\right]^2 - \frac{1}{2}\frac{\left(m - m_{\text{prior}}\right)^2}{\sigma_m^2}\right\}, \tag{5.73}$$

where b is a generic constant. If we assume the data variances σ_1^2 and σ_2^2 are known, then maximizing the posterior is equivalent to minimizing the following objective function:

$$O(m) = \frac{1}{2\sigma_1^2}\sum_{i=1}^{N_{d1}}\left[(y_m)_{i,1} - m\right]^2 + \frac{1}{2\sigma_2^2}\sum_{i=1}^{N_{d2}}\left[(y_m)_{i,2} - m\right]^2 + \frac{1}{2}\frac{\left(m - m_{\text{prior}}\right)^2}{\sigma_m^2}. \tag{5.74}$$

It can be shown that the estimate of m minimizing Equation (5.74) is given as

$$\tilde{m} = \frac{\frac{\sum_{i=1}^{N_{d1}}(y_m)_{i,1}}{\sigma_1^2} + \frac{\sum_{i=1}^{N_{d2}}(y_m)_{i,2}}{\sigma_2^2} + \frac{m_{\text{prior}}}{\sigma_m^2}}{\frac{N_{d1}}{\sigma_1^2} + \frac{N_{d2}}{\sigma_2^2} + \frac{1}{\sigma_m^2}} = \frac{\frac{N_{d1}(\bar{y}_m)_1}{\sigma_1^2} + \frac{N_{d2}(\bar{y}_m)_2}{\sigma_2^2} + \frac{m_{\text{prior}}}{\sigma_m^2}}{\frac{N_{d1}}{\sigma_1^2} + \frac{N_{d2}}{\sigma_2^2} + \frac{1}{\sigma_m^2}}, \tag{5.75}$$

where $\tilde{m}$ is referred to as the posterior mean. As can be seen from Equation (5.75), the Bayes estimate of m is a weighted combination of the prior mean m_{prior}, and the sample evidences about m, that is, $(\bar{y}_m)_j$ for $j = 1$ and 2. In addition, the weights are the reciprocals of the respective variances of the prior and sample evidences, that is, σ_m^2, σ_1^2/N_{d1}, and σ_2^2/N_{d2}.

It can also be shown that the variance of $\tilde{m}$ (also called the posterior variance of $\tilde{m}$) is given as

$$\sigma_{\tilde{m}}^2 = \frac{1}{\frac{N_{d1}}{\sigma_1^2} + \frac{N_{d2}}{\sigma_2^2} + \frac{1}{\sigma_m^2}}, \tag{5.76}$$

which indicates that the variance of $\tilde{m}$ is the reciprocal of the sum of the precision of the prior evidence $(1/\sigma_m^2)$ and the precisions of the observed sample evidences (N_{d1}/σ_1^2 and N_{d2}/σ_2^2). Note that as σ_m^2 tends to infinity in Equations (5.75) and (5.76), i.e., no prior information is available on m (all values of m are plausible), then, as expected, the posterior mean of m (Equation (5.75)) tends to the WLS estimate of m (Equation (5.41)) and the posterior variance of m (Equation (5.76)) tends to the corresponding WLS estimate of m (Equation (5.42)). For this specific case, both the Bayes and WLS estimates are identical and depend on only the information content and the precision of the observed data sets.

It can be deduced from Equations (5.75) and (5.76) that adding *a priori* information on m helps reducing the uncertainty in $\tilde{m}$ and that if the number of data available for estimation is sufficiently large, the bias on $\tilde{m}$ introduced by m_{prior} diminishes. The latter observation may also imply that in the cases where an incorrect m_{prior} is assigned, its effect on the estimated parameter may not be important, and one may still obtain an estimate of m that honors observed data sets well.

If we treat the variances of the data sets in addition to the model parameter m in the Bayesian setting for the example case considered above, then we may consider the minimization of the objective function given by Equation (5.68) (with $N_s = 2$ and $M = 1$) with respect to the model parameter m, which may be rewritten as

$$\tilde{O}(m) = \frac{1}{2}\sum_{j=1}^{2} N_{dj} \ln\left\{\sum_{i=1}^{N_{dj}} \left[(y_m)_{i,j} - m\right]^2\right\} + \frac{1}{2}\frac{\left(m - m_{\text{prior}}\right)^2}{\sigma_m^2}. \quad (5.77)$$

The estimate of m that minimizes the above objective function is given as

$$m = \frac{\dfrac{\sum_{i=1}^{N_{d1}} (y_m)_{i,1}}{\sigma_1^2(m)} + \dfrac{\sum_{i=1}^{N_{d2}} (y_m)_{i,2}}{\sigma_2^2(m)} + \dfrac{m_{\text{prior}}}{\sigma_m^2}}{\dfrac{N_{d1}}{\sigma_1^2(m)} + \dfrac{N_{d2}}{\sigma_2^2(m)} + \dfrac{1}{\sigma_m^2}} = \frac{\dfrac{N_{d1}(\bar{y}_m)_1}{\sigma_1^2(m)} + \dfrac{N_{d2}(\bar{y}_m)_2}{\sigma_2^2(m)} + \dfrac{m_{\text{prior}}}{\sigma_m^2}}{\dfrac{N_{d1}}{\sigma_1^2(m)} + \dfrac{N_{d2}}{\sigma_2^2(m)} + \dfrac{1}{\sigma_m^2}}, \quad (5.78)$$

where the data variances in Equation (5.78) are given as

$$\sigma_j^2(m) = \frac{1}{N_{dj}}\sum_{i=1}^{N_{dj}} \left[(y_m)_{i,j} - m\right]^2 \quad \text{for } j = 1 \text{ and } 2. \quad (5.79)$$

Like MLE without prior information (given by Equations (5.47) and (5.48)), MLE in the Bayesian setting for finding the solution $\tilde{m}$ from Equations (5.78) and (5.79) will require a nonlinear optimization method because Equations (5.78) and (5.79) are obviously nonlinear with respect to the model parameter m. Note that when σ_m^2 tends to infinity in Equation (5.78), i.e., no prior information is available on m and all values of m are plausible, then, as expected, the posterior mean of m (Equation (5.78)) tends to the ML estimate of m without prior that depends on only the information content of the observed data sets (Equation (5.47)).

5.8.2. An example application

We will use the same model and observed data sets previously considered in Section 5.7.2 to demonstrate the points made above for the Bayesian estimation. We consider the three data sets as given in Table 5.1 and the two-parameter linear model of Equation (5.51) to history match the data sets using Bayesian estimation. We will assume that both model parameters m_1 and m_2 (see Equation (5.51)) are independent *a priori* and that each prior model is a normal distribution with mean $m_{\text{prior},j}$ and variance $\sigma^2_{m_j}$, as given by

$$p_0(m_j) = \frac{1}{\sqrt{2\pi\sigma^2_{m_j}}} \exp\left[-\frac{1}{2}\frac{\left(m_j - m_{\text{prior}_j}\right)^2}{\sigma^2_{m_j}}\right] \tag{5.80}$$

for $j = 1$ and 2. The joint prior of the model parameters is then given as

$$p_0(m_1, m_2) = p_0(m_1)p_0(m_2) = \frac{1}{\sqrt{2\pi\sigma^2_{m_1}}\sqrt{2\pi\sigma^2_{m_2}}} \exp \times \left\{-\frac{1}{2}\left[\frac{\left(m_1 - m_{\text{prior}_1}\right)^2}{\sigma^2_{m_1}} + \frac{\left(m_2 - m_{\text{prior}_2}\right)^2}{\sigma^2_{m_2}}\right]\right\}. \tag{5.81}$$

The likelihood function L for this example can be written as

$$L\left(\mathbf{m}, \sigma_1^2, \sigma_2^2, \sigma_3^2 | \mathbf{y}_m\right) = \frac{1}{\sqrt{\left(2\pi\sigma_1^2\right)^{N_{d1}}}\sqrt{\left(2\pi\sigma_2^2\right)^{N_{d2}}}\sqrt{\left(2\pi\sigma_3^2\right)^{N_{d3}}}} \times \exp\left\{-\frac{1}{2\sigma_1^2}\sum_{i=1}^{N_{d1}}\left[(y_m)_{i,1} - f\left(\mathbf{m}, t_{i,1}\right)\right]^2 - \frac{1}{2\sigma_2^2}\sum_{i=1}^{N_{d2}}\left[(y_m)_{i,2} - f\left(\mathbf{m}, t_{i,2}\right)\right]^2 - \frac{1}{2\sigma_3^2}\sum_{i=1}^{N_{d3}}\left[(y_m)_{i,3} - f\left(\mathbf{m}, t_{i,3}\right)\right]^2\right\}, \tag{5.82}$$

where $\mathbf{m} = (m_1, m_2)^T$ and the model functions $f\left(\mathbf{m}, t_{i,j}\right)$ are defined by Equation (5.51) and the observed data $(y_m)_{i,j}$ are given in Table 5.1.

Combining the prior model and the likelihood function given by Equations (5.81) and (5.82) in Equation (5.58) and then considering the MLE in the Bayesian setting, we obtain the posterior function (Equation (5.64) with $N_s = 3$ and $M = 2$) as

$$\Phi(\mathbf{m}, \sigma_1^2, \sigma_2^2, \sigma_3^2) = -\frac{1}{2}\sum_{j=1}^{3} N_{dj}\ \ln(\sigma_j^2) - \frac{1}{2}\sum_{j=1}^{3}\frac{1}{\sigma_j^2}\sum_{i=1}^{N_{dj}}\left[(y_m)_{i,j} - f(\mathbf{m}, t_{i,j})\right]^2 - \frac{1}{2}\sum_{i=1}^{2}\left[\frac{m_i - m_{\text{prior}_i}}{\sigma_{m_i}}\right]^2, \tag{5.83}$$

which is to be maximized to obtain the ML estimates of the model

Table 5.3 True model parameters and four different cases of prior models for the two-parameter linear model, three sets of observed data example case

	m_1 or m_{prior_1}	m_2 or m_{prior_2}	$\sigma^2_{m_1}$	$\sigma^2_{m_2}$
True values	10	5	–	–
Prior model 1	10	5	0.1	0.1
Prior model 2	10	5	1	1
Prior model 3	10	5	10	10
Prior model 4	10	5	∞	∞

parameters and data variances. As shown previously, maximizing Equation (5.83) is equivalent to minimizing the following objective function (in the context of the stage-wise maximization discussed previously) which can be written as

$$\tilde{O}(\mathbf{m}) = \frac{1}{2}\sum_{j=1}^{3} N_{dj} \ln\left\{\sum_{i=1}^{N_{dj}} \left[(y_m)_{i,j} - f(\mathbf{m}, t_{i,j})\right]^2\right\} + \frac{1}{2}\sum_{j=i}^{2}\left(\frac{m_i - m_{\text{prior}_i}}{\sigma_{m_i}}\right)^2. \tag{5.84}$$

Table 5.3 gives the true values of the model parameters together with four different priors, each having the same values of prior means, but different variances for the model parameters m_1 and m_2, to investigate the effect of the use of prior model on the MLE estimation in the Bayesian setting. The true data error variances corresponding to three data sets are given in Table 5.4.

Table 5.4 also presents the estimates of m_1, m_2, σ_1^2, σ_2^2, and σ_3^2 obtained for the four different priors considered in Table 5.3. The results of Table 5.4 indicate that as the variances (i.e., the uncertainties) of the model parameters decrease ($\sigma^2_{m_1}$ and $\sigma^2_{m_2}$), the ML estimates of the model parameters progressively approach the mode of the prior distribution (m_{prior_1} and m_{prior_2}), and the ML estimates of the data error variances approach the corresponding true values. Note that as we increase the prior variances of the model parameters, the estimates approach to the estimates obtained without the *a priori* model, as expected (compare the values in the 3-rd row of Table 5.2 and the last row of Table 5.4).

5.9. Simultaneous vs. Sequential History Matching of Observed Data Sets

It may not be obvious why one should consider sequential incorporation of observed data sets because simultaneous history matching

Table 5.4 Summary of results for the synthetic example; the MLE in the Bayesian setting

	m_1	m_2	$\tilde{\sigma}_1^2$	$\tilde{\sigma}_2^2$	$\tilde{\sigma}_3^2$	$\hat{\sigma}_1^2$	$\hat{\sigma}_2^2$	$\hat{\sigma}_3^2$
True values	10	5	–	–	–	4	0.25	1×10^{-4}
Prior model 1	9.996	5.003	3.828[a]	0.235[a]	9.159×10^{-5}[a]	4.006[b]	0.251[b]	9.585×10^{-5}[b]
Prior model 2	9.985	5.009	3.828[a]	0.234[a]	8.639×10^{-5}[a]	4.006[b]	0.251[b]	9.041×10^{-5}[b]
Prior model 3	9.957	5.023	3.831[a]	0.234[a]	7.936×10^{-5}[a]	4.009[b]	0.251[b]	8.305×10^{-5}[b]
Prior model 4	9.945	5.030	3.833[a]	0.234[a]	7.863×10^{-5}[a]	4.011[b]	0.251[b]	8.228×10^{-5}[b]

[a] Computed from Equation (5.70).
[b] Computed from Equation (5.69).

(or inversion) of all the data is clearly a valid and preferred procedure for both linear and nonlinear models. Using sequential or simultaneous matching is related more to computational efficiency considerations, depending on the number of data sets available as well as the model considered for matching. For example, if we are going to use the gradient-based Levenberg-Marquardt or Gauss-Newton method, then at every iteration (see Section 5.11), we have to compute the sensitivity and the approximate Hessian matrices. In Section 5.13, we provide the sensitivity and the approximate Hessian matrices for the objective functions given by (5.61) and (5.67). The sizes of these matrices are directly related to the number of data sets and pressure data points contained in each set. If the number of data sets and conditioning data contained in each set are quite large, it may be difficult to store all these matrices and computationally perform matrix-matrix and matrix-vector multiplications for simultaneous matching of all data. By breaking the problem into a sequence of smaller problems and sequentially matching different data sets, we can substantially reduce the storage requirements and computational effort required by simultaneous matching of all the data.

As shown by Tarantola (2005) for linear models the model parameters estimated by sequential history matching of independent sets of observed data in the Bayesian setting are identical to those estimated by simultaneous matching of all the data sets. We present a verification of this important result for the case of a single-parameter linear model. In Section 5.8, we have shown that the estimates of the LS posterior mean and its variance, obtained by simultaneous history matching of two independent observed data sets based on a single-parameter linear model $[f(m) = m]$ with prior information, are given by Equations (5.75) and (5.76), respectively.

Now suppose that we sequentially history match these two data sets, although the order of data sets to be used in sequential history matching is immaterial; suppose that we first history match the data set 1 containing N_{d1} data points with zero mean and variance σ_1^2 and then the second data set 2 containing N_{d2} data points with zero mean and variance σ_2^2. As the readers may easily verify from the results of Section 5.8, the LS objective function including the prior information (see Equation (5.71)) to be minimized for history matching the first data set is given as

$$O(m) = \frac{1}{2\sigma_1^2}\sum_{i=1}^{N_{d1}}\left[(y_m)_{i,1} - m\right]^2 + \frac{1}{2}\frac{\left(m - m_{\text{prior}}\right)^2}{\sigma_m^2}, \tag{5.85}$$

which has a unique minimum for m given by

$$\tilde{m}_1 = \frac{\frac{N_{d1}(\bar{y}_m)_1}{\sigma_1^2} + \frac{m_{\text{prior}}}{\sigma_m^2}}{\frac{N_{d1}}{\sigma_1^2} + \frac{1}{\sigma_m^2}}, \tag{5.86}$$

where the subscript "1" on $\tilde{m}$ refers to the posterior mean of m computed after conditioning to the first data set. The variance of $\tilde{m}_1$ is given as

$$\sigma^2_{\tilde{m}_1} = \frac{1}{\frac{N_{d1}}{\sigma_1^2} + \frac{1}{\sigma_m^2}}. \tag{5.87}$$

Now, we use the estimates of Equations (5.86) and (5.87) as the "prior" mean and variance of m, respectively, that is, we set $m_{\text{prior}} = \tilde{m}_1$ and $\sigma_m^2 = \sigma^2_{\tilde{m}_1}$ before history matching the second data set. Then, the LS objective function including this prior information to be minimized for history matching the second data set is given by

$$O(m) = \frac{1}{2\sigma_2^2}\sum_{i=1}^{N_d}\left[(y_m)_{i,2} - m\right]^2 + \frac{1}{2}\frac{(m-\tilde{m}_1)^2}{\sigma^2_{\tilde{m}_1}}. \tag{5.88}$$

The estimate of m (denoted by $\tilde{m}_2$) that minimizes Equation (5.88) is given by

$$\tilde{m}_2 = \frac{\frac{N_{d2}(\bar{y}_m)_2}{\sigma_2^2} + \frac{\tilde{m}_1}{\sigma^2_{\tilde{m}_1}}}{\frac{N_{d2}}{\sigma_2^2} + \frac{1}{\sigma^2_{\tilde{m}_1}}} \tag{5.89}$$

and its variance (denoted by $\sigma^2_{\tilde{m}_2}$) is given by

$$\sigma^2_{\tilde{m}_2} = \frac{1}{\frac{N_{d2}}{\sigma_2^2} + \frac{1}{\sigma^2_{\tilde{m}_1}}}. \tag{5.90}$$

Using Equations (5.86) and (5.87) in Equations (5.89) and (5.90) gives

$$\tilde{m}_2 = \frac{\frac{N_{d1}(\bar{y}_m)_1}{\sigma_1^2} + \frac{N_{d2}(\bar{y}_m)_2}{\sigma_2^2} + \frac{m_{\text{prior}}}{\sigma_m^2}}{\frac{N_{d2}}{\sigma_2^2} + \frac{N_{d1}}{\sigma_1^2} + \frac{1}{\sigma_m^2}} \tag{5.91}$$

and

$$\sigma^2_{\tilde{m}_2} = \frac{1}{\frac{N_{d2}}{\sigma_2^2} + \frac{N_{d1}}{\sigma_1^2} + \frac{1}{\sigma_m^2}}. \tag{5.92}$$

Equations (5.91) and (5.92) are identical to those estimates given by Equations (5.75) and (5.76) which are obtained by simultaneously history matching both independent sets of data.

For nonlinear history matching problems based on a gradient-based optimization procedure, Oliver (1994) shows that sequential history matching of pressure data sets yields very similar, almost identical, estimates of the model parameters as those obtained by simultaneous inversion of all the data sets. In fact, these results seem to indicate that like the linear problems, it may be proper to use the posterior distribution obtained from nonlinear history matching of the already completed observed data sets as the prior distribution for the other independent observed data sets yet to be conducted. We should also note that recently, there is a great interest in research for exploring the use of sequential (or recursive) methods such as the Ensemble Kalman Filters for nonlinear history matching problems, because these methods do not require computation of sensitivities and, consequently, appear to be far more efficient than a gradient-based history matching procedure when the forward model is represented by a reservoir simulator considering heterogeneous geological reservoir models (Aanonsen, Nvda, Oliver, & Reynolds, 2009; Li & Reynolds, 2009; Oliver et al., 2008). For instance, Li, Han, Banerjee, and Reynolds (2009) presents a successful implementation of Ensemble Kalman Filters in integrating well-test pressure data sets from single-phase flow of slightly compressible fluids in a heterogeneous single and commingled layered reservoir models. Together, all these results seem to indicate that sequential history matching for nonlinear pressure data sets appears to work well. It is beyond the scope of this book to further consider and discuss Ensemble Kalman Filters methods for nonlinear history matching applications, and hence we refer the interested reader to the above given references for further details.

Let us consider IPTT data sets as an example to demonstrate the use of sequential history matching based on a gradient-based optimization procedure. Further, suppose that we have two IPTTs at two different locations along the wellbore, and each test is conducted with a dual-packer and single probe configuration. Thus, each IPTT data contains two sets of pressure data; one for dual-packer and the other for the observation probe, and hence we have a total of four spatial pressure data sets to be history matched. Assuming that the prior mean vector of model parameters ($\mathbf{m}_{\text{prior}}$) as well as their covariance matrix ($\mathbf{C}_M$) are available, then first, we consider only the two sets of the pressure data corresponding to one of the two IPTTs (say this is the first IPTT), and we minimize this objective function with $N_s = 2$ in Equation (5.61) or Equation (5.67). At the end of this minimization, we obtain the vector of posterior model parameters ($\mathbf{m}_\infty$) and the posterior model covariance matrix ($\mathbf{C}_{MP}$) conditional to the two sets of pressure data for the first IPTT. Here, the $\mathbf{C}_{MP}$ matrix will be given by the approximate Hessian matrix evaluated at $\mathbf{m}_\infty$ (Oliver et al., 2008; Tarantola, 2005). The approximate Hessian matrices are given in Section 5.11. Next, by setting $\mathbf{m}_{\text{prior}} = \mathbf{m}_\infty$ and $\mathbf{C}_M = \mathbf{C}_{MP}$ in Equation (5.61) or Equation (5.67), we minimize this objective function with $N_s = 2$ to history match the two sets of pressure data of the second IPTT. At the end of this minimization, we obtain the vector of posterior model parameters ($\mathbf{m}_\infty$) and the posterior model covariance matrix ($\mathbf{C}_{MP}$) conditional to all four sets of pressure data for the first and second IPTTs. Thus, based on the concepts and results discussed in the preceding paragraph, for nonlinear problems, we can also use sequential history matching of observed data sets instead of simultaneous matching of all the data sets.

The above proposed sequential matching of multiple IPTTs should produce similar parameters estimated by simultaneously matching all these IPTT data sets at once based on Equation (5.61) or Equation (5.67). In fact, Gok et al. (2005) have applied the sequential matching of multiple IPTT data sets by considering the objective function as given by Equation (5.61) and have shown that the above proposed sequential conditioning of IPTT data sets produces satisfactory results (also see Example 5.15.6 in Section 5.15).

5.10. Summary on MLE and LSE Methods

In the previous sections, we have provided detailed technical descriptions of the MLE and LSE methods and provided the objective functions to be minimized with respect to the model parameter vector **m** with and without prior information, under the assumptions that the observed data and the prior model parameter distribution are Gaussian distributions. Further, to make the MLE method feasible for pressure transient test applications, we have assumed that the error covariance matrix for each $(\mathbf{y}_m)_j$ is diagonal with all diagonal elements identical, $\sigma_{i,j}^2 = \sigma_j^2$ for $i = 1, \ldots, N_{dj}$ for a given j. Such a formulation allowed us to estimate the model parameter vector **m** and treat the variance of each observed data set as unknown in estimation. In the case of the presence of prior Gaussian distribution information for the model parameters and N_s different sets of observed data with each set containing N_{dj} observations, we have previously derived the following objective function to be minimized with respect to the model parameter vector **m** as

$$\tilde{O}(\mathbf{m}) = \frac{1}{2}\sum_{j=1}^{N_s} N_{dj} \ln\left\{\left[(\mathbf{y}_m)_j - (\mathbf{y}_c)_j(\mathbf{m})\right]^T \left[(\mathbf{y}_m)_j - (\mathbf{y}_c)_j(\mathbf{m})\right]\right\} + \frac{1}{2}\left(\mathbf{m} - \mathbf{m}_{\text{prior}}\right)^T \mathbf{C}_M^{-1}\left(\mathbf{m} - \mathbf{m}_{\text{prior}}\right), \quad (5.93)$$

where $(\mathbf{y}_c)_j(\mathbf{m}) = f(\mathbf{m}, \mathbf{t}_j) = f(\mathbf{m}, t_{i,j})$ for $i = 1, 2, \ldots, N_{dj}$, and $j = 1, 2, \ldots, N_s$.

If *a priori* information for model parameters is not used in the estimation, we can simply delete the second term in the right-hand-side of

Equation (5.93) or set $\mathbf{C}_M^{-1} \to \mathbf{0}$. This corresponds to a case where one will assume that there is a complete uncertainty in the model parameters. The objective function to be minimized for this case is given by

$$\tilde{O}(\mathbf{m}) = \frac{1}{2}\sum_{j=1}^{N_s} N_{dj} \ln \left\{ \left[(\mathbf{y}_m)_j - (\mathbf{y}_c)_j(\mathbf{m}) \right]^T \left[(\mathbf{y}_m)_j - (\mathbf{y}_c)_j(\mathbf{m}) \right] \right\}. \quad (5.94)$$

After obtaining the $\tilde{\mathbf{m}}$ that minimizes the $\tilde{O}(\mathbf{m})$ given by Equations (5.93) and (5.94), the biased and unbiased error variances for each data set by using this $\tilde{\mathbf{m}}$ may be estimated from:

$$\tilde{\sigma}_j^2 = \frac{1}{N_{dj}} \sum_{i=1}^{N_{dj}} \left[(y_m)_{i,j} - f(\tilde{\mathbf{m}}, t_{i,j}) \right]^2 \quad (5.95)$$

and

$$\hat{\sigma}_j^2 = \frac{1}{\left(N_{dj} - M/N_s\right)} \sum_{i=1}^{N_{dj}} \left[(y_m)_{i,j} - f(\tilde{\mathbf{m}}, t_{i,j}) \right]^2. \quad (5.96)$$

If N_{dj} is large enough such that $N_{dj} \gg M/N_s$, which is normally the case for pressure transient tests to be history matched with the models that are parameterized by a small number of model parameters, then both Equations (5.95) and (5.96) provide very similar estimates of the true, unknown variances. In cases where $M/N_s \gg N_{dj}$, which is normally the case for pressure transient tests to be history matched with the numerical simulators considering heterogeneous geological reservoir models, then we can simply use Equation (5.95).

If the data error covariance matrix, $\mathbf{C}_{D,j}$, is assumed to be known or estimated *a priori* for each data set, then the MLE is identical to the LSE. If N_s independent observed data sets are to be history matched and an *a priori* term is included, then the following least-squares objective function is minimized:

$$O(\mathbf{m}) = \sum_{j=1}^{N_s} \frac{1}{2} \left[(\mathbf{y}_m)_j - (\mathbf{y}_c)_j(\mathbf{m}) \right]^T \mathbf{C}_{D,j}^{-1} \left[(\mathbf{y}_m)_j - (\mathbf{y}_c)_j(\mathbf{m}) \right]$$
$$+ \frac{1}{2} \left(\mathbf{m} - \mathbf{m}_{\text{prior}}\right)^T \mathbf{C}_M^{-1} \left(\mathbf{m} - \mathbf{m}_{\text{prior}}\right). \quad (5.97)$$

If *a priori* information for model parameters is not used in the estimation, we can simply delete the second term in the right-hand-side of equation Equation (5.97) or set $\mathbf{C}_M^{-1}$ to an $M \times M$ dimensional null matrix, $\mathbf{C}_M^{-1} \to \mathbf{0}$, i.e., a matrix with all elements equal to zero, in which case Equation (5.97) reduces to:

$$O(\mathbf{m}) = \sum_{j=1}^{N_s} \frac{1}{2} \left[(\mathbf{y}_m)_j - (\mathbf{y}_c)_j(\mathbf{m})\right]^T \mathbf{C}_{D,j}^{-1} \left[(\mathbf{y}_m)_j - (\mathbf{y}_c)_j(\mathbf{m})\right]. \quad (5.98)$$

As discussed in the next section, the same optimization algorithms used to minimize least-squares objective functions (Equations (5.97) and (5.98)) can be used to minimize the objective functions for the maximum likelihood (see Equations (5.93) and (5.94)). For example, the well-known gradient based Levenberg-Marquardt algorithm can be used to minimize the objective functions (Equations (5.93) and (5.94)).

Finally, we should note that all objective functions defined above are dimensionless quantities because observed data and model parameters are normalized by their variances having the same physical units.

5.11. Minimization of MLE and LSE Objective Functions

In this section, we present details of minimization of the objective functions based on MLE (Equation (5.93)) and LSE (Equation (5.97)) by using gradient-based optimization algorithms such as the Levenberg-Marquardt (L-M) algorithm (Bard, 1974; Fletcher, 1986; Gill, Murray, & Wright, 1981; Levenberg, 1944; Marquardt, 1963; Oliver et al., 2008). An extensive review and a detailed treatment of various optimization methods available for nonlinear problems can be found in Oliver et al. (2008, Chap. 8).

Let us recall Equation (5.93), the MLE objective function for N_s different, independent sets of observed data with prior information, that is,

$$\tilde{O}(\mathbf{m}) = \frac{1}{2} \sum_{j=1}^{N_s} N_{dj} \ln\left\{\left[(\mathbf{y}_m)_j - (\mathbf{y}_c)_j(\mathbf{m})\right]^T \left[(\mathbf{y}_m)_j - (\mathbf{y}_c)_j(\mathbf{m})\right]\right\} + \frac{1}{2} \left(\mathbf{m} - \mathbf{m}_{\text{prior}}\right)^T \mathbf{C}_M^{-1} \left(\mathbf{m} - \mathbf{m}_{\text{prior}}\right). \quad (5.99)$$

Taking the first derivative of Equation (5.99) with respect to model parameter vector **m** gives the gradient vector of the objective function as

$$\nabla \tilde{O}(\mathbf{m}) = \sum_{j=1}^{N_s} \frac{N_{dj}}{S_j(\mathbf{m})} \left\{\nabla \left[\mathbf{r}_j(\mathbf{m})\right]^T\right\} \left[\mathbf{r}_j(\mathbf{m})\right] + \mathbf{C}_M^{-1} \left(\mathbf{m} - \mathbf{m}_{\text{prior}}\right), \quad (5.100)$$

where $\mathbf{r}_j$ is the vector of residuals for the data set j, that is

$$\mathbf{r}_j(\mathbf{m}) = (\mathbf{y}_m)_j - (\mathbf{y}_c)_j(\mathbf{m}), \tag{5.101}$$

$S_j(\mathbf{m})$ is the sum of the squares of the residuals for the observed data set j, given by

$$\begin{aligned} S_j(\mathbf{m}) &= \left[\mathbf{r}_j(\mathbf{m})\right]^T \left[\mathbf{r}_j(\mathbf{m})\right] \\ &= \left[(\mathbf{y}_m)_j - (\mathbf{y}_c)_j(\mathbf{m})\right]^T \left[(\mathbf{y}_m)_j - (\mathbf{y}_c)_j(\mathbf{m})\right] \\ &= \sum_{i=1}^{N_{dj}} \left[(y_m)_{i,j} - f(\mathbf{m}, t_{i,j})\right]^2, \end{aligned} \tag{5.102}$$

and $\nabla\left[\mathbf{r}_j(\mathbf{m})\right]^T$ is an $M \times N_{dj}$ matrix given by

$$\nabla\left[\mathbf{r}_j(\mathbf{m})\right]^T = \nabla\left[(\mathbf{y}_m)_j - (\mathbf{y}_c)_j(\mathbf{m})\right]^T = -\nabla\left[(\mathbf{y}_c)_j(\mathbf{m})\right]^T = -\mathbf{G}_j^T(\mathbf{m}). \tag{5.103}$$

Here, the $N_{dj} \times M$ $\mathbf{G}_j$ matrix is defined as the sensitivity coefficient matrix, that is

$$\mathbf{G}_j(\mathbf{m}) = \begin{bmatrix} \dfrac{\partial f_{1,j}(\mathbf{m})}{\partial m_1} & \dfrac{\partial f_{1,j}(\mathbf{m})}{\partial m_2} & \cdots & \dfrac{\partial f_{1,j}(\mathbf{m})}{\partial m_M} \\ \dfrac{\partial f_{2,j}(\mathbf{m})}{\partial m_1} & \dfrac{\partial f_{2,j}(\mathbf{m})}{\partial m_2} & \cdots & \dfrac{\partial f_{2,j}(\mathbf{m})}{\partial m_M} \\ \vdots & & \cdots & \\ \dfrac{\partial f_{N_{dj},j}(\mathbf{m})}{\partial m_1} & \dfrac{\partial f_{N_{dj},j}(\mathbf{m})}{\partial m_2} & \cdots & \dfrac{\partial f_{N_{dj},j}(\mathbf{m})}{\partial m_M} \end{bmatrix}, \tag{5.104}$$

where we have used $f_{i,j}(\mathbf{m}) = f(\mathbf{m}, t_{i,j})$ for simplicity. Each derivative in Equation (5.104) is called the sensitivity coefficient, which gives a measure of the magnitude of the change in the model response function ($f_{i,j}(\mathbf{m})$) that will result from a change in the model parameter. If this sensitivity is very small or zero, then we expect that the particular model parameter will not be determined reliably by the particular model function. As will be shown, the sensitivity coefficients (or equivalently the sensitivity coefficient matrix) form the approximate Hessian matrix used in the Levenberg-Marquardt algorithm and determine the search direction when minimizing the objective function. The sensitivity coefficient could be positive or negative. Furthermore, it can change sign with time. A positive sensitivity coefficient means that an increase in the value of model parameter results in an increase in the model response function.

The second derivative of the objective function is the $M \times M$ Hessian matrix ($\mathbf{H}$), that is,

$$\nabla\left(\nabla \tilde{O}(\mathbf{m})\right)^T = \mathbf{C}_M^{-1} + \sum_{j=1}^{N_s} \frac{N_{dj}}{S_j}\left(\nabla \mathbf{r}_j^T\right)\left(\nabla \mathbf{r}_j^T\right)^T - \sum_{j=1}^{N_s} \frac{N_{dj}}{S_j^2} \nabla \mathbf{r}_j^T \mathbf{r}_j \mathbf{r}_j^T \left(\nabla \mathbf{r}_j^T\right)^T + \sum_{j=1}^{N_s} \frac{N_{dj}}{S_j} \sum_{i=1}^{N_{dj}} r_{ij} \nabla^2 r_{ij}. \tag{5.105}$$

We consider the Gauss-Newton approximation to Newton's method. Hence, it is a common assumption to ignore the term involving second derivatives of residuals and the terms involving the square of the residuals in Equation (5.105). The justification for this is that these terms are small compared to the first derivatives because these terms are pre-multiplied by the residuals, which are small near the minimum (Bard, 1974; Fletcher, 1986; Gill et al., 1981). Thus, the Hessian given by Equation (5.105) can be approximated by

$$\mathbf{H}(\mathbf{m}) = C_M^{-1} + \sum_{j=1}^{N_s} \frac{N_{dj}}{S_j}\left(\nabla \mathbf{r}_j^T\right)\left(\nabla \mathbf{r}_j^T\right)^T. \tag{5.106}$$

Note that we can write the gradient (Equation (5.100)) and approximate Hessian (Equation (5.106)) in terms of the sensitivity coefficient matrix $\mathbf{G}$, respectively,

$$\nabla \tilde{O}(\mathbf{m}) = -\sum_{j=1}^{N_s} \frac{N_{dj}}{S_j} \mathbf{G}_j^T \mathbf{r}_j + \mathbf{C}_M^{-1}\left(\mathbf{m} - \mathbf{m}_{\text{prior}}\right) \tag{5.107}$$

and

$$\mathbf{H}(\mathbf{m}) = \mathbf{C}_M^{-1} + \sum_{j=1}^{N_s} \frac{N_{dj}}{S_j} \mathbf{G}_j^T \mathbf{G}_j. \tag{5.108}$$

Equations (5.107) and (5.108) are used in the Levenberg-Marquardt algorithm iteratively to find the minimum of Equation (5.99) as

$$\left[\mathbf{H}(\mathbf{m}^k) + \vartheta \mathbf{I}\right] \delta\mathbf{m}^{k+1} = -\nabla \tilde{O}(\mathbf{m}^k), \tag{5.109}$$

or

$$\delta\mathbf{m}^{k+1} = -\left[\mathbf{H}(\mathbf{m}^k) + \vartheta \mathbf{I}\right]^{-1} \nabla \tilde{O}(\mathbf{m}^k) \tag{5.110}$$

and update the model parameter vector by

$$\mathbf{m}^{k+1} = \mathbf{m}^k + \delta\mathbf{m}^{k+1}. \tag{5.111}$$

In Equations (5.109) and (5.110), $\mathbf{I}$ is an $M \times M$ identity matrix, and $\vartheta \geq 0$ is a nonnegative scalar which is also referred to as the L-M regularization parameter. The vector $\delta\mathbf{m}^{k+1}$ in Eqs. (5.109)–(5.111) gives the search direction at the $(k+1)$-st iteration.

An efficient algorithm should normally have a line search or restricted step to determine how far to step in a given search direction. Therefore, many different algorithms of the L-M method have been suggested (Fletcher, 1986; Gill et al., 1981). Some use line search algorithms based on a quadratic function and others use restricted step (or trust region) methods. These methods differ in calculating step length $\delta\mathbf{m}^{k+1}$ and updating the scalar ϑ. In our applications, we have successfully applied an algorithm based on a restricted step as described in Fletcher's book [see Fletcher (1986, Chap. 5)] to solve Equations (5.109) and (5.111).

In the case we do not have *a priori* information about the model parameters, we set $\mathbf{C}_M^{-1} \to \mathbf{0}$ in Equations (5.107) and (5.108). Thus, we use the L-M algorithm described by Equation (5.109) to Equation (5.111) with

$$\nabla \tilde{O}(\mathbf{m}) = -\sum_{j=1}^{N_s} \frac{N_{dj}}{S_j} \mathbf{G}_j^T \mathbf{r}_j \tag{5.112}$$

and

$$\mathbf{H}(\mathbf{m}) = \sum_{j=1}^{N_s} \frac{N_{d,j}}{S_j} \mathbf{G}_j^T \mathbf{G}_j \tag{5.113}$$

when minimizing the objective function without the prior term (i.e., Equation (5.99) with the second term deleted in the right-hand side).

In the cases that the LSE objective function given by Equation (5.97) is used for parameter estimation, the gradient vector, $\nabla \tilde{O}$, and approximate

Hessian matrix, **H**, in Equations (5.109) and (5.110) are replaced by the following equations, respectively.

$$\nabla \tilde{O}(\mathbf{m}) = -\sum_{j=1}^{N_s} \mathbf{G}_j^T \mathbf{C}_{D,j}^{-1} \left[(\mathbf{y}_m)_j - (\mathbf{y}_c)_j(\mathbf{m}) \right] + \mathbf{C}_M^{-1} \left[\mathbf{m} - \mathbf{m}_{\text{prior}} \right] \tag{5.114}$$

and

$$\mathbf{H}(\mathbf{m}) = \sum_{j=1}^{N_s} \mathbf{G}_j^T \mathbf{C}_{D,j}^{-1} \mathbf{G}_j + \mathbf{C}_M^{-1}. \tag{5.115}$$

Any computer program that implements an iterative algorithm like the L-M method described above must include instructions to stop computation at predetermined suitable termination criteria. In the nonlinear regression case, it is convenient to stop iterating based on the following criteria:

1. $\tilde{O}\left(\mathbf{m}^k\right) - \tilde{O}\left(\mathbf{m}^{k+1}\right) \leq \varepsilon_1$.
2. $\frac{\left\|\delta \mathbf{m}^{k+1}\right\|_2}{\left\|\mathbf{m}^{k+1}\right\|_2 + \varepsilon_3} \leq \varepsilon_2$, where $\|\bullet\|_2$ denotes the l_2 norm of a vector.
3. $\left\|\nabla \tilde{O}\left(\mathbf{m}^{k+1}\right)\right\|_2 \leq \varepsilon_4$.
4. An integer MAXNIT to terminate the iteration if any of the above criteria is not satisfied and the number of iterations exceeds user specified MAXNIT (e.g., MAXNIT = 50). Typical values are $\varepsilon_1 = 10^{-3}$, $\varepsilon_2 = \varepsilon_4 = 10^{-4}$, and $\varepsilon_3 = 10^{-14}$.

The Levenberg-Marquardt method described above is a very efficient and fast optimization tool, particularly if the number of parameters to be estimated is less than the number of observation points, because it has an advantage of using derivative information of the functions to be minimized. Because of these properties, it is the most widely used optimization method for non-quadratic, continuous functions in practice. However, it may get stuck at local minima if "smart" initial guesses close to the global minimum are not given, because it is a local optimization method. There are global optimization methods such as polytope and simulated annealing which do not require derivative information (Gill et al., 1981; Press, Teukolsyk, Vetterling, & Flannery, 1986). Because of their slow convergence to a global minimum, polytope and simulated annealing methods, however, do not seem computationally as attractive as the L-M method to use for the pressure-transient models and the objective functions considered here.

Mengen and Onur (2004) presented a comparison of the Levenberg-Marquardt, polytope, and simulated annealing optimization methods for simultaneously history matching active and interference pressure-rate data

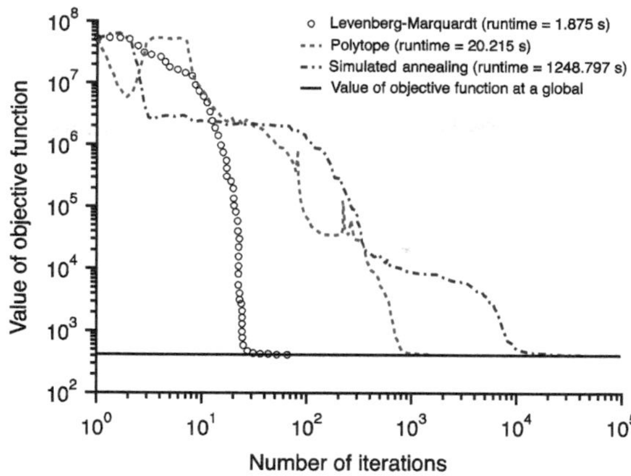

Figure 5.9 Comparison of performance of L-M, polytope, simulated annealing methods. *Source:* Reproduced from Mengen and Onur (2004).

sets from a single layer and commingled layered reservoirs based on the minimization of a weighted least-squares objective function (Equation (5.98)). Mengen and Onur's results indicate that although simulated annealing and polytope methods do converge to a global minimum regardless of initial guesses chosen, these methods do not appear to be a feasible parameter estimation method to be used in practice because they require a large number of iterations and more CPU times to converge to a global minimum compared to the L-M method. A comparison of the performance of Levenberg-Marquardt, polytope, and simulated annealing optimization methods is shown in Figure 5.9 for a history matching application based on a two-well interference test in a single layer system. For this application, there are four sets of observed data ($N_s = 4$); two sets of pressure data, and two sets of sandface rates at the active and observation wells, and each data set contains $N_{dj} = 100$, for $j = 1, 2, 3$, and 4 data points, and the number of unknown parameters is equal to $M = 4$.

5.12. Constraining Unknown Parameters In Minimization

Although the prior term $(1/2)\left(\mathbf{m} - \mathbf{m}_{\text{prior}}\right)^T \mathbf{C}_M^{-1}\left(\mathbf{m} - \mathbf{m}_{\text{prior}}\right)$ (also referred to as the model mismatch term) in the objective functions given by Equation (5.93) or Equation (5.97) can also be thought of as a constraint to keep the model parameters near their prior means during minimization, it is possible that some of the parameters during minimization can assume "unrealistic" values. It is also important to note that even in the cases where we do not consider a prior term (i.e., the model mismatch term deleted in

Equation (5.93) or Equation (5.97)), we will also need a constraint algorithm to keep the parameters within their user specified values. For instance, negative permeability values are not permitted during history matching of pressure transient data.

For the cases in which the number of the unknown model parameters is far less than the number of observed data points (i.e., $M \ll N_d$) we can use the so-called imaging procedure proposed by Carvalho, Thompson, Redner, and Reynolds (1996) to constrain parameters within an unconstrained minimization procedure because it has been shown to perform better than the constraint algorithms implementing penalty functions. Further details regarding implementation of this constraint algorithm can be found in Carvalho et al. (1996).

There exist also other proposed constraint methods such as damping and log transformation methods as proposed by Gao and Reynolds (2006). These constraint methods are particularly useful and quite efficient for the cases in which the unknown model parameters is greater than the number of observations (i.e., $M > N_d$), for instance, a more general numerical model considering heterogeneities based on a geostatistical model is used for history matching pressure-transient data sets.

5.13. Computation of Sensitivity Coefficients

As mentioned previously (Section 5.11), the use of the gradient-based Levenberg-Marquardt algorithm requires that the sensitivity matrix (or sensitivity coefficients) of computed (or model) pressures with respect to model parameters be computed at every iteration. The computational cost per iteration in the L-M method (or any gradient-based optimization algorithm) depends on the cost of computing the required sensitivity coefficients.

There are two different cases of parameter estimation that require specific treatment for the computation of the sensitivity coefficients in term of computational efficiency. **Case 1**: The number of unknown model parameters is less than the number of observed pressure data ($N_d \gg M$, an overdetermined system). **Case 2**: The number of unknown model parameters is greater than the number of observed data ($M \gg N_d$, an underdetermined system). As mentioned previously, the parameter estimation problem for **Case 1** arises in the cases where we consider simple reservoir models (e.g., a layer cake model where each layer is modeled as a homogeneous, anisotropic medium) which contain a fewer model parameters to history match the observed pressure-transient test data. On the other hand, **Case 2** arises in the cases where we consider more general reservoir models (e.g., a numerical reservoir model based on fully discretized difference equations where the permeability in each cell is treated as unknown) which contain many more unknown model parameters than the available observed data to be history matched.

For the cases where we deal with an overdetermined system ($N_d \gg M$) (**Case 1**), one can use an efficient numerical procedure based on a forward difference derivative which only requires $M + 1$ forward runs to compute the entries required in the sensitivity matrix in Equation (5.104), which can be given as

$$\frac{\partial f_{i,j}(\mathbf{m}^k)}{\partial m_l} \approx \frac{f_{i,j}\left(\mathbf{m}^k + \delta m_l \mathbf{e}_l\right) - f_{i,j}(\mathbf{m}^k)}{\delta m_l}, \tag{5.116}$$

for $i = 1, 2, \ldots, N_{dj}$, $j = 1, 2, \ldots, N_s$, and $l = 1, 2, \ldots, M$. In Equation (5.116), $\mathbf{e}_l$ is an M-dimensional unit column vector with its l-th entry equal to unity and all other entries equal to zero, δm_l is the perturbation of m_l^k. The vector $\mathbf{m}^k$ denotes the vector of estimated model parameters at the k-th iteration of the L-M algorithm. Here, $f_{i,j}(\mathbf{m})$ $[=f(\mathbf{m}, t_{i,j})]$ denotes computed (or model) pressure data. Based on our experience, it is usually sufficient to obtain accurate sensitivity coefficients if one uses $\delta m_l = 10^{-4} m_l^k$, except for the case when m_l denotes a skin factor which may be zero during iteration. In this case, where $m_l = 0$, we set $\delta m_l = 10^{-4}$ in Equation (5.116). Equation (5.116) is normally used for simple analytical pressure-transient models to history match the observed pressure data, and is referred to as a "direct" method for computing sensitivities. In these cases, we use a reservoir model based on fully discretized difference equations, but the discretized reservoir model has only a few parameters, e.g. some form of zonation is used so that $M < N_d$, then one can consider the gradient simulator method which has been introduced to petroleum engineering literature by Anterion, Karcher, and Eymard (1989).

For the cases where we deal with an undetermined system ($M \gg N_d$) (**Case 2**), the adjoint or optimal control method (Carter, Kemp, Pierce, & Williams, 1974; Chavent, Dupuy, & Lemmonnier, 1975; Chen, Gavalas, Seinfeld, & Wasserman, 1974; Li, Reynolds, & Oliver, 2003; Wu, Reynolds, & Oliver, 1999) has shown to be computationally more efficient than the gradient simulator or "direct" method because it requires only N_d+1 instead of $M + 1$ forward runs using a numerical reservoir model to obtain the required sensitivity coefficients in Equation (5.52) [see Ewing, Pilant, Wade, and Watson (1994) and Oliver et al. (2008)].

5.14. STATISTICAL INFERENCE

Once the most likely value of **m** (denoted by $\tilde{\mathbf{m}}$) that minimizes the ML objective function given by Equation (5.93) (or Equation (5.94)) is estimated, we can compute statistics of the match, e.g., variance (or standard deviation) of each observed data, confidence intervals for estimated parameters, overall RMS value for the fit, individual RMS values for each data set matched, and correlation coefficients between model parameters. Computing such statistics reveal important features of the match; for instance, we could answer the following important questions: How well the data fits to the model, and which of model parameters could be determined reliably, and independently?

The equations to be given below for computing such statistics are based on the assumption that the model function $\mathbf{y}_c$ behaves linearly near the minimum of the objective function given by Equation (5.93) or Equation (5.94) and are very similar to those used for the weighted least-squares regression for linear models. As we consider nonlinear models for history matching, the readers should be aware of the fact that the statistical measures given below are approximated from the linear model theory. A more rigorous treatment of statistical measures and concepts will not be given here in detail, if needed Bard (1974) and Barlow (1989) should be referred to.

The unbiased estimate of the variance for each observed data set is computed from Equation (5.96). An approximate 95% confidence interval for the i-th model parameter is computed from

$$\tilde{m}_i - 2\sqrt{\left(\mathbf{H}^{-1}\right)_{ii}} \le \breve{m}_i \le \tilde{m}_i + 2\sqrt{\left(\mathbf{H}^{-1}\right)_{ii}}, \tag{5.117}$$

where $\tilde{m}_i$ denotes the estimate of the i-th model parameter obtained by minimizing Equation (5.93) (or Equation (5.94)), $\breve{m}_i$ denotes the true, but unknown, value of the i-th model parameter, and $\left(\mathbf{H}^{-1}\right)_{ii}$ denotes the i-th diagonal element of the inverse of the Hessian matrix evaluated at $\tilde{\mathbf{m}}$. Recall that the approximate Hessian matrix at $\tilde{\mathbf{m}}$ is computed from

$$\mathbf{H}(\tilde{\mathbf{m}}) = \mathbf{C}_M^{-1} + \sum_{j=1}^{N_s} \frac{N_{dj}}{S_j(\tilde{\mathbf{m}})} \left[\mathbf{G}_j(\tilde{\mathbf{m}})\right]^T \mathbf{G}_j(\tilde{\mathbf{m}}). \tag{5.118}$$

The confidence intervals to be computed from Equation (5.117) provide a means of assessing and reporting the precision of a point estimate and determine which of various parameters are well determined. A confidence interval is a range of values that is normally used to describe the uncertainty around a point estimate of the parameter. Therefore confidence intervals are a measure of the variability in the data. Generally speaking, confidence intervals describe how much different the point estimate could have been if the underlying conditions stayed the same, but chance had led to a different set of data. Confidence intervals are calculated with a stated probability (e.g. 95%). Most confidence intervals are calculated as 95% confidence intervals because most statistical tests are done at the 0.05 level. It is important to note that the true value of a parameter is a constant, even though its value is unknown, but a confidence interval is a random quantity whose value depends on the random sample or data from which it is to be calculated. Therefore we describe a 95% (say) confidence interval as having a 95% probability of covering the true value, rather than saying that there is a 95% probability that the true value falls within the confidence interval. A small confidence interval indicates less uncertainty than a large confidence interval for a given parameter. If the data under consideration do not show significant sensitivity to some of the model parameters of interest, the variances[4] are expected to be large compared to those having strong influence on the data, and hence the confidence intervals for those model parameters are expected to be large, indicating that such model parameters cannot be reliably or well determined.

Now, we briefly discuss correlations among parameters on parameter estimation. Correlations among the model parameters can be described as the combined impact of model parameter changes on the model or system behavior. The correlation coefficient for the i-*th* and j-*th* model parameters is computed from

$$\rho_{m_i,m_j} = \frac{(\mathbf{H}^{-1})_{ij}}{\sqrt{(\mathbf{H}^{-1})_{ii}(\mathbf{H}^{-1})_{jj}}}. \tag{5.119}$$

As can be seen from Equations (5.117) and (5.119), the diagonal entries of the estimated covariance or Hessian matrix **H** determine the variance for each estimated parameter, while the off diagonal entries determines the correlation between estimated model parameters. ρ_{m_i,m_j} is a dimensionless scalar between -1 and $+1$ and is a measure of the degree of linear relationship between the two random model parameters or in general two random variables. If ρ_{m_i,m_j} is zero then m_i and m_j are uncorrelated. A positive correlation means that a particular m_i and m_j happens to be larger than their means so that a larger m_i will imply a larger m_j. For a negative ρ_{m_i,m_j}, a larger m_i will imply a smaller m_j. If ρ_{m_i,m_j} is exactly 1 (or -1), then m_i and m_j are completely correlated. For this case, the value of one parameter specifies precisely the value of the other. The correlation coefficients depend on the data available and also on the number of simultaneously estimated parameters. If correlations exist, the uncertainty of one model parameter affects the uncertainty of the other model parameter. Therefore, correlation coefficients could be used to identify which of the parameters are uniquely determined from the estimation. High correlation

[4] The diagonal elements of the Hessian, also called the posterior covariance matrix, of the estimated model parameters represent the variances of the estimated model parameters.

coefficients may indicate that two model parameters could not be resolved independently by the data set and hence may not be uniquely determined.

The overall goodness of the match (denoted by χ^2) is computed from

$$\chi^2 = \frac{2\tilde{O}(\tilde{\mathbf{m}})}{N_{dt}}, \tag{5.120}$$

where $N_{dt} = \sum_{j=1}^{N_s} N_{dj}$ and $O(\tilde{\mathbf{m}})$ are computed from

$$\tilde{O}(\tilde{\mathbf{m}}) = \frac{1}{2}\sum_{j=1}^{N_s} \frac{1}{(\tilde{\sigma}_j)^2} \left[\mathbf{r}_j(\tilde{\mathbf{m}})\right]^T \mathbf{r}_j(\tilde{\mathbf{m}}) + \frac{1}{2}\left(\tilde{\mathbf{m}} - \mathbf{m}_{\text{prior}}\right)^T \mathbf{C}_M^{-1}\left(\tilde{\mathbf{m}} - \mathbf{m}_{\text{prior}}\right). \tag{5.121}$$

It is important to note that the overall χ^2 (*chi-squared*) function given by Equation (5.120) is a dimensionless quantity. As can be shown [see Barlow (1989) and Tarantola (2005)], $2\tilde{O}(\tilde{\mathbf{m}})$ is a random variable having a χ^2 distribution with N_{dt} degrees of freedom. Its mean is N_{dt} and variance $2N_{dt}$. Thus, if we have an acceptable match, we would expect the quantity χ^2 given by Equation (5.120) to be close to unity. As proposed by Oliver et al. (2008), one may use the following interval of χ^2 to define an acceptable match satisfying the following inequality:

$$1 - 5\sqrt{\frac{2}{N_{dt}}} \le \chi^2 \le 1 + 5\sqrt{\frac{2}{N_{dt}}}. \tag{5.122}$$

If the estimated $\tilde{\mathbf{m}}$ from minimization does not satisfy Equation (5.122), one should suspect that we may have converged to a local minimum. As also stated by Oliver et al. (2008), this possibility could be investigated by starting with different initial guesses of the model parameter vector $\mathbf{m}$. If regardless of the initial guesses of $\mathbf{m}$, the optimization algorithm converges to a $\tilde{\mathbf{m}}$ value which does not satisfy Equation (5.122), then one should suspect that the prior model chosen may be inappropriate. Of course, if we use the LSE method (e.g., Equation (5.97)), which is based on the assumption that the error covariance matrix $\mathbf{C}_{D,j}$s are known, then one may also suspect that the covariance matrices used in history matching are incorrect.

The individual RMS values for each observed data set is computed from

$$\text{RMS}_j = \sqrt{\frac{2[\mathbf{r}_j(\tilde{\mathbf{m}})]^T \mathbf{r}_j(\tilde{\mathbf{m}})}{N_{dj}}}, \tag{5.123}$$

for $j = 1, 2, \ldots, N_s$.

In cases where the MLE objective function without a prior term, i.e., Equation (5.94) is considered, we set $\mathbf{C}_M^{-1} = \mathbf{0}$ in Equations (5.118) and (5.121). In cases where the least-squares method (e.g., Equation (5.97)) is used for parameter estimation, then the 95% percent confidence intervals and correlation coefficients for estimated parameters are computed from Equations (5.117) and (5.119), in which the Hessian matrix is defined by Equation (5.115).

5.15. Examples

The examples given in this chapter will cover the particular applications of nonlinear parameter estimation techniques without going through the whole interpretation procedure. Therefore, in this chapter, we will only examine and analyze derivative plots of buildup test pressures because throughout this book we focus on applications of nonlinear parameter estimation techniques rather than a full interpretation of a given test. For derivatives, we will use Bourdet, Ayoub, and Pirard (1989) derivatives with respect to an appropriate time function, such as time, Horner time, superposition time, etc. (see Chapter 2 for details). For simplicity, we will call them derivatives or pressure derivatives.

5.15.1. Example 1

A synthetic packer-probe test with a two-probe configuration in a three-layer system (Figure 5.10) is simulated by using the analytical solutions presented by Kuchuk and Onur (2003). The packer is located in the middle of the bottom (third) layer, and the probes 1 and 2 are located in the middles of the second and first layers, respectively. The input parameters of interest are given in Table 5.5.

The simulated test consists of a 2000-second drawdown period with a constant production rate of 16 B/D, followed by a 2000-second buildup. Normal random errors with mean zero and specified variances were added to corrupt the true packer, probe 1 and probe 2 pressure data sets generated from the analytical solution. The variance used to corrupt packer pressure data is 1 psi^2 (or standard deviation of 1 psi), while the variance for both probe pressure data sets is the same and equal to 0.01 psi^2 (or standard deviation of 0.1 psi). Each pressure data set contains 193 data points. Figures 5.11 and 5.12 show measured (corrupted) packer and probe pressure data sets. The corrupted data sets as measured pressures will be used for the history matching in nonlinear parameter estimation. Packer pressure data are intentionally made much noisier than probe pressure data because in practice, packer pressure data during the drawdown period often show larger noise due to pump-out effects.

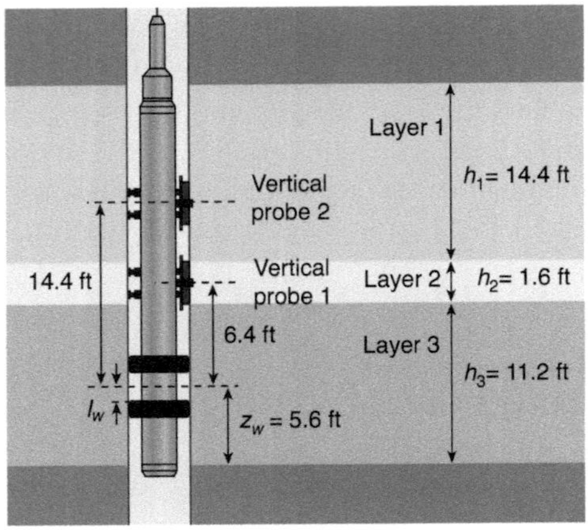

Figure 5.10 A schematic representation of a two-probe configuration of a packer-probe wireline formation tester in a three-layer system for Example 1.

Table 5.5 Formation and fluid properties for Example 1

$\phi_1 = \phi_2 = \phi_3$	fraction	0.15
k_{h1}	md	5.0
k_{h2}	md	0.5
k_{h3}	md	50
k_{v1}	md	1.0
k_{v2}	md	10
k_{v3}	md	15
$c_{t1} = c_{t2} = c_{t3}$	psi^{-1}	6×10^{-6}
$\mu_1 = \mu_2 = \mu_3$	cp	1.0
S	unitless	0.5
C	B/psi	2×10^{-6}
r_w	ft	0.354
l_w	ft	1.6

Figure 5.13 presents buildup derivatives corresponding to the packer and probe pressure buildup tests. As can be seen in this figure, after a transition period, the packer pressure derivative exhibits a negative half-slope ($m = -1/2$) on a log-log plot, indicating a spherical flow regime with the spherical slope (Culham, 1974), given as

$$m_{sp} = \frac{2453 q \mu \sqrt{\mu \phi c_t}}{k_s^{3/2}} = \frac{2453 q \mu \sqrt{\mu \phi c_t}}{k_h \sqrt{k_v}}, \tag{5.124}$$

where m_{sp} is in psi hr$^{1/2}$ and k_s is referred to as the spherical permeability

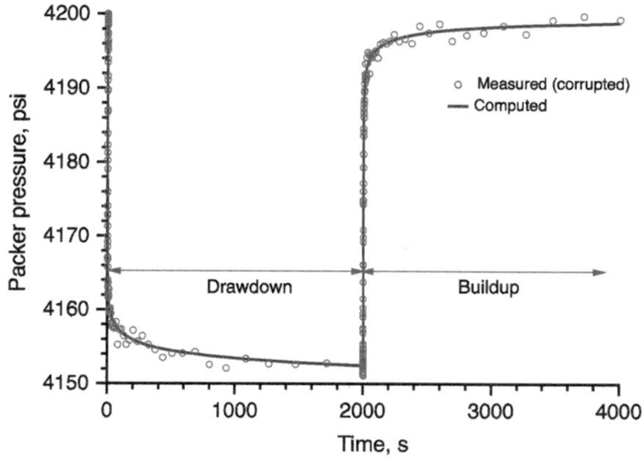

Figure 5.11 Comparison of the packer measured (corrupted) and computed pressures for Example 1.

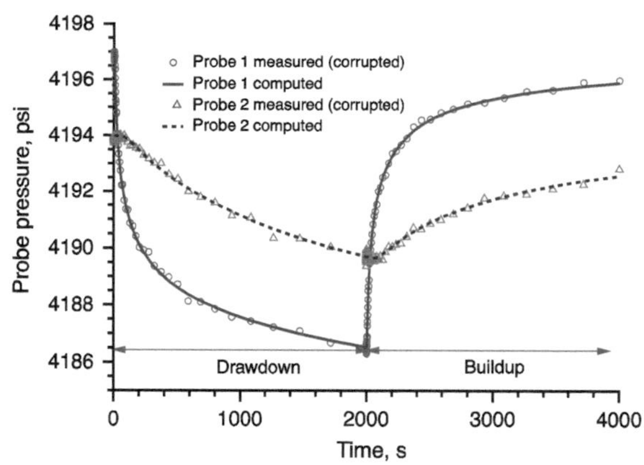

Figure 5.12 Comparison of the probe measured (corrupted) and computed pressures for Example 1.

and is given by

$$k_s = \left(k_h^2 k_v\right)^{1/3}. \tag{5.125}$$

A spherical permeability, k_s, is computed from Equation (5.124) and the derivative given in Figure 5.13 to be 30.5 md, which compares well with the true spherical permeability 33.5 md of the third layer. Notice from Figure 5.13, this spherical flow regime is very short because of the close proximity of the non-permeable lower boundary and the less permeable first layer (see Table 5.5 and Figure 5.10). Just after spherical flow, both packer and probe buildup pressure derivatives become almost flat, indicating that

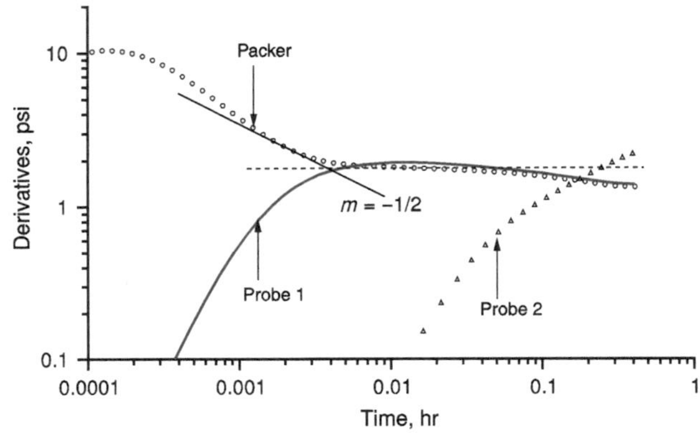

Figure 5.13 Packer, and Probe 1 and 2 buildup derivative for the synthetic interval test in a three-layer system.

the flow is predominately radial, with a radial slope, m_r (in psi/log-cycle) given as

$$m_r = \frac{162.6 q \mu}{k_h h}. \tag{5.126}$$

A horizontal permeability, k_h, is computed from Equation (5.126) and the derivative given in Equation (5.125) to be 56 md, which compares well with the true horizontal permeability 50 md of the third layer. Using Equation (5.125) and the values of the horizontal and spherical permeability, the vertical permeability is obtained to be 9 md, which is close to the true vertical permeability 15 md of the third layer. However, it should be stated that these flow regimes do not fully develop because of the considerable downward flow from the first and second layers, and strictly speaking, Equations (5.124) and (5.126) are valid for a single-layer system. When they are applied to a layered system, k_h, k_v, and ϕc_t in Equations (5.124) and (5.126) should represent some average properties. As shown in Figure 5.13, the probe 1 derivative also exhibits the same radial flow regime in the time interval from 0.005 to 0.03 hr as the packer, but the probe 2 does not. Both packer and probe 1 derivatives slightly bend downward after 0.03 hr because the second layer vertical permeability is 20 times higher than its horizontal permeability (Table 5.5). Therefore, the flow in the second layer is predominately vertical and the derivatives go downward (double permeability model, see Bourdet (2002)). The probe 1 derivative starts flattening at the same time as the packer derivative. This is an indication that both packer and probe derivatives are the signature of the same model, but the probe 2 has not seen the same model yet. In other words, the effect of the bottom no-flow was not felt by the probe 2. Therefore, the pseudo-cylindrically radial flow regime for the whole formation due the no-flow top and bottom boundaries is not established during the simulated test duration because all three pressure derivatives have not converged to the same radial flow regime. The above analysis based on pressure derivatives provides very useful information about a certain range of the layer properties. Basically the pressure derivative analyses provide:

1. Improving the layer definitions and features obtained from geoscience data, and
2. Improving the initial guesses for the layer parameters to be used in the nonlinear regression.

Next, nonlinear regression analysis is performed for this example with the objective of estimating layer horizontal and vertical permeabilities as well as the skin in the producing interval by simultaneously matching packer, probe 1, and probe 2 pressure data sets using the maximum likelihood method. Here, we use the MLE objective function given by Equation (5.37) (with $N_d = 193$, $M = 7$, and $N_s = 3$) as each data set contains the same number of pressure measurements, and does not consider a prior term (i.e., we are assuming an infinite variance for the layer properties and skin factor).

Table 5.6 summarizes the results obtained from the MLE procedure for this synthetic example. The 95% confidence intervals for estimated parameters are given in the same table. The numbers given in parenthesis represent the 95% confidence intervals computed from Equation (5.117). Figures 5.11 and 5.12 show the history matches obtained for packer and probe pressures. The results from the history matches are given in Table 5.6. As can be seen from Figures 5.11 and 5.12, and Table 5.6, the ML method provides very good matches of pressure data sets and estimates of model parameters as well as the standard deviations of errors for the pressure data sets (computed from Equation (5.38)). The estimated layer permeabilities and skin factor, and standard deviation of errors in packer, probe 1 and 2 data sets compare very well with the corresponding true layer permeabilities, skin factor and standard deviation of errors. Although not shown here, the weighted least-squares estimation (Equation (5.52)) was tried to simultaneously match the pressure data sets. However, its use is cumbersome, because as noted previously, the WLSE requires us to perform a trial-and-error procedure to determine the appropriate weights (inverse variances) to be used in regression. On the other hand, the MLE method provides a significant advantage over the weighted least-squares method as it avoids the trial-and-procedure for determining the appropriate relative weights.

5.15.2. Example 2

This is a real field IPTT example acquired with a packer/probe tool configuration from a vertical well in a three-layer system (Figure 5.14).

Table 5.6 Summary of results from nonlinear regression analysis based on the MLE method, Example 1

	k_{h1} (md)	k_{h2} (md)	k_{h3} (md)	k_{v1} (md)	k_{v2} (md)	k_{v3} (md)	S	σ_p (psi)	σ_{v1} (psi)	σ_{v2} (psi)
True values	5	0.5	50	1	10	15	0.5	1.0	0.1	0.1
Initial guess	1	0.1	56	0.1	1	9	1.0	–	–	–
MLE method	5.7	0.12	49.7	1.1	4	17	0.54	1.06	0.11	0.11
	(±1.3)	(±0.08)	(±1.2)	(±0.1)	(±2)	(±3)	(±0.06)			

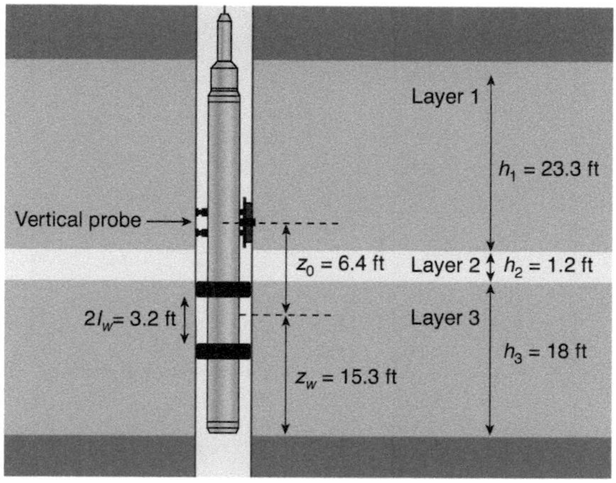

Figure 5.14 A schematic representation of packer/probe test in a three-layer system, three-layer field test data.

Table 5.7 Formation and fluid properties for Example 2

ϕ_1	fraction	0.245
ϕ_2	fraction	0.110
ϕ_3	fraction	0.205
h_1	ft	23.3
h_2	ft	1.2
h_3	ft	18
$c_{t1} = c_{t2} = c_{t3}$	psi^{-1}	4.5×10^{-5}
$\mu_1 = \mu_2 = \mu_3$	cp	1.0
r_w	ft	0.354
z_w	ft	15.3
z_o	ft	6.4
l_w	ft	1.6

Table 5.7 and Figure 5.14 show the layer definitions, porosities and compressibilities per layer, that were determined from open-hole logs with the local geology knowledge, and PVT and core data. This initial model setup was used for this IPTT design and was crucial to build a suitable model to determine the optimal packer and probe locations and test duration to attain a reasonable confidence in parameter estimation. The model-building step is very important for interpreting interval tests, particularly the probe pressure (high accuracy and resolution observation data), which is more sensitive to model attributes than packer module data, which can be affected by production noise, filtrate cleanup, etc.

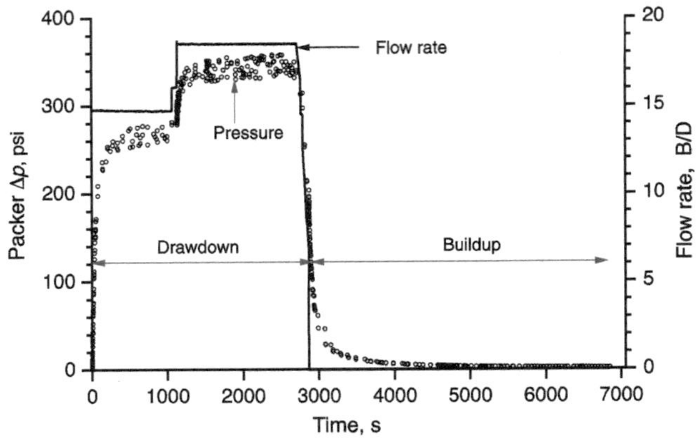

Figure 5.15 Packer pressure changes and flow-rate history for the three-layer field example.

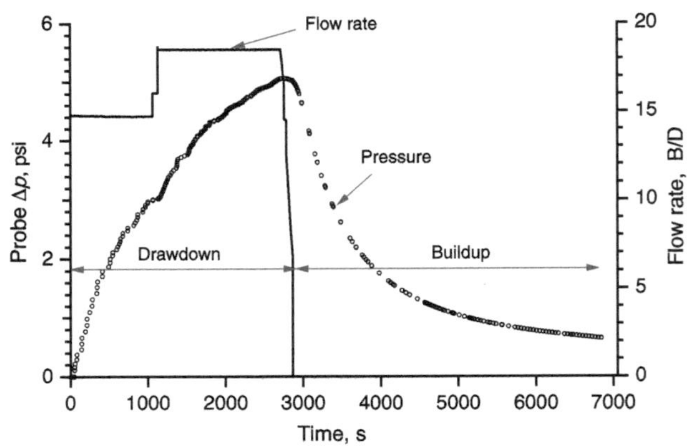

Figure 5.16 Probe pressure changes and flow-rate history for the three-layer field example.

For this IPTT, the packer and vertical probe were located in the third and the first layers, respectively. Figures 5.15 and 5.16 present the measured packer and probe pressure changes and rate history of the test. Each pressure data set contains 500 data points. Note that packer pressure-change data are approximately an order of magnitude larger and much noisier than vertical-observation probe pressure-change data.

Figure 5.17 shows the log-log plots of buildup pressure-derivatives for the packer and probe. Here, all derivatives were taken with respect to multirate superposition time (see Chapter 2). As can be seen in this figure, both derivatives exhibit spherical flow regimes towards the end of the buildup period, although the durations of the spherical flow regimes are short. A large separation in derivative values during the spherical flow regime indicates that both packer and probe derivatives are affected by different

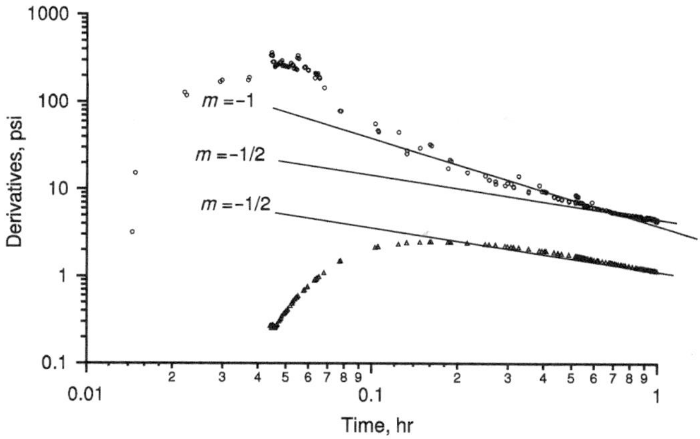

Figure 5.17 Packer and probe pressure-derivatives for the buildup period for the three-layer field example.

sectors of the model. Furthermore, any effect of the no-flow top and/or bottom boundaries of the system is not observed from both derivatives during the buildup period.

Performing spherical flow-regime analysis using the derivative values during the spherical flow regimes given in Figure 5.17 and the ϕc_t values given in Table 5.7 by Equation (5.124) yields $k_s = 10$ md for the packer region (should be dominated by Layer 3) and $k_s = 25$ md for the probe region (should be dominated Layer 1). This indicates that the third layer, where the packer is located, is less permeable than the first layer, where the probe is located. However, the individual values of the horizontal and vertical permeabilities cannot be estimated for these layers because, as shown in Figure 5.17, a pseudo-cylindrically radial flow regime for the total system is not established. Next, a nonlinear regression analysis is performed in an attempt to determine individual layer permeabilities as well as the skin factor and wellbore storage coefficient for this IPTT.

The objective of the nonlinear regression analysis here is to estimate for each layer horizontal and vertical permeabilities in addition to S and C using the MLE (Equation (5.37)) by regressing on packer and probe pressure data simultaneously. For this IPTT, the total number of unknown model parameters is $M = 8$, the number of data sets is $N_s = 2$, and the number of data points for each data set is $N_d = 500$. For comparison purposes, we also performed UWLS regression based on Equation (5.26). No prior term is used for both MLE and UWLSE methods.

The model fits of the packer pressure changes from the UWLS and ML regressions are shown in Figure 5.18. As can be seen from this figure, the packer pressure changes estimated from UWLS and ML regressions match the measured data very well. The difference between the computed pressure changes from UWLSE and MLE is very small and cannot be seen very well in

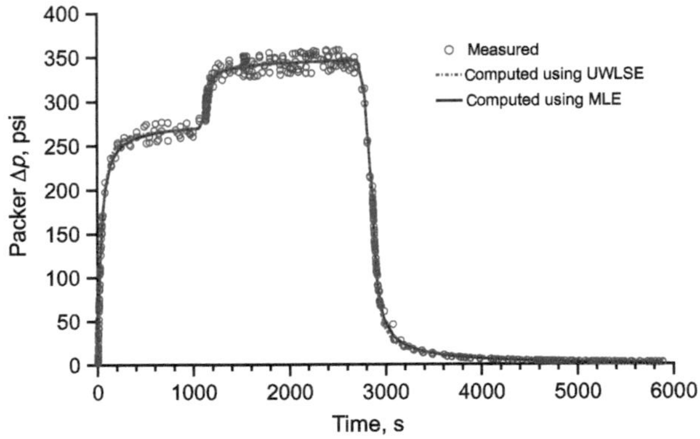

Figure 5.18 Comparison of the measured and computed packer pressure changes using UWLSE and MLE for Example 2.

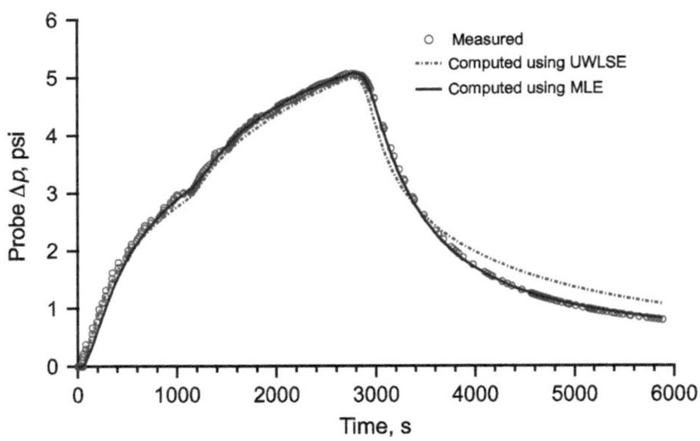

Figure 5.19 Comparison of the measured and computed probe pressure changes using UWLSE and MLE for Example 2.

Figure 5.18. Although not shown here, the match of the computed buildup packer pressure-derivative with the measured buildup packer derivative from the MLE was also excellent. The estimate of the variance for the errors in packer pressure change data obtained from UWLSE and MLE is the same and is equal to 29 psi^2 (a standard deviation of 5.4 psi).

The model matches of vertical probe pressure change data from the UWLS and ML regressions are shown in Figure 5.19. As can be seen in Figure 5.19, the computed pressure changes from MLE matches very well the probe measured data, while the computed data from UWLSE do not match the probe buildup data well. The estimated variance for the errors in probe pressure change obtained by the UWLSE is 0.04 psi^2 (a standard deviation of 0.2 psi), and by the MLE is 0.0031 psi^2 (a standard deviation of 0.055 psi). As in the packer case, the match of the computed buildup probe pressure-derivative with the measured one was excellent, although not shown here.

We have also tried several different initial guesses for the unknown model parameters from those given in Table 5.8 for the UWLS estimation, but we were not able to obtain a better quality of matches with these initial guesses than the one shown in Figure 5.19. This is, in fact, expected because it is evident that the measured packer data and probe data exhibit different levels of noise, and hence the UWLSE is not an appropriate method of estimation for a simultaneous match of the field packer and probe data sets.

On the other hand, the MLE provides very good matches of both packer and probe pressure data (Figures 5.18 and 5.19) for both the drawdown and buildup periods. The error variances given above indicate that packer pressure data are much noisier than vertical probe pressure data.

The parameter estimates and their associated 95% confidence intervals from UWLSE and MLE are given in Table 5.8. (The confidence intervals are the numbers given in parentheses in Table 5.8.) Clearly, UWLSE and MLE yield quite different parameter estimates for the top layer (Layer 1), where the probe is located, and also the confidence intervals for the estimated parameters for the UWLSE are much higher than those for the MLE. The reason may be because the UWLSE gives more emphasis to the data with a larger noise (packer data) and hence ignores the information content of the less noisy probe data located in the first layer, and the MLE method properly accounts for both noise levels in packer and probe data sets.

As mentioned previously, it is very useful to inspect the confidence intervals for the estimated parameters because the 95% confidence intervals are determined by the noise level in data and the sensitivity of the data to a given model parameter. A high value of the absolute (or relative) confidence interval for an estimated parameter indicates "larger" uncertainty and, hence, it can be stated that the parameter in question is not well determined from nonlinear regression. An inspection of the 95% confidence intervals for the parameters estimated from the MLE (Table 5.8) indicates that all parameters, except the horizontal permeability of the second layer, are well determined from the MLE. The relative confidence interval (defined as the ratio of a 95% absolute confidence interval to the estimated value) for the horizontal permeability of the second layer is more than 100%, which indicates that the permeability is not well determined. Therefore, one cannot trust the optimized value of 0.003 md for the horizontal permeability of the second layer. This is further verified by the results given in Figure 5.20, which shows that the sensitivity of packer and probe pressure changes to k_{h2} is very small. To generate the results shown in Figure 5.20, all model parameters except k_{h2} were fixed at their estimated values given in the last row of Table 5.8 for the MLE.

As can be seen from Figure 5.20, increasing k_{h2} about a 10^5-fold (from 3×10^{-5} to 3 md) has almost no effect on the packer response. On the other

Table 5.8 Results of nonlinear regression analysis of the packer/probe IPTT for Example 2

	k_{h1} (md)	k_{h2} (md)	k_{h3} (md)	k_{v1} (md)	k_{v2} (md)	k_{v3} (md)	C (B/psi)	S
Initial guess	50	5	10	20	3	2	$1. \times 10^{-5}$	0
UWLSE	8.9	1.0×10^{-3}	6.1	9.6	1.1	37.6	2.9×10^{-5}	1.0
	(±7)	(±11.0)	(±30)	(±10)	(±0.8)	(±5)	$(\pm 1 \times 10^{-6})$	(±0.8)
MLE	39.6	3.0×10^{-3}	8.7	16.6	0.85	2.9	2.6×10^{-5}	1.3
	(±3)	(±0.7)	(±1.5)	(±7)	(±0.3)	(±1.5)	$(\pm 1 \times 10^{-6})$	(±0.3)

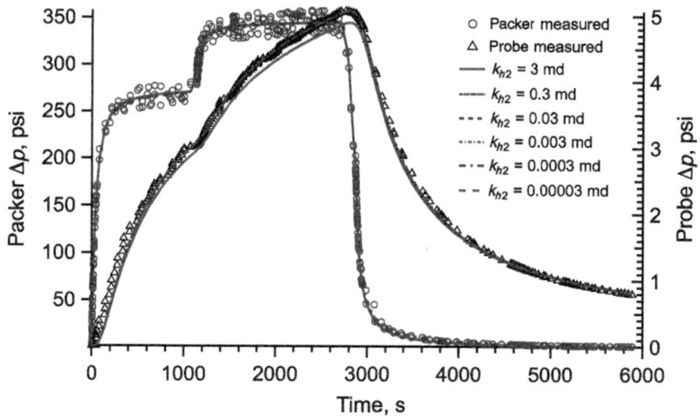

Figure 5.20 Sensitivity of the packer model and the probe model to the horizontal permeability of the second layer (k_{h2}).

hand, the probe data show some sensitivity to k_{h2}, but only indicates that k_{h2} cannot exceed a certain value of k_{h2}, roughly equal to 1.0 md (obtained after performing a refined sensitivity analysis between the curves corresponding to 0.3 and 3.0 md, as shown in Figure 5.20). So, this result indicates that the observation probe data for this test configuration provides a useful constraint for the value of k_{h2}, but cannot provide a reliable value for k_{h2}, as also indicated by the estimated 95% confidence interval for this parameter in Table 5.8, which also indicates that k_{h2} may take any value between 0^+ and 0.7 md. Therefore, inspecting confidence intervals is a useful and nice feature of nonlinear regression analysis for identifying which parameters are or are not well determined by nonlinear regression.

If another observation probe were to be placed in the second layer, in addition to the observation probe at the first layer, as in the synthetic three-layer model considered in Example 1, k_{h2} would have been determined well by simultaneously history matching packer and two sets of observation probe pressures by the MLE method.

5.15.3. Example 3

The example application pertains to a nonlinear regression analysis of an IPTT with a dual packer and single probe configuration in a slanted well with a deviation angle of $\theta_w = 45$ degrees (from the vertical) in a 54.5-ft thick oil zone (Figure 5.21). The formation and fluid properties of this example are given in Table 5.9.

This IPTT consisted of an 1158-second production (pumpout) period, during which about 40 liters of fluid were produced, followed by a 3639-second buildup test. The packer and the probe pressure data were recorded with strain gauges. The pressure and flow-rate history of the IPTT are shown in Figures 5.22 and 5.23.

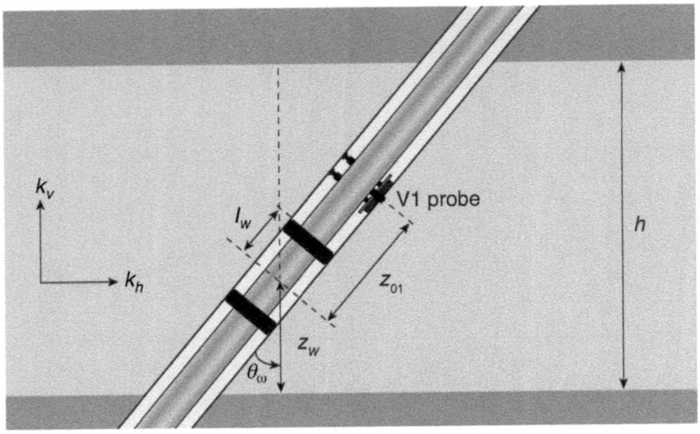

Figure 5.21 A schematic representation of a packer-probe IPTT in a slanted well for Example 3.

Table 5.9 Formation and fluid properties for Example 3

ϕc_t	psi^{-1}	3×10^{-6}
μ	cp	0.387
C	B/psi	2×10^{-6}
r_w	ft	0.354
θ_w	degrees	45
h	ft	54.5
z_w	ft	21.3
z_o	ft	6.4
l_w	ft	1.6

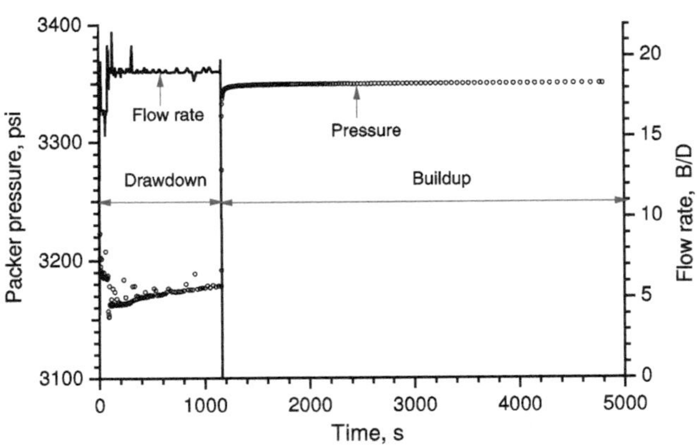

Figure 5.22 Measured packer pressure and flow-rate history for Example 3.

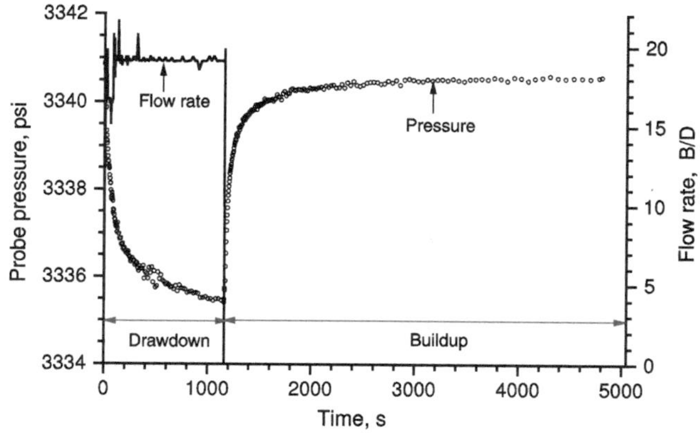

Figure 5.23 Measured probe pressure and flow-rate history for Example 3.

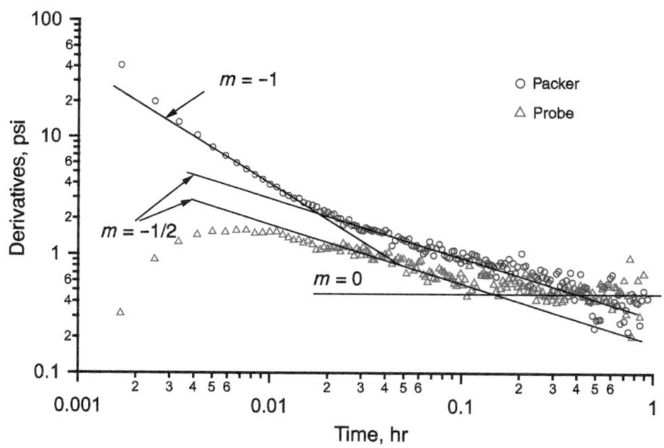

Figure 5.24 Packer and probe pressure-derivatives for the buildup tests for Example 3.

Both packer and probe buildup pressure derivatives are shown in Figure 5.24. Both derivatives look quite noisy, most likely due to the poor resolution of the strain gauges. The packer buildup derivative displays a negative unit slope $m = -1$ period due to the tool flowline storage. After the storage dominated flow period, the packer derivative exhibits a spherical flow regime $m = -1/2$ from 0.02 to 0.4 hr (80 to 1400 seconds). This should be expected because the packer interval is located close to the middle of the thick zone. The probe derivative also exhibits a spherical flow regime $m = -1/2$ from 0.02 to 0.1 hr (80 to 360 seconds). Notice from Figure 5.24 that the probe derivative values are slightly smaller than those of the packer and the duration of the probe-derivative spherical flow regime is much

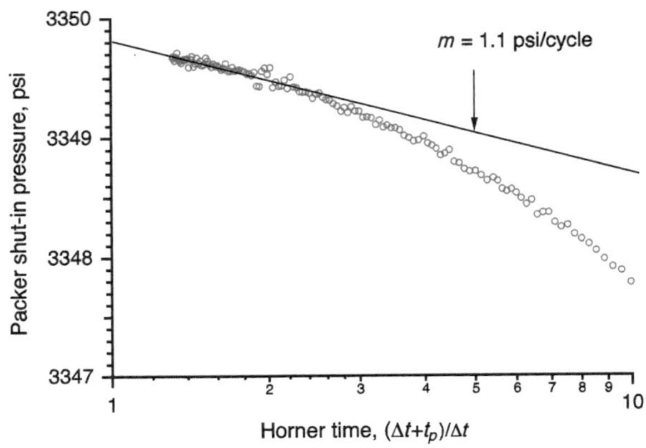

Figure 5.25 Horner plot of the packer buildup pressure for Example 3.

shorter. This is most likely due to weak heterogeneity in the formation. In other words, one should not expect the horizontal and vertical permeabilities to be exactly the same in a 54.5-ft thick formation.

Finally, both derivatives start to flatten, the $m = 0$ line in Figure 5.24, at 0.5 hr (1800 seconds), indicating a radial flow regime is most likely due to the whole formation that is bounded by no-flow top and bottom boundaries. Although they are noisy, both derivatives converge to a single semilog straight line, i.e. they represent an average value of k_h of the whole formation. Overall, the derivative signatures are consistent with geological and log data not shown here. Thus, a single-layer slanted-well model will be used for the flow regime analysis and the nonlinear regression.

The Horner plot of the packer buildup pressure given in Figure 5.25 exhibits a well-defined semilog straight line from 0.3 hr (a Horner time of 2.0) to the end of the buildup with a slope $m = 1.1$ psi/cycle. The Horner semilog straight line given in Figure 5.25 starts slightly earlier than the one from the packer derivative given in Figure 5.24. Although it is not shown here, the Horner plot of the probe buildup pressure also exhibits a semilog straight line with a slope $m = 0.97$ psi/cycle. Using these slopes of semilog straight lines and the flow rate of 19 B/D (Figure 5.22), the average horizontal permeabilities (k_h) from Equation (5.126) are obtained to be 19.6 and 23 md for the packer and probe buildup tests, respectively. A total skin of 175 is also computed from the packer data; this skin is a result of both the mechanical skin and the pseudo-skins due to partial completion created by the dual-packer and inclined angle geometry of the well (Onur et al., 2004a).

The Horner extrapolated pressures determined for the packer (see Figure 5.25) and probe buildup tests are 3349.80 and 3340.70 psi. The difference between the packer and probe extrapolated (initial) pressures is higher than expected from the well inclination angle and hydrostatic pressure, but it should be considered reasonable given the accuracy of the strain pressure gauges used for this IPTT.

The spherical permeability is obtained to be $k_s = 11$ md from the packer derivative given in Figure 5.24 and $k_s = 15$ md from the probe derivative given in Figure 5.24, using Equation (5.124) and the other parameters given in Table 5.9. Using these spherical permeabilities, k_h values estimated from Horner analysis, and Equation (5.125), we obtain $k_v = 3$ md for the packer buildup test and $k_v = 6.5$ md for the probe. Note that spherical permeabilities estimated are somewhat different, which may indicate some kind of heterogeneity, as mentioned above, in vertical permeability between the packer and probe locations or due to effects of the filtrate invasion zone (Gok, Onur, Hegeman, & Kuchuk, 2006; Gok et al., 2005). Note the mechanical skin factor can also be estimated from the packer's spherical-flow period data, based on the equation presented by Onur et al. (2004a). This analysis gives a mechanical skin to be $S = 8.0$.

The next step is to apply a nonlinear regression analysis based on all the data available, accounting for the full-rate history and considering a general slanted-well model (Abbaszadeh and Hegeman, 1990). This is to check the consistency of the parameter estimates and to further refine the parameter obtained from the Horner and spherical flow analyses of the packer and probe buildup tests.

Before the nonlinear regression analysis of this IPTT, as pointed out by (Onur et al., 2004a), it should be stated that it is often very difficult to estimate uniquely the individual values of k_h, k_v, C, and S (assuming ϕc_t and the inclination angle θ_w are known) by regressing on packer interval pressures alone, even if both spherical flow and late-radial flow are observed from the derivative or Horner plots (Figures 5.24 and 5.25). This is because the nonlinear regression often gets trapped in local minima due to strong correlation existing between k_h, k_v, and S (Onur et al., 2004a). Actually, this emphasizes the estimation of k_h, k_v, and S based on spherical and/or late-radial flow, if they prevail during the test duration, analyses of packer-interval pressure data and then the use of these estimated parameters in nonlinear regression. However, matching observation probe pressures alone often provides unique estimates of k_h and k_v, and even the inclination angle, θ_w, without requiring the late-radial flow period. This is because observation probe pressures are essentially not affected by tool storage and skin effects at the packer interval, and show significant sensitivity to permeabilities and well inclination angle (Onur et al., 2004a). Hence, simultaneous matching of both packer and probe(s) pressures helps to better resolve all parameters reliably, including skin and wellbore storage at the packer interval.

Next, a nonlinear regression parameter estimation based on the MLE method (Equation (5.37)) will be performed to match the packer and/or probe pressures, using the flow-rate history shown in Figure 5.22. In matching, only the buildup portion of the test (times greater than 1158 seconds) was considered, but accounting for the entire flow-rate

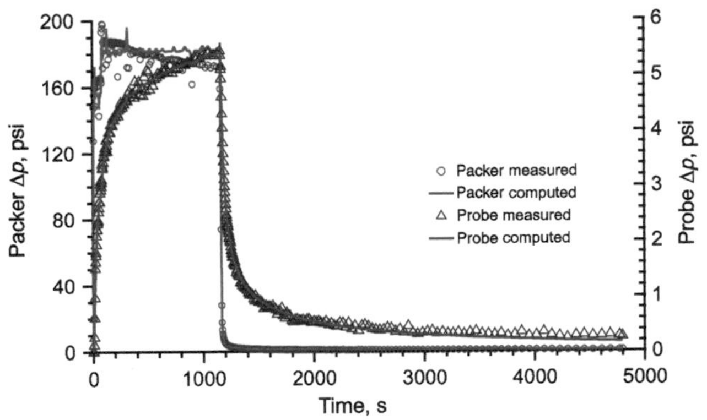

Figure 5.26 History matches of the packer and observation probe pressure changes with corresponding computed pressure changes for Example 3.

history, because data before these times were affected by tool opening and cleanup effects in this test. We wish to determine k_h, k_v, S (mechanical skin), C (tool storage), as well as the initial pressures at packer and probe locations (denoted as p_{op} and p_{ov}), while keeping all other tool and formation parameters as known in regression. Initial guesses for parameters used in regression are given in Table 5.10, which were determined from the radial and spherical flow analysis of the buildup data discussed previously.

Table 5.10 summarizes the results obtained from three different matching options for the MLE regression as follows:

- Option 1 using packer pressure alone,
- Option 2 using probe pressures alone, and
- Option 3 simultaneous matching of packer and probe pressures.

The numbers given in parentheses in the table represent the 95% absolute confidence intervals for the estimated parameters. When matching packer pressures alone (Option 1), the horizontal permeability at the value of 22 md determined from the Horner semilog straight line (radial flow regime) is fixed. As discussed previously, including k_h, k_v, and S as unknowns does not estimate unique values of these parameters due to strong correlation among parameters. Results given in the third row of Table 5.10 indicate that estimated parameters do not change significantly and are in good agreement with those obtained from radial and spherical flow regime analyses. Also the narrow confidence intervals for the estimated parameters (and standard deviation of measurement errors packer and probe pressure) verify that they were estimated quite reliably.

Figure 5.26 shows the history matches of measured packer and probe pressure changes with the corresponding model (or computed) pressure changes for the regression of Option 3 given in Table 5.10. As can be observed from this figure, the matches of measured packer and probe pressure

Table 5.10 Nonlinear regression analysis results for Example 3

Parameters	k_h (md)	k_v (md)	S	C (B/psi)	p_{op} (psi)	p_{ov} (psi)	θ_w °	σ^a (psi)
Initial guess	22	3	8	2×10^{-6}	3349.79	3340.70	45	–
Option 1	22^b	4 (±1.1)	8 (±0.2)	3.7×10^{-6} ($\pm 1.7 \times 10^{-7}$)	3349.71 (±0.26)	–	45^b	1.06
Option 2	16 (±0.3)	10 (±0.7)	8^b	3.7×10^{-6b}	3349.79^b	3340.83 (±0.01)	45^b	0.05
Option 3	17 (±2.5)	7 (±2.2)	6 (±1.2)	3.7×10^{-6} ($\pm 1.2 \times 10^{-7}$)	3349.76 (±0.18)	3340.86 (±0.20)	45^b	1.06^p 0.07^{pr}

^{p} for packer and ^{pr} for probe.

[a] Standard deviation of errors (or residuals) in pressure data.

[b] Parameter fixed in regression.

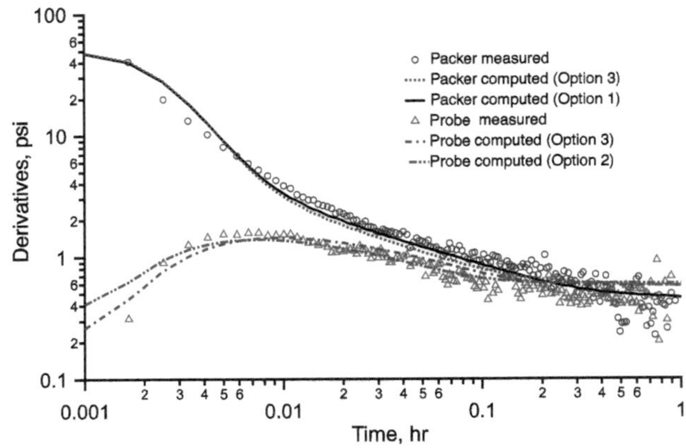

Figure 5.27 History matches of field packer and observation probe buildup derivatives with corresponding computed buildup derivatives.

changes with the computed ones are very good, although a slight mismatch with the measured packer pressure change can be observed during the drawdown period. The measured probe pressure changes match well for the entire test duration.

Figure 5.27 shows the history matches of the packer and probe buildup derivatives with the corresponding model derivatives based Options 1 and 2, and Option 3 regression parameters given in Table 5.10. As can be seen from Figure 5.27, the computed packer and probe buildup derivatives from Options 1 and 2, and Option 3 match reasonable well with those of from the measured data, while slight mismatches are observed towards the end of the buildup tests. As mentioned previously, these mismatches are likely to be due to some heterogeneity between packer and probe locations and/or filtrate invasion.

5.15.4. Example 4

This is a packer-probe field IPTT conducted in a vertical well in an 11-ft thick water zone. As shown in Figure 5.28, the single vertical probe is mounted 6.4 ft above the middle of the packer interval. The other pertinent parameters are $z_w = 2$ ft, $l_w = 1.6$ ft, $z_o = 6.4$ ft, $c_t = 8 \times 10^{-6}$ psi^{-1}, $\mu = 1$ cp, $r_w = 0.354$ ft, $\phi = 0.22$, and the estimated tool storage coefficient (C) is about 2.0×10^{-6} B/psi. The test consists of an 1862-second production (pumpout) period, during which about 34 liters of water was produced, followed by a 3826-second buildup. The packer and probe pressures were acquired with high-resolution quartz gauges. The packer and probe pressures, as well as flow-rate history computed from the pump displacement (which will be called measured), are shown in Figures 5.29

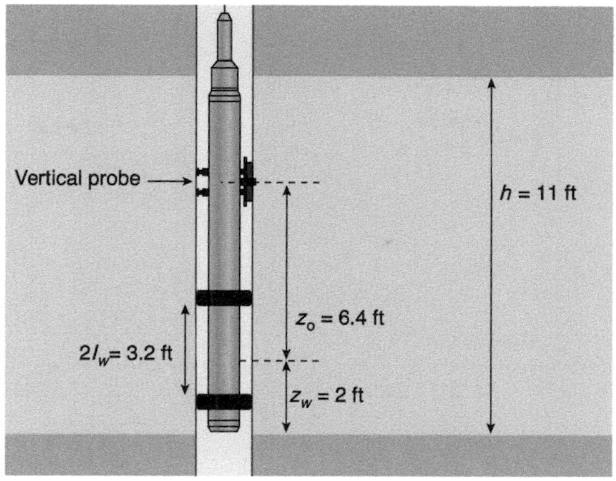

Figure 5.28 A schematic representation of a packer/probe configuration in a single-layer system for Example 4.

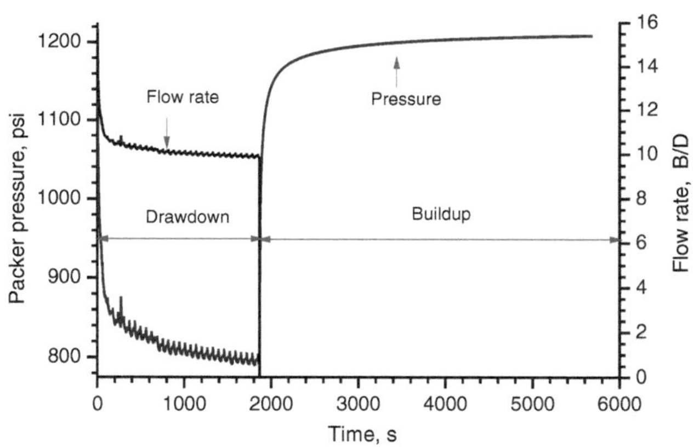

Figure 5.29 Measured packer pressure and flow-rate data for Example 4.

and 5.30. It should be pointed out that there could be some uncertainty in measured flow rates.

As discussed in detail in Chapter 3, in the cases where flow-rate data are not reliable or are uncertain, which is a possibility for the example test here, the need for the flow rate to be used in the pressure-rate (p-r) convolution for estimating permeability from wireline formation tester multiprobe and packer-probe pressure data sets can be eliminated by using the pressure-pressure (p-p) convolution (Goode, Pop, & Murphy, 1991; Onur et al., 2004b). The p-p convolution is based on the use of different pressure data sets recorded at two different spatial locations and, thus, enables one to perform parameter estimation without flow rate. Because pressure

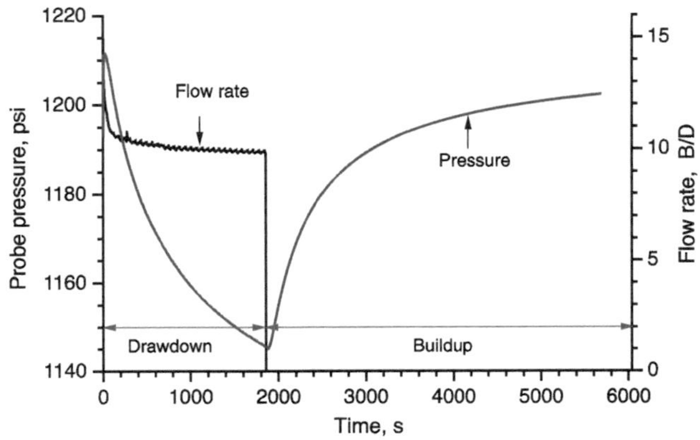

Figure 5.30 Measured probe pressure and flow-rate data for Example 4.

measurements at the probes are acquired accurately using high-resolution quartz pressure gauges, the *p*-*p* convolution becomes quite attractive for model (flow regime) identification as well as parameter estimation without flow rate measurements. In Chapter 3, various *p*-*p* convolution formulations for different probe-probe and packer-probe configurations for understanding non-uniqueness issues and correlations between different parameters are presented [see also Onur et al. (2004b)]. Furthermore, Onur et al. (2004b) have shown which *p*-*p* convolution formulation should be used to perform nonlinear optimization to obtain the maximum number of parameters uniquely. For this example packer-probe IPTT, shown in Figures 5.29 and 5.30, we present nonlinear regression analyses based on the *p*-*r* and *p*-*p* convolutions and discuss the estimates from both convolutions.

The buildup pressure derivatives for the packer and probe are shown in Figure 5.31. As can be seen from this figure the pressure-derivatives for both packer probe buildup tests exhibit radial flow regimes (in fact the derivatives merge into a single $m = 0$ slope line) from 0.2 hr (720 seconds) to the end of the buildup tests. In the time interval from 0.01 to 0.08 hr, the packer derivative exhibits a well-defined spherical or hemispherical flow ($m = -1/2$). For the given values of h, z_w and l_w, as shown in Figure 5.28, the packer interval is very close to the bottom of the non-permeable zone. Therefore, we would expect that the spherical flow around the packer interval should be very short and most likely is dominated by the tool storage. So, the $-1/2$ slope observed should be that of the hemispherical flow regime. The observed hemispherical flow regime is due to the flow period until the effect of the top non-permeable boundary is felt by the both packer and the probe. The probe derivative does not exhibit the hemispherical flow regime that is observed by the packer because the probe location is very close to the top non-permeable boundary. Overall, this derivative signature is typical of a single-layer formation with sealed top and bottom boundaries,

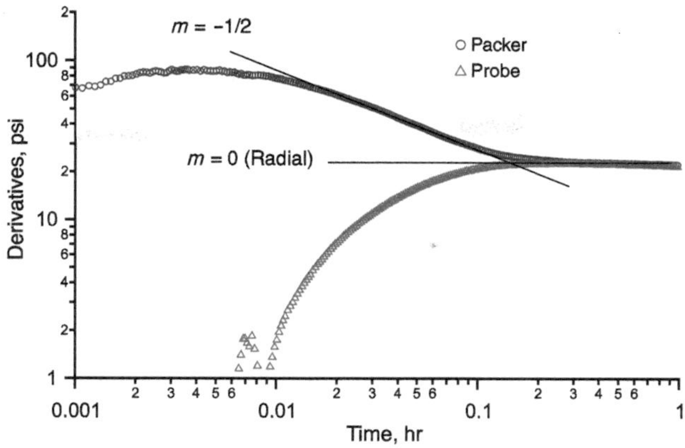

Figure 5.31 Pressure derivatives for the packer and probe buildup tests for Example 4.

and this is consistent with geological and log data. Thus, we use a single-layer model for the analysis.

The horizontal mobility, λ_h, is estimated to be 2.9 md/cp from the derivative value given at the $m = 0$ slope line using the flow rate of 10 B/D and the other required parameters given above, due to very well-defined a radial flow regime, as shown in Figure 5.31. The total skin (S_t) is estimated to be 4.5, which is composed of the mechanical (damage) skin and the pseudo-skin due to limited-entry completion created by the dual-packer geometry.

From the hemispherical flow regime ($m = -1/2$) shown in Figure 5.31, the spherical mobility λ_s is calculated to be 4 md/cp using the flow rate of 10 B/D and the other required parameters in Equation (5.124), in which the constant 4906 (hemispherical flow regime) is used instead of 2453. Thus, using the above values of the horizontal and spherical permeabilities in Equation (5.125), the vertical λ_v is calculated to be 2 md/cp. The mechanical skin factor is calculated to be almost zero using the formula $S = 2l_w/h(S_t - S_p)$, where S_t is the total skin given above and the pseudo-skin $S_p = 4.6$ from the formula given by Papatzacos (1987).

Next, let us assume that the measured flow rate that was computed from the pump displacement is not available or is only approximate, we then use the p-p convolution analysis because it does not require rate data. A parameter estimation scheme based on the MLE method was performed by matching the probe pressure data and the model that is expressed as the convolution of the packer pressure with the unit-response of the system (see Chapter 3). In matching, we have not used the probe pressure data earlier than 23 seconds as they were effected by tool opening. In the MLE using the p-p convolution, λ_h, λ_v, and S (mechanical skin) are estimated while keeping all other tool and formation parameters as known.

We tried many different sets of initial guesses for nonlinear regression applications based on the p-p convolution, but to illustrate our point, we

Table 5.11 Nonlinear regression analysis results for the packer-probe field Example 4

No.	Description	λ_h (md/cp)	λ_v (md/cp)	S	σ^a (psi)
1	Initial guess 1	2.81	2.0	0.0	–
2	Initial guess 2	2.81	2.81	0.5	
3	v-p p-p (using initial guess 1)	6.26 (±0.486)	2.55 (±0.01)	−0.002 (±0.04)	0.158
4	v-p p-p (using initial guess 2)	3.08 (±0.155)	2.58 (±0.007)	0.365 (±0.03)	0.165
5	v-p p-p (λ_h is fixed)	2.81	2.58 (±0.007)	0.411 (±0.004)	0.165

[a] Standard deviation of measurement errors.

present the regression results for two of the initial guess sets (No. 1 and 2) in Table 5.11. The results (No. 3 and 4 in Table 5.11) show that different initial guesses give different local minima, as is seen from the variation in the estimated values of λ_h and S. The correlation coefficient between λ_h and S for both sets of initial guesses is close to -1 (-0.996 and -0.991, for initial guesses 1 and 2, respectively). These results indicate that λ_h and S cannot be determined independently and hence uniquely, even though the computed 95% confidence intervals are narrow, and the standard deviation of probe pressure residuals are small. On the other hand, we note that vertical mobility is more reliably determined, because the confidence interval associated with λ_v is the smallest and both initial guesses (No. 1 and 2) yield essentially an identical value of λ_v.

The results noted above are not surprising given the theoretical finding of Onur et al. (2004b), who show that λ_h and S are correlated for the probe-packer p-p convolution matching in the vertical well and thus cannot be determined uniquely (see also Chapter 2). Next, by fixing $\lambda_h = 2.81$ md/cp obtained from the radial flow regime of the derivative plot given in Figure 5.31, we performed a nonlinear regression to estimate λ_v and S. These new results are given in the last row of Table 5.11, and the probe pressure match is shown in Figure 5.32. As can be seen from this figure, the pressure matches are excellent, and the measured pressures are indistinguishable from the computed ones, which are computed from three regression estimations (No. 3-5) shown in Table 5.11.

Figure 5.33 presents the estimated flow rate (based on a rate extraction method given by Onur et al. (2004b)) using the measured flow rate with the packer pressure data and the estimated parameters from all three regression estimations. Figure 5.33 compares the measured flow rate with computed flow rates. As shown in this figure, the measured flow rate does not match well with the computed flow rate obtained from the parameters determined

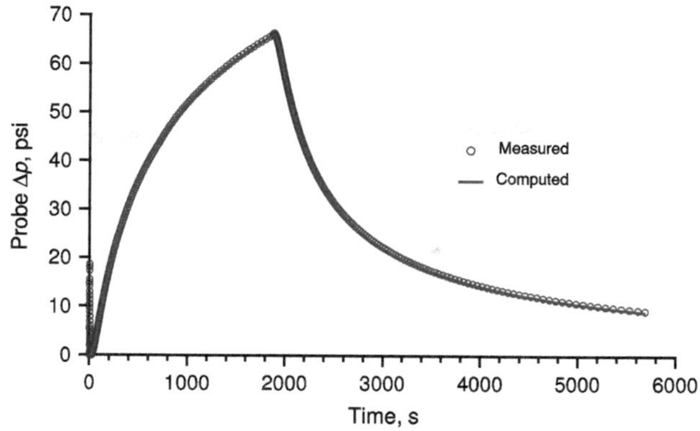

Figure 5.32 Pressure match at the vertical probe for Example 4.

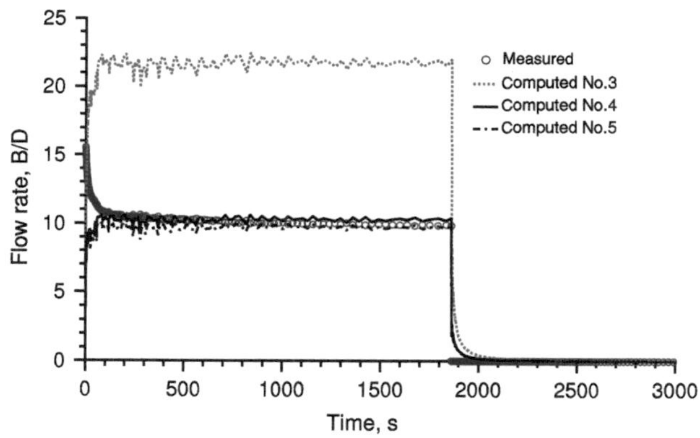

Figure 5.33 Flow rate match at the packer-interval for Example 4.

from an initial guess 1 (No. 3 in Table 5.11). In other words, estimated λ_h and S values cannot be accepted. However, the computed flow-rate data using the estimated parameters given in the last two rows (No. 4 and 5) of Table 5.11 do match well the measured flow rate. Therefore, the estimated λ_h and S values are acceptable. Notice that these λ_h, S as well as λ_v are also in close agreement with those estimated previously from the radial and hemispherical flow regimes. Although there is a difference between computed flow and pumpout flow rates at very early times, the overall standard deviation of pressure residuals is quite small and the confidence intervals are quite narrow, so we have high confidence in the analysis.

5.15.5. Example 5

Here, we revisit the field IPTT in Example 4, but this time we wish to simultaneously estimate the formation parameters and flow-rate history by using pressure-rate convolution because the measured flow rate shown in Figure 5.29 could have some uncertainty.

Therefore, in this example application, we take into account the uncertainty in the flow-rate history (as represented by step-wise constant function) by considering a prior term for the flow rate in the MLE nonlinear regression. Based on the theoretical development presented in Section 5.8 of this chapter, the MLE objective function considering a prior term for the flow rate can be expressed as

$$\tilde{O}(\mathbf{m}, \mathbf{q}_c) = \frac{1}{2}\sum_{j=1}^{N_s} N_{dj} \ln \left\{ \sum_{i=1}^{N_{dj}} \left[(p_m)_{i,j} - p_c(\mathbf{m}, \mathbf{q}_c, t_{i,j})\right]^2 \right\}^2 + \frac{1}{2}\sum_{i=1}^{N_q} \left[\frac{(q_c)_i - (q_c)_{\text{prior}_i}}{\sigma_{(q_c)_i}}\right], \tag{5.127}$$

where $\mathbf{m}$ denotes the vector of the unknown model parameters ($\mathbf{m} = [p_{op}, p_{ov}, \lambda_h, \lambda_v, S, C]^T$), p_{op} and p_{ov} are the initial formation pressures at the packer and probe locations, respectively. $\mathbf{q}_c$ is the N_q-dimensional vector of flow rates to be estimated (or adjusted) given as $\mathbf{q}_c = [(q_c)_1, (q_c)_2, \ldots, (q_c)_{N_q}]^T$, where N_q denotes the total number of flow rates to be treated as unknown. Also $(p_m)_{i,j}$ denotes the measured pressure and p_c denotes the corresponding model pressure to be computed at time $t_{i,j}$ for the data set j. In this example, we have two sets of pressure data; measured at the packer interval and observation probe. Both sets of pressure data will be used in nonlinear regression, and $N_s = 2$ in Equation (5.127). The total number of pressure data points for each data set for this IPTT are the same and $N_{d1} = N_{d2} = 513$ in Equation (5.127). In Equation (5.127), $(q_c)_{\text{prior}_i}$ and $\sigma_{(q_c)_i}$ denote the prior mean and the variance of the flow rate $(q_c)_i$, respectively. We will set $(q_c)_{\text{prior}_i}$ equal to the flow rate shown in Figure 5.29. Normally, it is more difficult to have a good prior knowledge of error variance in flow rate than its mean. The magnitude of noise in measured flow-rate data could typically range from 1 to 15%. So, we will consider different values of error variance $\sigma^2_{(q_c)_i}$ in Equation (5.127) to investigate the impact of $\sigma^2_{(q_c)_i}$ on parameter estimation.

It is important to note that the flow-rate schedule represented by the discrete q_i (if unknown it is called q_c) values plays a dual role in Equation (5.127). Some of the q_i values can be treated as unknown $[(q_c)]$, and others as known in the estimation procedure. In other words, q_i values can be partitioned into two subsections as known and unknown subsections. In principle, there is no limit on the number of partitions. For instance, in typical pressure transient well tests, the drawdown flow rates can be treated as unknown, while the buildup rates are treated as known and their values

Table 5.12 Error levels considered for the flow-rate history shown in Figure 5.29 for Example 5

Case No.	Error level (%)
	$\eta_q\%$
1	0
2	5
3	8
4	15
5	50

are normally set to zero. For this IPTT, the buildup starts at 1862 seconds, and hence flow-rate data after 1862 seconds will be treated as known and the drawdown (production period) flow rates will be treated as unknown. Thus, the total number of flow rates to be estimated is $N_q = 257$ in Equation (5.127). The second usage of the flow-rate data, as shown in Equation (5.127), is that $\mathbf{q}_c = [(q_c)_1, (q_c)_2, \ldots, (q_c)_{Nq})^T$ are used with $\mathbf{m}$ to compute the model response p_c.

Table 5.12 presents the error levels considered in flow rates. It is assumed that each computed flow rate contains a constant relative error of $\eta_q\%$. Thus, the standard deviation of error for the computed flow-rate data is calculated by means of $\sigma_{q,i} \approx \left(\eta_q \times q_{obs,i}\right)/100$, where q_{obs} denotes the measured flow rate shown in Figure 5.29. Case 1 shown in Table 5.12 corresponds to a test during which flow-rate data are accurately measured and used directly in the nonlinear estimation. Cases 2 to 5 correspond to different error levels in measured flow-rate data and they are treated as unknown in Equation (5.127). As can be seen in Table 5.12, we also consider a case (Case 5) where the flow-rate data have a quite large uncertainty (50%), which may not represent a realistic case.

Table 5.13 presents the regression results for all five cases given in Table 5.12. For all cases in the nonlinear regression, we used the same initial guesses for the unknown parameters (see 2nd row of Table 5.13). Considering the regression option yielding the narrowest 95% confidence intervals for the estimated parameters as a criterion, we may say that the regression results with a prior rate distribution considering a 5% error in rate data (Case No. 2 in Table 5.12) is more representative of the true, unknown parameters. In addition, the parameter estimates k_h, k_v, and S are also very similar to those obtained from the p-p convolution (for which rate data are not used) of the same test considered in Example 4 (see No. 4 and 5 of Table 5.11).

Figure 5.34 shows comparison of the optimized flow rates (by the use of Equation (5.127)) for Cases 2-5 of Table 5.12 with the measured rate shown in Figure 5.29. The results of Figure 5.34 indicate that the parameter

Table 5.13 Nonlinear regression analysis results for Example 5

Case No.	Description	k_h (md)	k_v (md)	S	C (B/psi)	p_{op} (psi)	p_{ov} (psi)	σ_v[b] (psi)	σ_p[a] (psi)	σ_q[c] (B/d)
	Initial guess	2.81	2.0	0.0	2×10^{-6}	1218.58	1211.40	–	–	–
1	0% error in flow rate	2.814 (±0.01)	2.384 (±0.02)	0.178 (±0.02)	5.5×10^{-6} ($\pm 2.1 \times 10^{-7}$)	1218.24 (±1.72)	1211.89 (±0.08)	11.8	0.45	–
2	5% error in flow rate	2.916 (±0.09)	2.594 (±0.01)	0.385 (±0.02)	4.1×10^{-6} ($\pm 1.4 \times 10^{-7}$)	1218.76 (±0.560)	1211.60 (±0.05)	3.12	0.11	1.37
3	8% error in flow rate	2.867 (±0.134)	2.594 (±0.01)	0.394 (±0.03)	4.0×10^{-6} ($\pm 7.5 \times 10^{-8}$)	1218.74 (±0.546)	1211.60 (±0.05)	2.95	0.11	1.54
4	15% error in flow rate	2.732 (±0.134)	2.598 (±0.01)	0.420 (±0.03)	3.8×10^{-6} ($\pm 7.5 \times 10^{-8}$)	1218.78 (±0.546)	1211.60 (±0.05)	2.81	0.10	2.02
5	50% error in flow rate	2.214 (±0.402)	2.616 (±0.01)	0.537 (±0.09)	3.2×10^{-6} ($\pm 5.2 \times 10^{-7}$)	1219.13 (±0.546)	1211.60 (±0.06)	2.75	0.10	3.77

[a] Standard deviation of errors (or residuals) in packer pressure data.
[b] Standard deviation of errors (or residuals) in observation probe pressure data.
[c] Standard deviation of errors (or residuals) in flow-rate data.

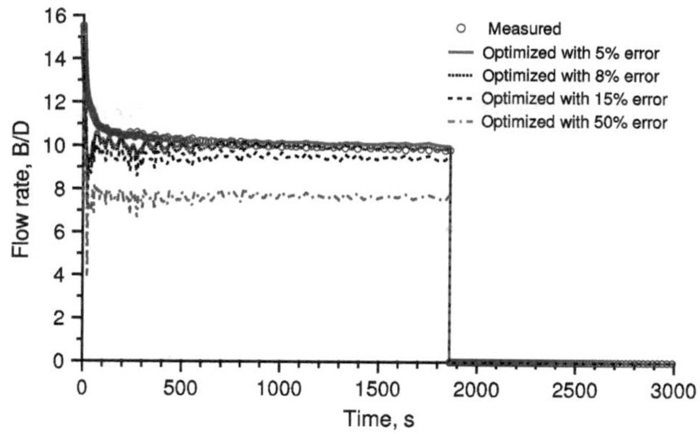

Figure 5.34 Comparison of optimized flow rates with the measured rate for four different cases shown in Table 5.13 with different error levels in rate data.

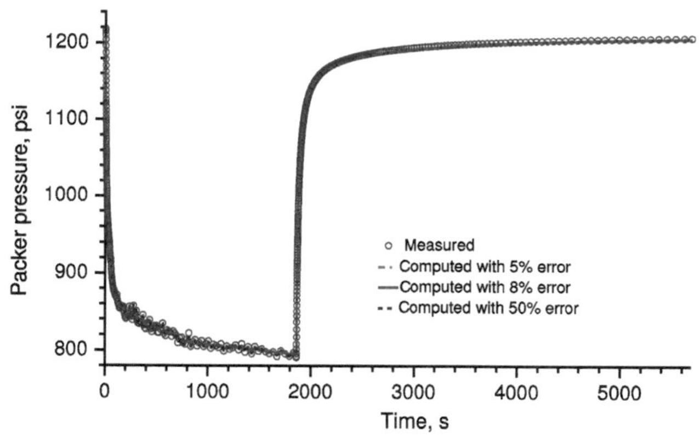

Figure 5.35 Pressure matches at the packer for Example 5.

estimates, as well as the optimized rates corresponding to the cases assuming 8%, 15%, and 50% error in flow-rate history, may not be accepted because the optimized flow rates do not honor the measured flow-rate data. We should point out that the regression adjusts the input flow-rate history more at the early times of the production period, indicating that either the input flow-rate history for that portion may not be not accurate or the model we use in convolution may not be consistent with the actual well/reservoir system (e.g., cleanup effects).

Figures 5.35 and 5.36 present the history matches of packer and observation probe data for Cases 2, 3, and 5 of Table 5.13. As shown in these figures, history matches of observed and computed packer and observation probe pressures for all cases are excellent. Figures 5.37 and 5.38 compare derivatives of the packer and probe buildup pressures with the computed

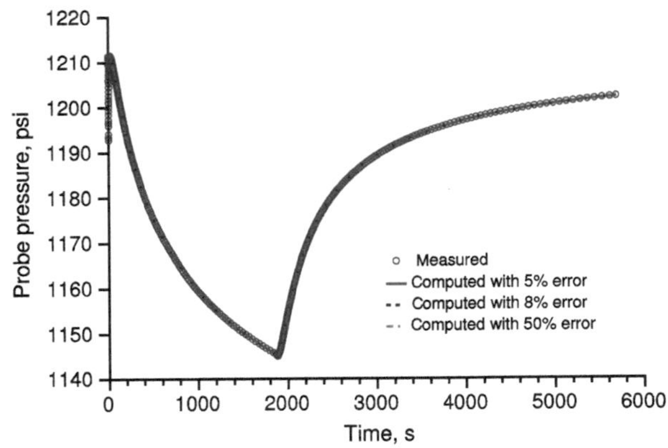

Figure 5.36 Pressure matches at the vertical probe for Example 5.

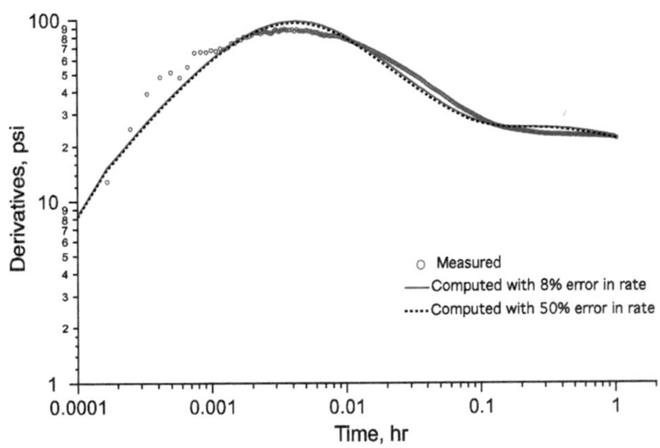

Figure 5.37 Comparison of measured and computed derivatives of the packer buildup pressure.

derivatives for Cases 3 and 5 of Table 5.13. These derivatives are taken with respect to the superposition time function using the corresponding flow-rate history (measured or computed flow rates). As in the pressure case, the computed derivatives match very well with the measured derivatives, as can be seen from these figures.

The results shown in Table 5.13 and Figures 5.34–5.38 reveal an important fact that flow-rate optimization with formation parameters simultaneously should be used with caution because such an optimization procedure cannot provide unique results for flow rates and formation parameters. This is because flow rates, and particularly some parameters (e.g., horizontal permeability and skin factor; see the values of these parameters for different cases in Table 5.13), are adjusted in optimization for the sake

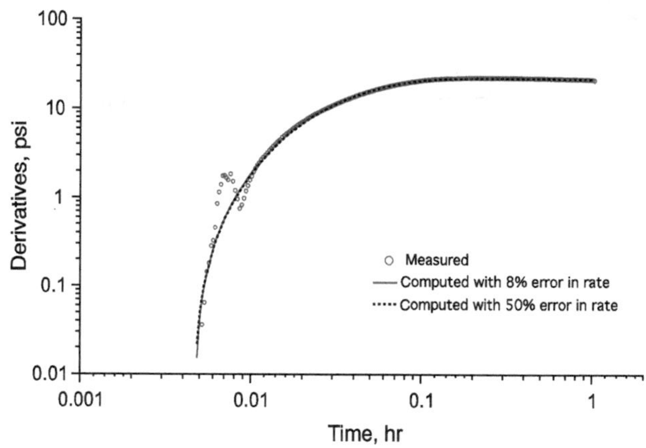

Figure 5.38 Comparison of measured and computed derivatives of the probe buildup pressure.

of "honoring" measured packer and probe pressures. The same problem also occurs in attempting to correct flow-rate history by deconvolution (see Chapter 4). For instance, although the optimized flow-rate histories of the cases assuming 15% and 50% uncertainties in rate data are different from the input flow-rate data, the computed packer and probe pressures and their derivatives are almost identical (Figures 5.35 and 5.38). So, we should have additional information for choosing the correct optimized rate history and formation parameters. For example, we should check whether the optimized flow-rate history or histories agree with the cumulative produced fluid during the test—for the example test considered it is 34 liters and only the optimized flow rates corresponding to Cases 1-3 do satisfy this amount—and/or we should have a good knowledge of the spread of the error (or expected error margins) in rate data. Moreover, we should check and identify whether we have consistent data with the convolution model. Otherwise, flow-rate optimization should not be considered as a tool for correcting flow-rate history by optimizing the formation parameters from pressure measurements alone.

5.15.6. Example 6

This is a synthetic IPTT example that consists of two tests conducted at different locations along the wellbore with a packer/probe tool combination in a heterogenous formation. The packer and probe locations in a vertical well with a 54.5 ft formation thickness for the first and second IPTTs are shown in Figure 5.39. The outer radius of the reservoir in the r-direction (r_e) is 1000 ft. Other pertinent input data for the reservoir model is given in Table 5.14. The reservoir is simulated with a $r - \theta - z$ numerical simulator with a total of 1550 grid blocks ($N_r = 25$, $N_\theta = 1$, $N_z = 62$).

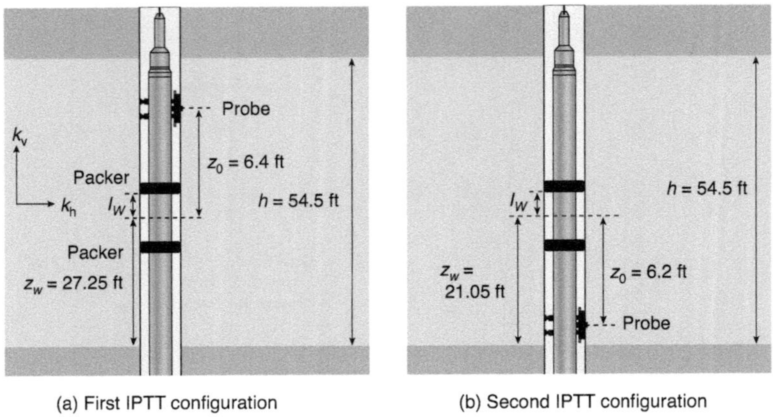

Figure 5.39 Schematic representations of the first (a) and second (b) IPTT packer-probe configurations in a heterogeneous formation shown in Figure 5.40.

Table 5.14 Formation and fluid properties for Example 6

c_t	psi^{-1}	2.5×10^{-5}
μ	cp	0.387
C	B/psi	4×10^{-6}
S	unitless	5
r_w	ft	0.354
r_e	ft	1000
h	ft	54.5
z_w	ft	21.3
z_o	ft	6.4
l_w	ft	1.6
q	B/D	20
p_o	psi	3500

The MLE Bayesian methodology for performing history matching of the synthetic IPTT data sets will be applied to this heterogeneous reservoir model, for which horizontal and vertical permeability, and porosity fields were generated by using geostatistical methods (Journel & Huijbregts, 1978). This example has been given in detail by Gok et al. (2005). Unlike all the other examples presented above, the number of unknown parameters far exceed the number of observed data (i.e., an undetermined system with a prior geostatistical model) in the MLE for this example.

The formation rock property fields are assigned as follows: A log-normal horizontal permeability $\ln(k_h)$ distribution with log-variance of 1.0 and log-mean equal to 2.99 (20 md) and a log-normal vertical permeability

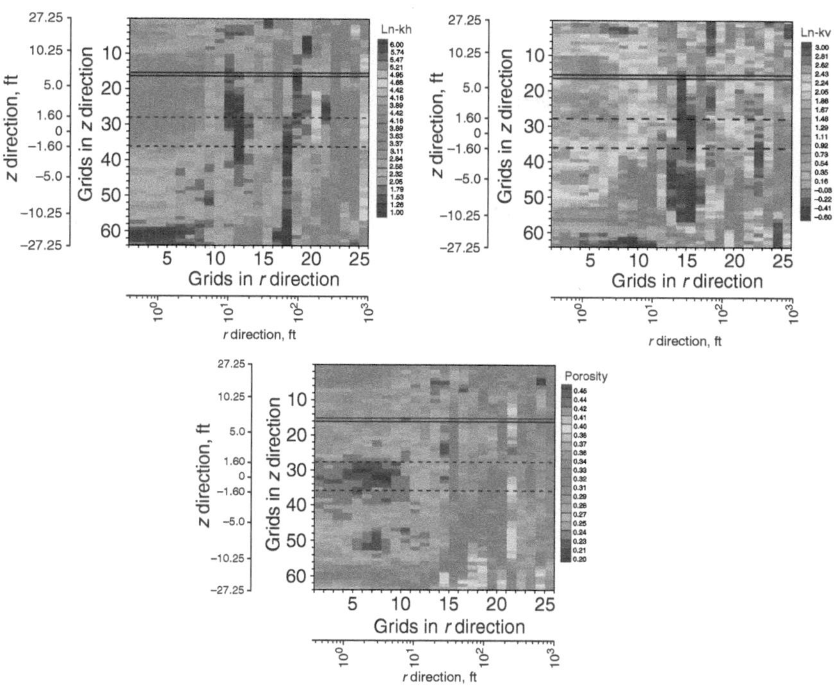

Figure 5.40 True log-horizontal permeability ($\ln k_h$) field (top left), true log-vertical permeability $\ln(k_v)$ field (top right), and true porosity (ϕ) field (bottom), Example 6 (see color plate 1).

$\ln(k_v)$ distribution with log-variance of 0.5 and log-mean equal to 1.79 (6 md). Both permeability distributions are described by isotropic spherical variogram models. The ranges (or correlation lengths) of $\ln(k_h)$ and $\ln(k_v)$ are equal to 50 and 27.5 ft, respectively. The porosity distribution is assumed to be normal with mean equal to 0.32 and variance equal to 0.0025 and is represented by an isotropic spherical variogram with a range of 50 ft. It is also assumed that there is no correlation between log-horizontal and log-vertical permeabilities, log-horizontal permeability and porosity, or log-vertical permeability and porosity. Shown in Figure 5.40 are unconditional realizations of $\ln(k_h)$, $\ln(k_v)$ and ϕ grid block values in the r-z directions, generated from Cholesky decomposition (Alabert, 1987).

For the inverse problem, we refer to the distributions shown in Figure 5.40 as the "true" distributions because they are used by the simulator to generate the observed IPTT data sets for history matching. The true values of other parameters (skin, storage, etc.) are as given in Table 5.14. The dashed horizontal lines in Figure 5.40 are used to show the boundaries of the dual-packer interval, while the solid horizontal lines show the grid block boundaries where the observation probe is located (for the first IPTT). Note that we consider a radially symmetric case from the wellbore and the

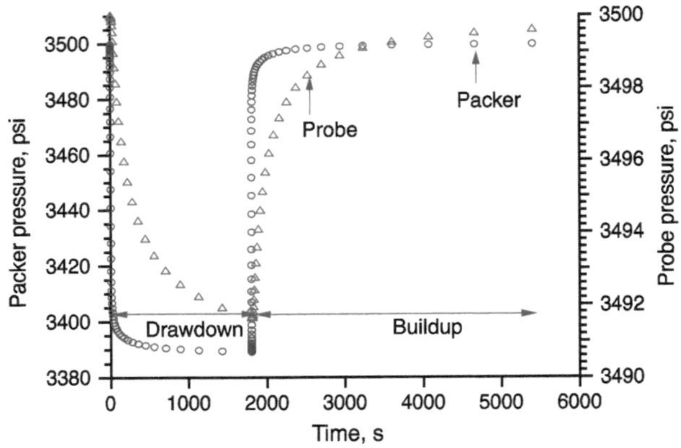

Figure 5.41 Simulated packer and probe pressure histories for the first IPTT of Example 6.

first grid block left-boundary in the r-direction corresponds to the wellbore ($r_w = 0.354$ ft). The last grid block right-boundary in the r-direction is at $r_e = 1000$ ft. A no-flow boundary at $r_e = 1000$ ft is considered. The r- and z-direction axes shown in Figure 5.40 are used to facilitate relating grid indexing to distances in r- and z-directions. The distance in the z-direction is measured as positive from the middle of the dual-packer interval to the top no-flow boundary and as negative from the middle of the dual-packer to the bottom no-flow boundary of the formation.

Figures 5.41 and 5.42 show the simulated pressure data for the first and second IPTTs (Figure 5.39) by using the rock property fields shown in Figure 5.40 and other input model parameters given in Table 5.14. Each IPTT is simulated independently from the initial reservoir pressure; i.e., no transient effects from the first IPTT to the second one. In history matching given here, we work with corrupted (noise added) data sets of the true packer and probe pressures. For both IPTTs, the corrupted packer pressures contain the same normal random noise with a zero mean and standard deviation of 0.7 psi, whereas the corrupted probe pressures contain the same normal random noise with a zero mean and standard deviation of 0.1 psi. Each pressure data set contains 118 data points. As in the previous synthetic examples, the corrupted (noise added) data are called the measured data as if they are real.

Figure 5.43 presents the buildup pressure derivatives (based on the true pressures, i.e., no noise added) for the packer and probe for both IPTTs. As shown in this figure, both packer IPTT buildups derivatives exhibit three distinct flow periods: (1) A positive unit-slope ($m = 1$) storage-dominated flow regime at very early time, (2) A transition period, and (3) A spherical flow regime ($m = -1/2$). Note that all four derivatives merge into a single curve, regardless of the packer and probe locations, after 0.1 hr. However,

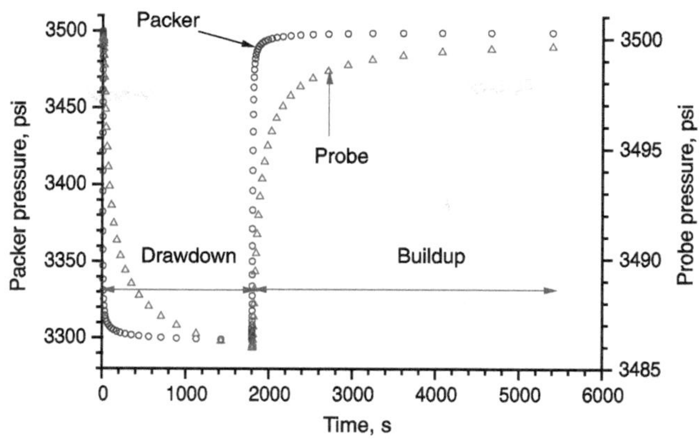

Figure 5.42 Simulated packer and probe pressure histories for the second IPTT of Example 6.

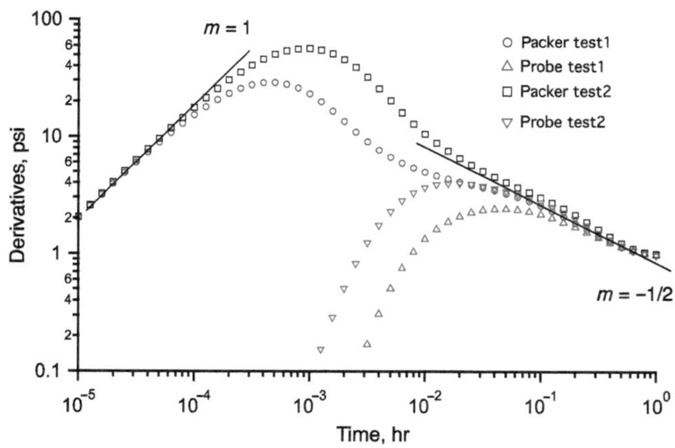

Figure 5.43 Packer and probe pressure derivatives for the first and second IPTTs for Example 6.

the start of the spherical flow regime for each test is different due to different packer/probe locations.

Next, we history match the simulated IPTT data sets in the Bayesian framework by minimizing the objective function given in Equation (5.97) (Section 5.10) using the Levenberg-Marquardt algorithm. For this purpose, the sequential history matching procedure described in Section 5.9 will be used. More specifically, first we perform history matching of packer and probe pressure data (simultaneously) for the first IPTT, starting with the prior means ($\mathbf{m}_{\text{prior}}$) and covariance matrix ($\mathbf{C}_M$) in Equation (5.97). After the first IPTT, a history matching will be performed for the packer and probe pressure data (simultaneously) of the second IPTT by starting with

$\mathbf{m}_{\text{prior}} = \mathbf{m}_{\infty}$ and $\mathbf{C}_M = \mathbf{C}_{MP}$ in Equation (5.97). Here $\mathbf{m}_{\infty}$ and $\mathbf{C}_{MP}$ denote the a maximum posterior estimate of the model parameters and maximum posterior covariance matrix obtained after history matching the first IPTT packer and probe data sets, respectively. In both history matchings, we match the entire drawdown and buildup pressure data. Only horizontal and vertical permeabilities in each grid block are to be treated as unknown and to be estimated from simultaneous history matching of packer and probe pressures using the LSE objective function given by Equation (5.97). The porosity, wellbore storage, and skin are treated as known at their true values in regression.

For each nonlinear estimation, the number of unknowns to be estimated from a total number of 236 pressure measurements are equal to 3100, which far exceed the observed pressure data points. The sensitivity coefficients required in minimizing the objective function of Equation (5.97) by Levenberg-Marquardt method are computed from a fully discrete adjoint method (Gok et al. (2005)). When performing simultaneous history matching of the packer and probe pressure data for the first IPTT data set, we use the prior means of $\ln(k_h)$ and $\ln(k_v)$ as initial guesses, i.e., we assign each grid block the values of prior means of $\ln(k_h) = 2.99$ and $\ln(k_v) = 1.79$, and the prior covariance matrix $\mathbf{C}_M$ in Equation (5.97) based on the analytical spherical variogram models as described previously. The error data covariance matrix $\mathbf{C}_{D,j}$ for each data set in Equation (5.97) is diagonal, with the diagonal elements equal to the given variances of the packer and probe data sets; i.e., 0.49 psi^2 and 0.01 psi^2, respectively. The Levenberg-Marquardt algorithm started with an initial value of the objective function (Equation (5.97)) equal to 84,055 and converged to a solution in 96 iterations with a final value of the objective function equal to 133, which is a nearly thousand-fold reduction in the objective function.

Figure 5.44 shows the maximum *a posteriori* (MAP) estimates of $\ln(k_h)$ and $\ln(k_v)$ and their maps of the normalized posterior variances. In the Bayesian framework, the most probable model (also called the maximum *a posteriori* estimate and denoted by MAP) is referred to the parameter estimates obtained by minimizing the objective function starting with prior information (prior means, $\mathbf{m}_{\text{prior}}$ and prior covariance matrix, $\mathbf{C}_M$ in Equation (5.97)(Oliver et al., 2008)). History matches (not shown here) of the entire drawdown and buildup packer pressure and probe data with the corresponding model responses are nearly perfect, yielding an RMS value of 0.77 psi and 0.11, which are very close to the input standard deviation of noise in the packer and probe data; 0.7 and 0.1 psi, respectively. Here, only the history matches of the buildup portion of the first IPTT packer and probe data are presented in Figure 5.45. As shown, the pressure-derivative matches with the measured ones are quite good. The scattering of both the measured packer and probe derivative values, as shown in Figure 5.45, is due to the noise added to the simulated test data.

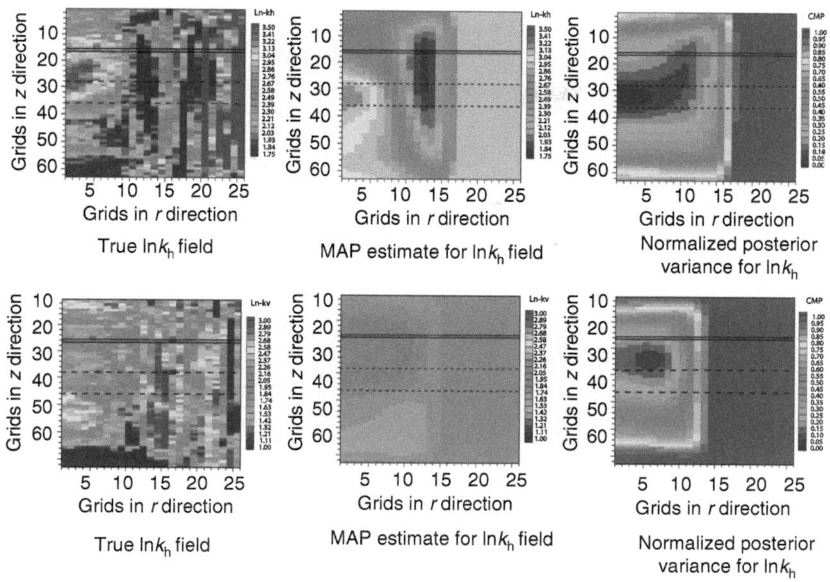

Figure 5.44 True, MAP estimates and normalized posterior variances for $\ln(k_h)$ and $\ln(k_v)$, packer and probe pressure matching simultaneously (see color plate 2).

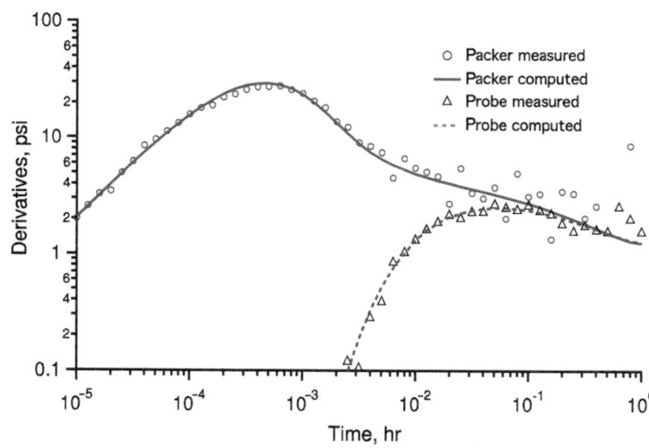

Figure 5.45 A comparison of model (computed) pressure change and derivative responses (solid and dashed lines in the figure represents the model responses computed at the MAP estimates) with the corresponding noisy responses for the buildup period for the first IPTT data set; Example 6.

As can be seen in Figure 5.44, the MAP estimates capture the major trends of the true permeability distributions, but they are quite smooth. Here, we will restrict our presentation to the MAP estimates, but we note that the MAP estimates conditioned to the pressure data are too smooth to be a credible model. To generate a realization conditional to the pressure data, but plausible too, one could use the randomized maximum likelihood approach (Kitanidis, 1986; Oliver et al., 1996; 2008) based on minimization of the same objective function of Equation (5.97), but by starting with $\mathbf{m}_{\text{prior}}$ equal to rough initial guesses of rock property fields generated from an unconditional simulation and a realization of the observed pressure data sets.

The posterior normalized variances shown in Figure 5.44 are defined by a ratio of the posterior variances of the estimated parameters to the prior variances of the model parameters; i.e., $\sigma^2_{m\infty}/\sigma^2_{m_{\text{prior}}}$. If this normalized posterior variance is close to one, then the conditioning of the IPTT data does not help much to reduce the uncertainty in this model parameter, while if it is less than one, then the conditioning of the IPTT data does indeed help to reduce the uncertainty. If the normalized posterior variance is zero, the conditioning of the data completely reduced the uncertainty for those particular model parameters.

An inspection of the results shown in Figure 5.44 indicates that the MAP estimates conditioned to the packer and probe data best resolve the true values of $\ln(k_h)$ and $\ln(k_v)$ in the near wellbore grid blocks within the open interval and in the grid blocks away from the wellbore between the upper boundary of the open interval and the probe location, as supported by the normalized posterior variances of $\ln(k_h)$ and $\ln(k_v)$. Beyond those regions, the MAP essentially equals to prior variances of the model parameters, indicating conditioning to both packer and probe data cannot help to resolve the true unknown $\ln(k_h)$ and $\ln(k_v)$. The locations where the MAP estimates of $\ln(k_h)$ and $\ln(k_v)$ are close to their corresponding true values are related to the sensitivities of the packer and probe pressure data. The sensitivities represent derivatives of packer and probe pressures (not pressure changes) with respect to $\ln(k_h)$ and $\ln(k_v)$ at a given time, as shown in Figures 5.46 (for the packer sensitivities) and 5.47 (for the observation probe sensitivities) for the drawdown period of the first IPTT (Figure 5.39, figure on the left) at two different drawdown times; $t = 5 \times 10^{-3}$ and $t = 0.5$ hr.

The sensitivities shown in Figures 5.46 and 5.47 have the units of psi because we look at the sensitivity of pressure to the log-permeability distribution. Here, we present only the sensitivity plots for the drawdown period, but we should note that the sensitivity behaviors of the packer and probe pressure buildup data are different from those for the drawdown pressure data (Gok et al., 2005).

As can be seen from Figure 5.46 (top two figures), the packer pressure is more sensitive to $\ln(k_h)$ values of the grid blocks in the vicinity of the open interval. Note also that the sensitivity is not uniform in these grid blocks; the grid blocks at the tips of the open interval show the greater sensitivity. Note that the region of sensitivity resembles a "cone". As time increases, the conical region of sensitivity grows as it moves deeper into the reservoir. For instance, at the end of drawdown period ($t = 0.5$ hr), the outer edge of the sensitivity region (the contour having a constant sensitivity about

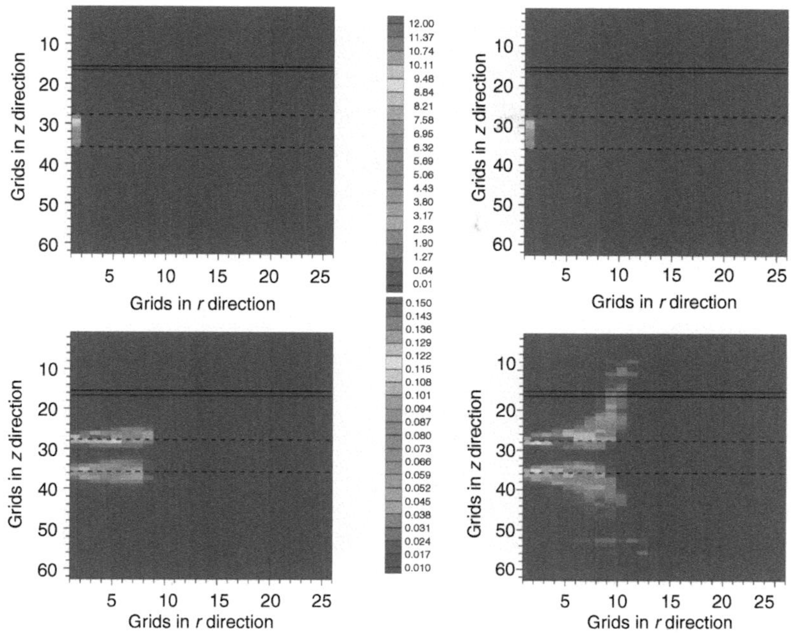

(a) $t = 0.005$ h of drawdown period (b) $t = 0.5$ h of drawdown period

Figure 5.46 Sensitivity of packer pressure to $\ln(k_h)$ (top two figures) and $\ln(k_v)$ (bottom two figures). The figures on the left are for $t = 0.005$ hr, and the ones on the right are for $t = 0.5$ hr of drawdown period (see color plate 3).
Source: Reproduced from Gok et al. (2005).

equal to 0.01 psi) reaches the 14th grid block (about 25 ft from origin) in the r-direction, and the 6th grid block (15 ft) in the positive z-direction and the 58th grid block (20 ft) in the negative z-direction. However, the sensitivity in the grid blocks further away from the open interval is very small compared to the sensitivity in the grid blocks close to the open interval. Interestingly, the packer pressure does not show any sensitivity at all to the horizontal permeabilities of grid blocks near wellbore above and below the open interval.

Regarding sensitivities of the packer pressure to $\ln(k_v)$ (see bottom two figures in Figure 5.46) the packer pressure is more sensitive to $\ln(k_v)$ values of the grid blocks lying through an imaginary horizontal line passing through the upper and lower boundaries of the open interval, to a certain distance in the r-direction from the wellbore. For all times during the drawdown, this distance is at the 8th grid block (about 3.5 ft from the wellbore). The sensitivity at the grid blocks lying along the upper boundary of the open interval is slightly larger than that at grid blocks along the lower boundary. This is because the vertical permeabilities along the upper boundary of the open interval are smaller than those along the lower boundary (Figure 5.40).

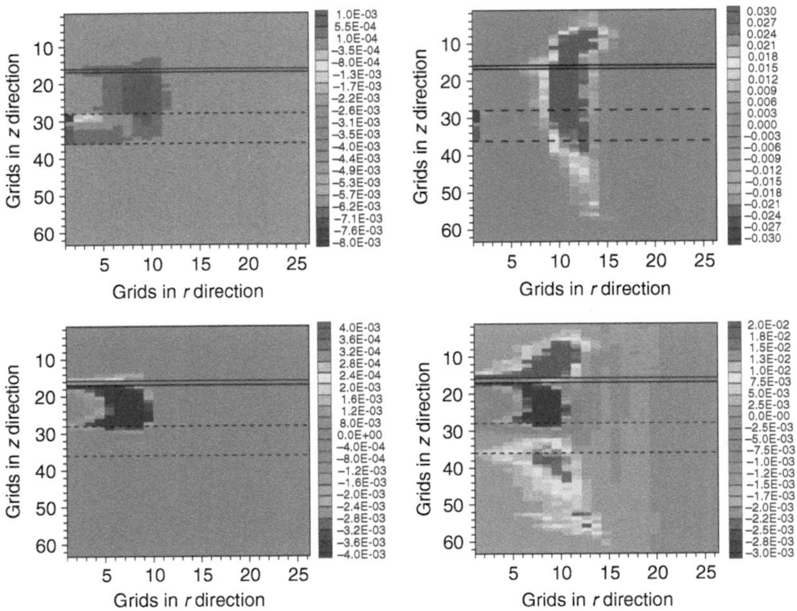

Figure 5.47 Sensitivity of probe pressure to $\ln(k_h)$ (top two figures) and $\ln(k_v)$ (bottom two figures). The figures on the left are for $t = 0.005$ hr, and the ones on the right are for $t = 0.5$ hr of drawdown period, reproduced from Gok et al. (2005) (see color plate 4).

As time increases, the sensitivity tends to increase from these grid blocks towards the grid blocks between these lines (horizontal lines described above) and no-flow boundaries of the reservoir. Note that the packer pressure shows much less sensitivity to the grid blocks within the open interval as well as the grid blocks, near the wellbore, above and below the open interval.

As can be seen from Figure 5.47 (top two figures), the observation probe pressure shows more sensitivity to $\ln(k_h)$ in the grid blocks adjacent to the wellbore near the lower tip of the open interval. The sensitivity at these grid blocks is positive. In addition, the probe pressure shows negative sensitivity at the grid blocks adjacent to the wellbore near the upper tip of the open interval. Note that a negative sensitivity means that decreasing horizontal permeability at these grid blocks increases probe pressure. At early times of the drawdown, the probe pressure also shows positive sensitivity to the horizontal permeability of the grid blocks located between the upper boundary of the packer interval and the observation probe, but confined into a small region away from the wellbore in the r-direction. For $t = 0.005$ hr, this region is confined between the 5th and 11th grid blocks (a radial ring between 1.3 and 10 ft) in the r-direction. As time increases, this region grows both in horizontal and vertical directions, but more in the vertical direction, towards the top and bottom boundaries of the formation. For example, at the end of drawdown ($t = 0.5$ hr) the boundary of sensitivity region in the r-direction reaches the 15th grid block, which is about 30 ft away from the wellbore. The vertical thickness of this region at this time reaches the top boundary (27.25 ft) and the 58th grid block towards the bottom boundary (almost 19 ft). Regarding the sensitivity of the probe pressure to $\ln(k_v)$ (bottom two figures in Figure 5.47), at early times, the probe pressure is more sensitive to the grid blocks between the upper boundary of the packer interval and observation probe and between the probe and top no-flow boundary. Notice also that the values of sensitivity at the grid blocks between the probe and upper boundary of the packer are negative, whereas the sensitivity values between the probe and upper boundary of the layer are positive. As time increases, the probe pressure also shows a significant (positive) sensitivity to the vertical permeability of the grid blocks lying between the lower boundary of the open interval and the bottom boundary. The sensitivity region grows towards the top and bottom boundaries and the magnitudes of sensitivity increase with time. The sensitivity region of the probe pressure to the vertical permeability spreads more in the vertical direction than the radial direction. It is also interesting to note that the probe pressure shows almost no sensitivity to the vertical permeabilities of the grid blocks within the upper and lower boundaries of the open interval.

As can be seen from the sensitivity graphs given in Figures 5.46 and 5.47, the influence of nonuniform permeability anisotropy and porosity distributions on the packer and probe pressure behaviors is quite complicated. These plots also indicate that the packer and probe pressure data (at a given time) are associated with a particular area of influence, depending on the location of the packer interval, the location of the observation probe, the flow-rate history (drawdown or buildup) and the formation flow properties. These sensitivity graphs also identify the regions where the formation properties will have more influence on the packer and probe pressure behaviors. It is clear that these pressure behaviors can show sensitivities to different parameters in different parts of the formation, or the influential areas and averaging process of the packer pressure are different from that of the probe. Hence, by simultaneously history matching both packer and probe pressure data, we can resolve $\ln(k_h)$ and $\ln(k_v)$ at different sections of the formation. As the information content of both pressure data sets are well reflected by the MAP estimates, as shown in Figure 5.44, the interpretation of this example may suggest that if we use a heterogeneous reservoir model, it may be always better to match packer and probe pressure data sets simultaneously because packer and probe pressures reflect heterogeneities in the formation properties at different sections of the reservoir. These results also suggest that if a single-layer homogeneous model were used to match packer and probe pressure data from a heterogeneous formation, it would be better to perform regression on packer and probe pressure data separately, not simultaneously.

Although the concept of sensitivity coefficients discussed above is quite useful to identify the regions in the reservoir and which model parameters show sensitivity to IPTT data, this does not necessarily mean that just by examining sensitivity coefficients we can precisely predict how well we resolve the model parameters by history matching. The resolution of model parameters by history matching of the packer and/or probe pressure data and geostatistical model is highly dependent on:

1. Observability of a parameter (or sensitivity),
2. The prior variances (level of uncertainty) of model parameters in the prior model,
3. Correlation between model parameters contained in the prior as well as in the posterior model, and
4. The accuracy and resolution of pressure measurements.

It should be stated that the sensitivity (observability) of a parameter in a region fundamentally depends on the pulse duration and its magnitude, the number of spatial observation points, and the number of pressure and rate data points at each location as a function of time in diffusive transient tests. If the pressure diffusion becomes steady-state or pseudosteady-state in the region, due to constant pressure or no-flow boundary conditions, or steady-state due to the nature spherical flow geometry in thick formations because of small produced volumes with a probe or a dual-packer, then the effect of a parameter is lumped with all the other parameters of the region. Therefore, the parameter in question cannot be observable (determined). This has two implications for the well test interpretation and parameter estimation:

1. The sensitivity of a region with its parameters increases and then decreases with the pulse duration and its magnitude (for instance, sensitivities for drawdown and buildup portions are different due to change in pulse, flow rate, magnitude and duration), and
2. The measurability of pressure diffusion and sensitivity depends on the accuracy and resolution of the measurements.

Next, we perform a history matching of the second IPTT data for which the middle point of the dual-packer interval was positioned just 6.2 ft below the middle of the dual-packer interval in the first IPTT (Figure 5.39). Note that in the second test, an observation probe is placed below the dual-packer interval. As in the first test, the second test also consists of a 0.5-hr drawdown test followed by a 1-hr buildup test (Figures 5.41 and 5.43). The main objective of performing the second IPTT is to explore whether we can resolve k_h and k_v values, especially for the lower section of the formation because, as discussed above, observability of these parameter (or sensitivity) was insignificant from the pressure data of the first IPTT.

For the history matching of both packer and probe pressure data for the second test, as discussed previously, we use the maximum *a posteriori* model (MAP) for $\ln(k_h)$ and $\ln(k_v)$ (Figure 5.44) conditional on the packer and

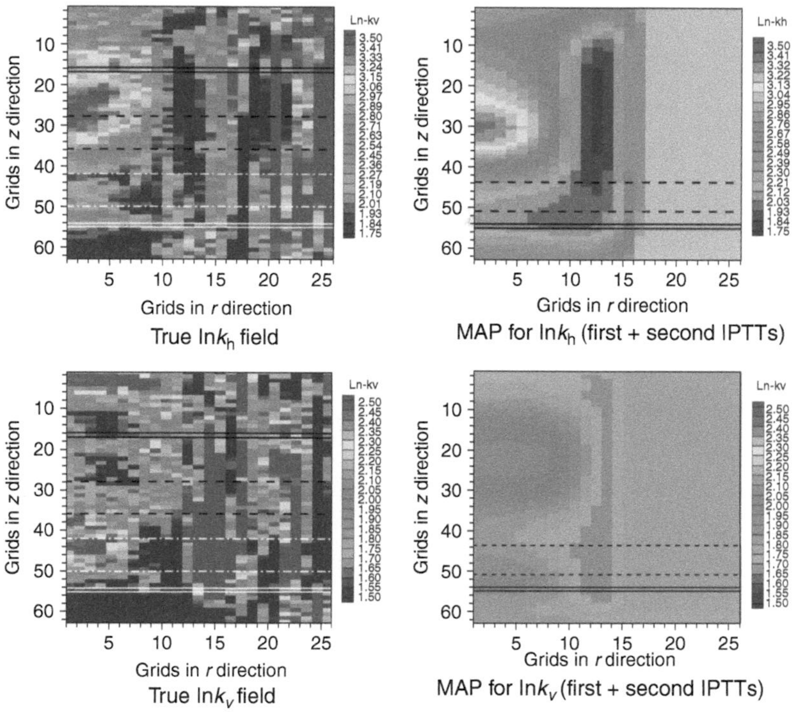

Figure 5.48 MAP estimates of log-horizontal and -vertical permeabilities conditioned to both packer and probe pressures of first and second IPTTs (see color plate 5).

probe pressure data for the first test (Figure 5.39) as the "prior model" when minimizing Equation (5.97). It means that we use the posterior covariance matrix of this estimate as the prior covariance matrix and the posterior (MAP) estimates as the prior means of the model parameters. Minimizing Equation (5.97) with such a prior model gives a maximum posterior estimate conditional to both pressure data sets of two tests. Note that this does not mean the values of parameters obtained from the first test are fixed. If pressure data of the second test have sensitivity to the grid blocks where the pressure data of the first test had sensitivity too, then the uncertainty at these grid blocks will be further reduced by conditioning to the pressure data of the second test. Actually, as discussed in Section 5.8, this is an important feature of the Bayesian estimation because it allows us to sequentially update the prior model as new data become available as in this example. This may also be true for most reservoirs, where IPTTs are normally conducted in open-holes before any DST or conventional pressure transient tests.

The MAP estimates of $\ln(k_h)$ and $\ln(k_v)$ conditional to both sets of IPTTs are shown in Figure 5.48. As seen from these figures the second IPTT helped us to resolve the k_h and k_v values at the grid blocks which the

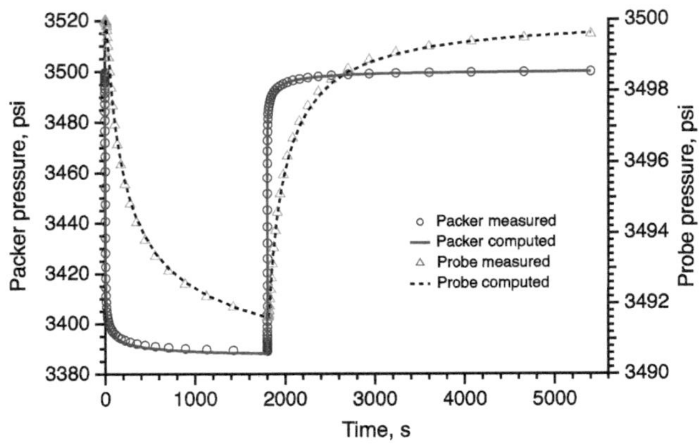

Figure 5.49 History matches of packer and probe pressure data for the first IPTT.

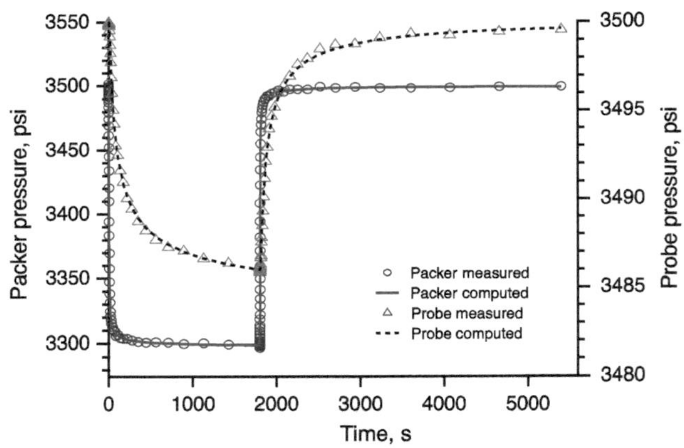

Figure 5.50 History matches of packer and probe pressure data for the second IPTT.

pressure data of the first test did not resolve; mainly the lower part of the layer below the packer interval location of the first test. Note that conditioning the pressure data for the second test did not significantly change the estimated values of $\ln(k_h)$ and $\ln(k_v)$ at the grid blocks which were well resolved by conditioning only the first IPTT data. This means that for this example, the second IPTT packer and probe data did not show significant sensitivity to those grid blocks which were resolved well previously by the first IPTT data. In that sense, the estimated values from the first test were preserved during the nonlinear parameter estimation for the second IPTT.

Figures 5.49 and 5.50 clearly show that we have perfect history matches of the data for both tests. Thus, this example demonstrates that this sequential conditioning method discussed in Section 5.9 is useful in cases where we wish to condition more than one interval pressure transient test data at different locations along the wellbore. Consequently, if this interpretation methodology is used in real-time as we conduct IPTTs, the test locations and the number of tests can be optimized such that we obtain more information for a given formation.

CHAPTER 6

Pressure Transient Test Design and Interpretation

Contents

6.1. Introduction

Pressure transient testing is one of the essential exploration tools as well as being a subset of the reservoir monitoring and surveillance that are an integral part of reservoir management. Reservoir monitoring and surveillance consist of acquisition of data (pressure, saturation, production profile, etc.) and samples (cores, fluid samples, etc.) continuously and/or periodically, interpretation of data, upscaling, downscaling, averaging, utilizing, and decision making. It is a dynamic and continuous decision making process: how, what, when, and where to acquire and utilize data for managing reservoir and wells. In this chapter we will cover the general principles of pressure transient test design and the interpretation process and present two examples.

In general, reservoirs are heterogeneous, but we make many simplifying assumptions to idealize them as homogeneous or inhomogeneous, such as

Developments in Petroleum Science, Volume 57
ISSN 0376-7361, DOI: 10.1016/S0376-7361(10)05712-2

layer-cake and double porosity models for well test interpretation. This is somewhat justified under drilling, logging, and completion requirements, where punctuality is vital because of safety and cost justification. Punctuality (computational speed) is also important for real-time interpretation as data are acquired.

As we discussed in Chapter 1, pressure transient testing provides dynamic data to assess reserves, reservoir pressure, well condition and productivity, formation transmissibility, and inhomogeneities, such as faults and fractures, heterogeneities, and horizontal and vertical permeability distributions, etc.

Pressure transient testing is now applied for a wide variety purposes including obtaining formation fluid gradients and samples, reservoir pressure, well productivity, reservoir characterization, well drainage volume and boundaries, perforation and stimulation evaluation, etc. These testing objectives are achieved employing a wide variety of hardware and acquisition systems deployed with wireline, slickline, coil tubing, drill pipes, and permanently installed pressure and rate systems, and using interpretation techniques usually via computer software.

Since the 1930s, analytical techniques have been used for interpretation of pressure transient test data for obtaining well and reservoir parameters and their characteristics, mostly from buildup tests and WFTs (Wireline Formation Tests). Since the 1990s, more diverse transient data such as flow rate, density, temperature, spatially measured observation pressure, and particularly continuous pressure datastreams from permanent downhole systems are also progressively used in pressure transient test interpretation.

Analytical reservoir or formation models, although based on single-phase fluid flow physics, usually cannot capture well/reservoir behavior because they do not take account of most of the inhomogeneity or heterogeneity. Since the 1990s, there has been tendency towards using single-phase or multiphase numerical models to capture the realistic behavior of reservoirs or formations because of multi-source data, improved data quality and their extent in space and time, advances in computational methods both in forward and inverse solutions, and computing power. Here, we do not mean that analytical reservoir or formation models will soon be irrelevant. As we stated above, analytical simplicity may still be required for interpretation of many transient tests under realistic oilfield conditions, however, the numerical option should also be available for the interpreter to better handle complicated well geometries, heterogeneity, and/or multiphase flow.

Formation and well pressure transient testing objectives can be broadly categorized (Kamal, Fryder, & Murray, 1995) as:

1. Obtain reservoir pressure and temperature;
2. Estimate formation permeability and skin (productivity or injectivity);
3. Estimate relative permeabilities;
4. Estimate storativity ($\varphi = \phi c_t$) usually from interference tests or multiprobe WFTs;

5. Obtain formation parting pressure;
6. Estimate mobility of fluid banks and their locations;
7. Estimate radius of investigation;
8. Obtain reservoir pressure gradients and fluid contacts;
9. Determine well deliverability;
10. Obtain formation fluid samples;
11. Determine fluid types for each flow unit;
12. Determine a number of fluid properties such as bubble point pressure, viscosity, density, etc. in situ;
13. Determine net to gross ratio;
14. Determine lateral and vertical continuity and hydraulic communication between wells;
15. Evaluate completion and stimulation (acidizing or fracturing) efficiency;
16. Estimate fracture length and conductivity;
17. Determine reservoir type such as fractured or layered;
18. Determine reservoir inhomogeneities, such as faults, fractures, etc.;
19. Assess reservoir heterogeneities;
20. Assess reservoir extent (reserve) and outer boundary types such as no flow, pressure support;
21. Facilitate and assess near-wellbore and hydraulic fracture clean up.

It should be pointed out that this list is much more extensive than that given by Kamal et al. (1995), but it is not an ultimate list.

Principally, many of the above geological, fluid, and formation parameters may also be obtained from other sources, for instance permeability from openhole cores and logs. Therefore, the objective of formation and well pressure transient testing is to acquire transient data sets to reduce uncertainty in these parameters and the reservoir model.

6.2. Pressure Transient Test Design and Interpretation Workflow

Well test interpretation as an inverse problem such as deconvolution and nonlinear parameter estimation is founded in mathematics and statistics, although inverse problems are typically ill-posed (Hadamard, 1953) compared to well-posed problems of transient pressure testing forward models: analytical or numerical.

The fundamental problem of the inverse pressure transient testing is the lack of sufficient useful data over the large 4D reservoir domain. In fact, a meta-theorem (Colton, Ewing, & Rundell, 1990) states that for a unique solution of an inverse problem, the overposed data (measurements) and the unknown property function (say permeability distribution) should lie in the same direction. More explicitly, it means that measuring pressure in the wellbore for days, months, and years without spatially distributed pressure measurements over the domain will not yield a unique permeability distribution in a reservoir. Nevertheless, a basic pressure transient test interpretation procedure (Ljung, 1987) can be described as:

1. Data (for instance, the shut-in pressure vs. time),
2. A possible set of reservoir models that may fit the data,
3. A set of rules that are used to identify the most likely candidate reservoir models (for instance, a linear flow regime followed by an infinite acting radial flow regime may indicate an infinite-conductivity fracture in an infinite-acting system),
4. Parameter estimation using the most likely reservoir model (for instance, using the Horner method to obtain the reservoir pressure, permeability, and skin),
5. Model validation: assessing uncertainty in each estimated parameter and model.

These steps in the interpretation procedure are essential to follow as a pressure transient test inverse problem, although highly ill-posed but it is quite analytical.

On the other hand, pressure transient test interpretation requires an integration of multi-domain knowledge using a variety of software tools and multi-source data on multiple scales (in space, time, or both): seismic and well log data. It is very difficult to formulate the integration of multi-domain knowledge and measurements within the Bayesian inference framework. Therefore, unlike other chapters, this chapter will be more descriptive and less analytical, and therefore subjective. In the petroleum literature, many papers have presented integrated formation and well test interpretation workflows (Ehlig-Economides, Joseph, Ambrose, & Norwood, 1990; Kuchuk, 2009b; Xian, Carnegie, Al Raisi, Petricola, & Chen, 2004). The multi-source and multi-domain interpretation process is called *geoengineering* by Corbett (2009) and Corbett et al. (2010), as shown in Figure 6.1. Corbett (2009) further describes geoengineering or petroleum geoengineering as: "The vision for petroleum geoengineering is of an all-encompassing discipline (Figure 6.2) that contains the functions and skill sets of geologists, engineers, petrophysicists, and geophysicists."

The interpretation workflow given by Figure 6.1 is excellent for the postmortem transient test interpretation but it does not include the whole process of pressure transient testing. The multi-source and multi-domain integration or geoengineering must have an impact on test design and its execution: the pulse duration and its magnitude, gauge resolution, etc. Otherwise, we end up with a transient test without any information content about the reservoir. Therefore, the test design and interpretation process given in Figure 6.3 should start when the reservoir monitoring and surveillance group, asset and/or exploration teams, and/or production and/or reservoir engineering decide to conduct a formation and/or well

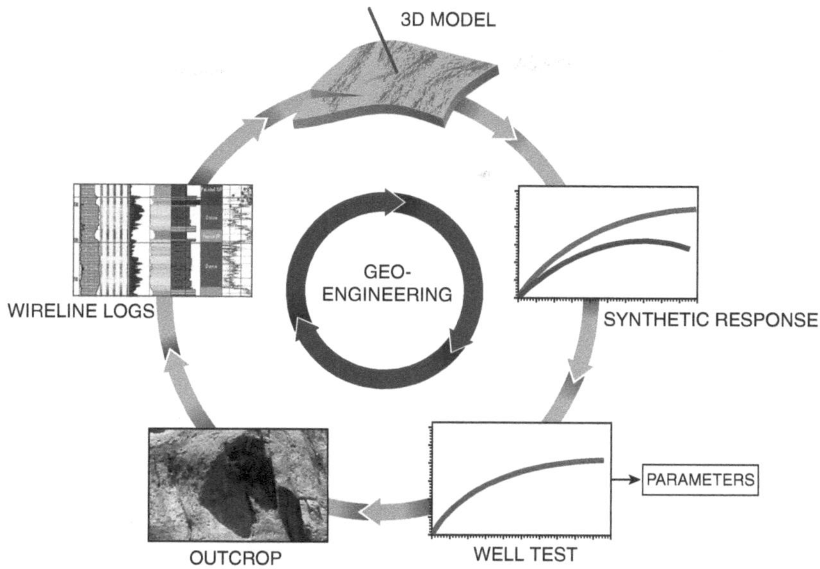

Figure 6.1 Pressure transient geoengineering workflow after Corbett et al. (2010).

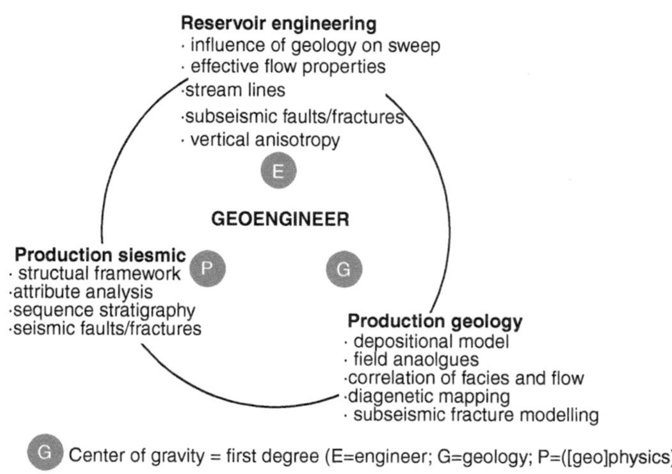

Figure 6.2 Geoengineering definition after Corbett (2009).

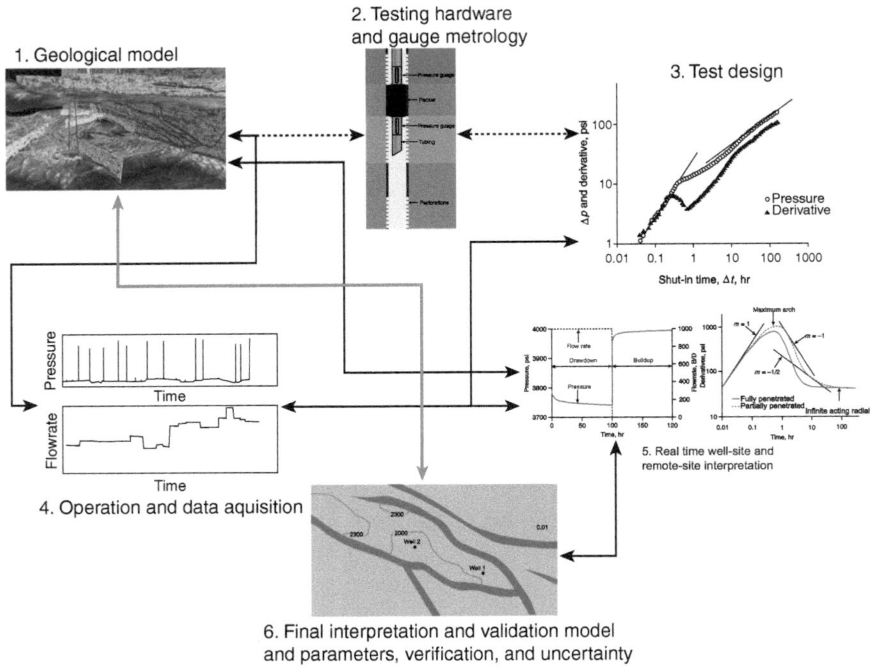

Figure 6.3 Pressure transient test design, execution, and interpretation (outer loop) workflow.

transient test. The process continues iteratively until the test objectives completely or partially are achieved. As shown in Figure 6.3, the test design, its execution, and interpretation of the acquired transient data can be divided into six continuous intertwined steps as follows:

1. Development of Geological and Reservoir Model,
2. Selection of Testing Hardware and Gauges,
3. Test Design,
4. Operation of Test and Data Acquisition,
5. Real-Time Interpretation (Well-site and/or Remote-site),
6. Final Interpretation and Validation.

The workflow shown in Figure 6.3, which is a highly iterative process, will be called the *outer loop*, and its steps will briefly be described next. It should be noticed that the outer loop workflow is almost the same as the workflow given by Corbett et al. (2010) in Figure 6.1, except the

Selection of Testing Hardware and Gauges, Operation of Test and Data Acquisition, and Test Design steps. The Development of Geological and Reservoir Model step in Figure 6.3 is given as outcrop, wireline logs, and 3D model steps by Corbett et al. (2010), as shown in Figure 6.1.

6.2.1. Development of the geological and reservoir model

Geological and reservoir model building and conditioning to pressure transient test and production data are very broad subjects. The readers can referred to the papers by Ballin, Journel, and Aziz (1992), Ballin, Aziz, Journel, and Zuccolo (1993), Corbett, Couples, and Gardiner (2004), Corbett, Ellabad, and Mohammed (2001), Farmer (1988), Landa and Horne (1997), Landa, Horne, Kamal, and Jenkins (2000), Li, Han, Banerjee, and Reynolds (2009), Oliver (1994), Oliver, He, and Reynolds (1996), Reynolds, He, Chu, and Oliver (1996), Suzuki, Caumon, and Caers (2008), Weber (1986), and Zheng and Corbett (2009) and the books by Caers (2005) and Oliver, Reynolds, and Liu (2008). The purpose of this brief subsection is to provide a very elementary and qualitative description of the geological and reservoir model building.

When a decision is made to conduct a transient (DST, WFT, or production) test to achieve one or more test objectives given above, we start with a well in a given formation, for which the well planning, trajectory, drilling with or without geosteering have already been done using a detailed or even crude geological and/or reservoir model. Therefore, before a formation or well transient test, there is usually a reservoir model (earth model) that is used for drilling and formation evaluation, etc., although it could be highly uncertain. Typically, the reservoir model is built from geology, seismic, well logs, and data from other wells. For test design purposes, the reservoir model has to be re-built over a much small volume: a well drainage area, including a few neighboring wells if any. Compared to a full-field reservoir model, the transient test reservoir model should have much higher resolution vertically and laterally, while keeping the number of grid blocks in the model manageable, particularly for the inverse transient test problem (see Example 6 in Chapter 5). It is important that the transient test formation or reservoir model should not build directly from an upscaled full-field reservoir model. It should be built from a detailed geological model, where small scale properties are kept with a minimum degree of upscaling.

A static transient test geological model is a numerical representation of the actual reservoir with distribution of rock properties (porosity, permeability, relative permeability, capillary pressure, etc.). Building a static transient test model is an integrated multi-domain process as described in detail by Corbett (2009) and as shown in Figure 6.2, where petrophysicists, geologists, geophysicists, and reservoir engineers work together. For transient test models, the use of reservoir-related geological seismic interpretation, particularly for faults, fractures, flow barriers, partial pressure connections, etc. is quite important.

The reservoir properties determined from the petrophysical analysis and geological interpretation of seismic data, preferably 3D, should be distributed in each grid block directly or geostatistically. The objective is to distribute each reservoir property (net thickness, porosity, water saturation and, if possible, permeability) in each grid block of the model. In general, the true location of geological features between wells may be unknown but the model should take account of their possible presence by using geostatistical techniques.

To transform the static geological model to a dynamic (reservoir) model, the fluid properties from PVT data, well completion data such as perforation intervals, well radius, etc., and the past production data, if any, has to be incorporated to complete the reservoir model. This pressure transient reservoir model is called the *formation* or *reservoir model,* which will be continuously updated during the data acquisition if surface-readout and/or real-time data transmission and real-time interpretation are available. Depending on the complexity of the model, the reservoir model can be simplified and solved analytically; perhaps its analytical solution is available in the literature or from commercial testing software. An analytical approach for real-time interpretation in many cases is perhaps unavoidable. Both analytical and numerical model options should be retained for test design, and the real-time and final interpretation.

6.2.2. Testing hardware and gauge selection

In Chapter 1, the most common testing hardware and gauge metrology were given. The reservoir model can also be used for hardware selection and gauge specifications. Selecting a proper test type and measurement system according to the test objective is one of the most important steps in testing. A downhole and/or surface data acquisition system should deliver the data at specified accuracy, resolution, and drift to achieve the test objective. Surface and downhole conditions considerably affect pressure gauge and rate measuring device performance. Therefore, it is important to assess these conditions when selecting gauges and a data acquisition system. The reservoir model can also be used to assess effects of accuracy, resolution, and drift on the test objective and interpretation. For instance, using the reservoir model, we can determine what the minimum pressure gauge resolution and drift should be in order to estimate the fault conductivity and distance to an active well at a reasonable testing duration from an interference test.

As we discussed in Chapter 4, pressure gauge accuracy and resolution affect deconvolution of pressure and rate data considerably, as shown in Figure 4.32, where the deconvolution derivative indicates totally incorrect geological features due to low gauge resolution and an incorrect initial pressure due to low gauge accuracy.

Once the testing hardware and acquisition system with gauges are selected and set downhole, current testing hardware systems do not allow us to change or modify them. Therefore, they have to be evaluated and selected carefully because it will be very costly to change the downhole hardware after setting it in the wellbore.

6.2.3. Test design

In actuality, Steps 1, 2, and 3 shown in Figure 6.3 can also be called Test Design. However, given the reservoir model (updated and modified until the end of interpretation), and hardware and acquisition systems (fixed and cannot be changed after setting downhole), we define the test design as determining the test type, its duration, and the magnitude of the rate pulse to achieve the test objective. Second, given the test type, duration, and magnitude of the rate pulse, sensitivities of model parameters to gauge accuracy, resolution, and drift should also be investigated in the test design step. The effects of the expected geological features and well geometry (limited entry, hydraulic fracture, etc.) should be simulated and their diagnostic plots should be examined to determine whether essential features will appear. These features could be: the end of wellbore storage effects, linear and infinite acting radial flow regimes, etc. The test design step continues iteratively until an optimal test type, its duration, and an optimal flow-rate schedule are achieved.

6.2.4. Operation of test and data acquisition

After the testing hardware and acquisition system are placed downhole, the test commences according to the design (planned), the data stream starts to flow into the well-site and/or remote-site computers. As data are acquired, pressure and rate measurements should be processed, diagnostic plots should be displayed and examined, and nonlinear parameter estimation should be performed simultaneously and continuously in real time. When the first pressure transient test interpretation results are obtained, the test sequence, test design, and reservoir model should be re-examined continuously and modified if needed. If available, other data such as temperature, density, and fluid samples should be processed and interpreted. In the mean time, data QA/QC (Quality Assurance/Quality Control) should be continuously performed to ensure that transient data are good quality (Xian et al., 2004). The data QA/QC is one of the important functions of the data acquisition system and of the objectives of well-site interpretation.

The data acquisition and processing system (could also be a part of an interpretation software) should provide a flexible methodology for interpreters to assess data quality, testing sequences and events. The purpose of the QA/QC process is to diagnose and understand the wellbore, formation and/or tool related issues. The QA/QC process will help us to optimize the operating procedure and testing sequences, and diagnose well and formation/reservoir related issues in order not to end up with an uninterpretable data set.

Proper data processing is one of the important steps to minimize interpretation and computer execution times, but the processed data set should still capture the dynamic formation or reservoir and well behavior completely. The data processing tasks can include (Xian et al., 2004): (1) To remove superfluous data, (2) To identify data with high frequency noise, (3) To define missing and non-transient data sections (sometimes the gauge may move unintentionally due to external factors), and (4) To decimate.

In order to minimize the computer execution time, the actual data points to be used in the nonlinear parameter estimation software should be reduced to a manageable number. Giving current computing capabilities, it is generally accepted that about one thousand dynamic data points are sufficient to perform parameter estimation without missing any formation or reservoir and well information. However, while a few hundred data points would be sufficient for some tests, perhaps a few thousand data points may not be sufficient for some other tests. Therefore, as we interpret data, it becomes obvious how many data points should be used for interpretation. The decimation procedure should be based on some advanced signal processing algorithms (Houze, Kikani, & Horne, 2009) and retain the essential character of the data stream. The diagnostic plots (for instance, log-log pressure and deconvolved derivatives) can be used to check the quality of data processing and the degree of data decimating. After pressure decimating, flow-rate data must be synchronized with pressure data. The rate synchronization with pressure data is critical for deconvolution, as discussed in Chapter 4, if they are not measured simultaneously at the sandface.

6.2.5. Real-time interpretation

As the data stream flows into the well-site and remote-site computers, formation or well test interpretation begins. This step can be performed at the well-site and/or remote-sites in parallel or sequentially. The procedure described here can be applied to both real-time and postmortem interpretation. The difference is: for real-time interpretation we may not have sufficient time to include Step 6, which is the Final Interpretation and Validation step. As can be seen from Figure 6.3, the interpretation in Step 6 is a part of the outer loop, where the geology and other geoscience information may be modified and re-interpreted after the transient data are acquired.

In addition to the data QA/QC, the main objective of the Real-Time Interpretation step is to achieve the test objective while the testing system is in place, by continuously updating the geological and reservoir model and modifying the test design. In other words, to avoid short tests with little information content, to optimize the test time, and to minimize

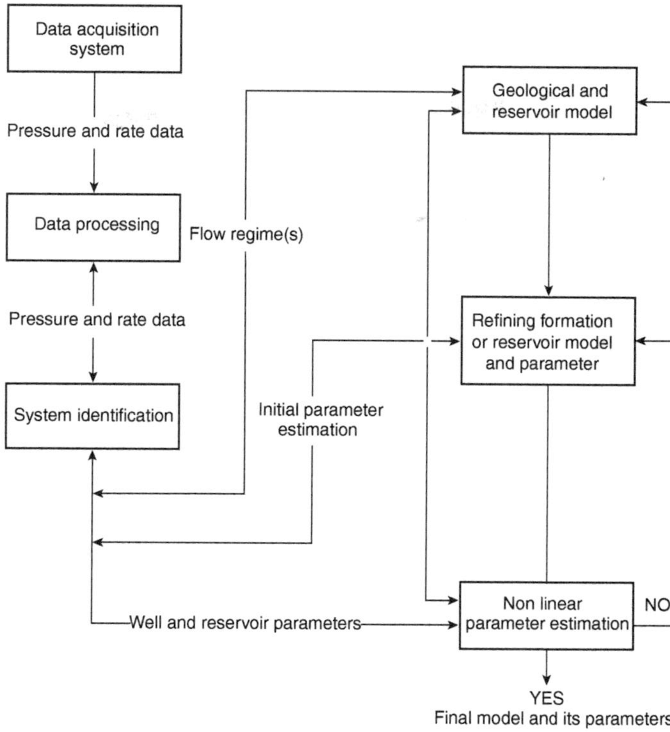

Figure 6.4 Pressure transient test inner loop interpretation workflow, modified after Xian et al. (2004).

expenditure and well-site exposure. Decision making on the termination of the downhole testing operation and data acquisition is an important and critical part of the real-time interpretation.

The interpretation in Step 5 can be defined as an inner loop and consists of the following steps, as shown in Figure 6.4:

1. Data Processing;
2. Updating the Geological and Reservoir Model;
3. System (Model) Identification;
   (a) Deconvolution,
   (b) Diagnostic plots such as derivative and semilog, and flow regime analysis,
4. Parameter Estimation;
   (a) Flow regime analysis,
   (b) Nonlinear parameter estimation (type-curve matching),
   (c) Confidence interval and statistical analysis;
5. Validation.

We have described the Data processing and Geological and Reservoir Model building steps in the previous sections. Here, we will briefly describe the remaining steps of the inner loop of the pressure transient test interpretation, as shown in Figure 6.4. It should be stated that all these steps can also be performed postmortem (after data acquisition and testing operation in the field).

6.2.5.1. System identification

As described in Chapter 4, deconvolution (determination of the constant rate pressure behavior of the system or unit response of influence function) of measured pressure and flow-rate data is the first step in the System Identification, if applicable. After deconvolution, all diagnostic derivatives, such as deconvolution and Horner derivatives, as shown in Chapter 4, should be plotted together on the same plot to identify reservoir or formation features and estimate well and reservoir parameters. It is also important to compare the deconvolution derivatives to assess their quality because they could be incorrect due to many external factors and measurement errors, as shown in Figure 4.32 (Chapter 4). Along with derivatives plots, other flow regime plots, for instance Horner semilog, should also be examined and interpreted.

Normally, wellbore effects such as wellbore storage and/or geometry complicate system identification considerably if sandface flow-rate data are not used for deconvolution and computing superposition times. In principle, model identification is defined as obtaining the "signature" of the formation or reservoir behavior from a library of available drawdown type curves or synthetic model responses. Many system identification techniques are given in well testing books (Bourdet, 2002; Earlougher, 1977; Horne, 1995; Kamal et al., 2009; Lee, Rollins, & Spivey, 2003; Matthews & Russell, 1967) and papers (Bourdet, Ayoub, & Pirard, 1989; Gringarten, 2008; Ramey, 1970; 1976b). We have also presented the most common flow regimes that are encountered in pressure transient tests in Chapter 2 and many examples in Chapters 4 and 5.

6.2.5.2. Parameter estimation

As shown in Chapter 2, flow regimes identified from derivatives and diagnostic plots are analyzed to estimate a number of parameters. For instance, permeability and skin may be obtained from the Horner semilog plot (see Figure 2.5) with the aid of corresponding derivative plots (deconvolution or the Horner superposition time) or directly from the derivative plots. The reservoir pressure is usually obtained from the Horner semilog plot.

The parameters obtained from the flow regime analyses are refined and/or additional parameters are estimated during the nonlinear parameter estimation step. As we stated above, besides the formation or reservoir model, downhole hardware, tool configuration, and completion information are also needed to compete the model to be used for nonlinear estimation (type-curve or history matching). Finally, a set of initial guesses for the unknown parameters should be obtained from the flow regime analyses and other available sources. The details of nonlinear parameter estimation and several examples are given in Chapter 5.

Nonlinear parameter estimation using analytical and/or numerical models can be performed employing weighted least square (WLS), for which weights are assumed to be known, or maximum likelihood estimation (MLE), for which the weights may be unknown. As discussed in detail in Chapter 5, we prefer to use the MLE method to better handle transient formation or well test data sets, in which the error variances (weights) of observed data are uncertain, and to estimate error variances in data sets along with the unknown formation parameters. It is well known that if the UWLSE (unweighted least-squares estimation) is used for data sets having disparate orders of magnitude, noise or both, then those data sets with large magnitudes and/or noise will dominate those having small magnitudes and/or noise in estimation (Onur & Kuchuk, 2000). Thus, information contained in data sets with small magnitudes and/or noise will be lost.

Besides the estimated parameters, statistics of these parameters, including confidence intervals and correlation coefficients should also be computed, which are very useful for identifying which parameters could be determined reliably from the available data and the reservoir model. The computational efficiency and reliability of the reservoir model and nonlinear regression algorithm considerably affect the results from nonlinear parameter estimation. In general, the nonlinear optimization is constrained by defining lower and upper limits for the parameters, as discussed in Chapter 5.

When the estimated parameters and model are unacceptable, the geological model and other input parameters must be reexamined and modified, and nonlinear parameter estimation should be performed again. In any inverse problem, it is important to understand: (1) The range of uncertainties in estimated parameters and (2) How these uncertainties can affect the match between computed and measured (observed) data. Uncertainties can arise from the measurement errors in pressure and rate data as well as in input formation and fluid parameters such as porosity, viscosity, compressibility, layering, and heterogeneity of the formation.

The Validation given in the Real-Time Interpretation section is the same as the Final Interpretation and Validation Step 6. Therefore, it will be given next as the final step of interpretation.

6.2.6. Final interpretation and validation

In principle, the final Step 6 is a continuation of the Real-Time Interpretation step as shown in Figure 6.3, and in detail in Figure 6.4, but it is performed after the data acquisition is completed and the testing operation is terminated. It is a postmortem interpretation. As we discussed above, an important feature of transient well test interpretation is: it is a sequential process whether it is real-time or postmortem. This interpretation task cannot be performed as parallel subtasks via computer software. For instance, first geological and reservoir models have to be built, then flow regimes have to be identified and analyzed before nonlinear parameter estimation. We cannot continuously iterate on geology before a few cycles of parameter estimations.

All interpretation steps described in the Real-Time Interpretation section can be performed again during this final step using analytical and/or numerical models if the results are not satisfactory or more parameters are need to be estimated. The final interpretation step can be subdivided for convenience as follows:

1. Updating the geological and reservoir model,
2. Updating some of the parameters with values obtained from the regression.
3. Examine non-uniqueness of the estimated parameters and model,
4. Determining degree of consistency of estimated parameters with other dynamic and static data, and
5. Determining uncertainty or confidence in the estimated parameters and model.

At the end of these steps the measured pressure and rate histories should be compared with values computed from the final model. If the match, the estimated parameters, and model are satisfactory and acceptable, the interpretation process is completed. If they are unacceptable, the interpretation process given in Figure 6.4 (the inner loop) will be followed until a satisfactory and acceptable match, estimated parameters, and model are obtained with the acceptable confidence intervals.

Since the 1990s, there have been many publications in the petroleum engineering literature on the integrated multidisciplinary pressure transient test interpretation (Corbett, Zheng, Pinisetti, Mesmari, & Stewart, 1998; He, Oliver, & Reynolds, 2000; Landa, 2009; Landa et al., 2000; Landa & Horne, 1997; Oliver, 1994). One recent paper by Dubost, Zheng, and Corbett (2004) presented a detailed inner loop interpretation for multiprobe wireline tester pressure transient tests in a thinly-bedded turbidities formation, where surface and subsurface geological information were utilized well. This is an excellent example for the inner loop interpretation process (Figure 6.4). Dubost et al. (2004) looked at most possible geological models that matched the observed transient data. The only missing part of the interpretation procedure of Dubost et al. (2004) was confidence in the estimated parameters and the model. In this chapter, we will present two examples to demonstrate the full interpretation with the statistics of the estimates. The first example is a synthetic one, where the geology is assumed to be known but which can be estimated with confidence. The second example is a field case, where openhole logs are available but with a limited geological knowledge.

At the onset of any interpretation process, we have to start with an assumption that all sedimentary formations such as sandstone, carbonate, shales, etc. are usually heterogeneous. On the other hand, the complexity of a model should not go beyond the support of the data. Perhaps, a single model with bulk properties is an over simplification. However, there is no universal and agreed interpretation methodology on what is an acceptable complexity of the geological and reservoir model for pressure transient test interpretation. We hope that in this book we have provided some insights to interpretation based on our own experience. Complexity or ultimate simplicity is not the objective of our transient test interpretation methodology; common sense, integration of multiple disciplines, the use of multi-source data and all available tools, and careful test design that provides a data set to support the complexity of the model are the main elements of our methodology.

As a summary of this interpretation and test design chapter, the objectives of transient tests are to estimate well and reservoir parameters, and define an accurate reservoir model. To achieve these objectives, a pressure transient test has first to be designed and conducted to have the maximum possible information content in transient data so that all critical parameters of the well, reservoir and the model are determined with high certainty. However, given oil field conditions such as cost, safety, environmental, etc., we will never have enough data spatially distributed for long time periods to solve the well test inverse problem uniquely. Therefore, the integration of multiple disciplines and multi-source data are absolutely necessary for test design, execution, and interpretation to obtain the well/reservoir parameters and define an accurate reservoir model with an assessed uncertainty. The overall interpretation procedure can simply be summarized as: Data processing, system identification and flow regime analysis via deconvolution and derivatives, the integration of multiple disciplines and multi-source data, a Bayesian approach for the nonlinear estimation with uncertainty statistics for the estimated parameters and underlying geological model.

6.3. Multiwell Interference Test Example

Here we consider a synthetic interference test involving one active and two observation wells in a rectangular, homogeneous, and isotropic single-layer reservoir, as shown in Figure 6.5. All the boundaries of the rectangle are no-flow (closed). The total length of the reservoir in the East-West (E-W) direction is 4200.00 ft, and the total length in the South-North (S-N) direction is 900.00 ft. The distance between observation well 1 and the active well is 900.00 ft, and the distance between observation well 2 and the active

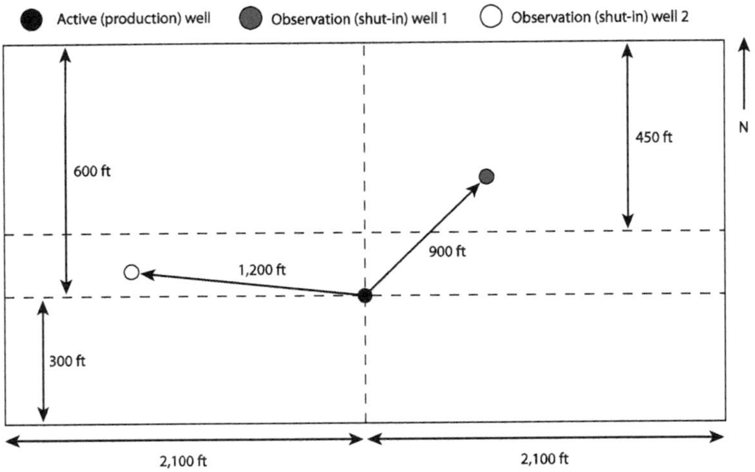

Figure 6.5 Schematic of the well/reservoir configuration for the simulated interference test in a closed rectangular, homogeneous, and isotropic single-layer reservoir.

well is 1200.00 ft. The input rock and fluid properties, relevant reservoir dimensions, and well locations are listed in Table 6.1. All wells are fully penetrating and vertical, and have identical radii. The initial reservoir static pressure at each well is the same and equal to 5000.00 psi. The objective of this test is to show the full interpretation procedure with the estimation statistics, as we discussed above.

The simulated active and observation well pressures, as well as the flow-rate history at the active well are displayed in Figures 6.6–6.8. The test sequence consists of two drawdown and two buildup periods. The total test duration is 125 hr, as can be seen from Figure 6.6. The durations of the first drawdown (DD1) and buildup (BU1) periods are 15 hr and 20 hr, respectively. The durations of the second drawdown (DD2) and buildup (BU2) periods are 40 hr and 50 hr, respectively. The surface flow rates (converted to reservoir barrel per day) during the first and second drawdown periods are 1500 and 2000 B/D, respectively. For simplicity, we assume that wellbore storage and skin effects at both observation wells are negligible.

We have added different levels of noise to active and observation well pressures. The noise for each well pressure is of Gaussian with zero mean and specified standard deviation (or variance). The standard deviations of noise for the active well pressures are 0.25 psi for the drawdown tests and 0.05 psi for the buildup tests. The standard deviations of noise for both observation well pressures are the same and equal to 0.01 psi for the entire interference tests. Figure 6.9 presents the logarithmic derivatives (based on the superposition time) for BU1, DD2, and BU2 periods for the active well, while Figures 6.10 and 6.11 present the logarithmic derivatives of DD1 and DD2 periods for both observation wells. All derivatives are rate-normalized

Table 6.1 Formation and fluid properties, relevant reservoir dimensions, and well locations

ϕ	fraction	0.20
k	md	200.0
h	ft	60.0
c_t	psi^{-1}	1.0×10^{-5}
μ	cp	1.0
C_w (at the active well)	B/psi	1.0×10^{-2}
S (at the active well)		4
r_w	ft	0.354
p_o	psi	5000.00
L_x (length of the reservoir in the E-W direction)	ft	4200.00
L_y (length of the reservoir in the S-N direction)	ft	900.00
X_a (x-coordinate of the active well from the lower-left)	ft	2100.00
Y_a (y-coordinate of the active well from the lower-left)	ft	300.00
X_{o1} (x-coordinate of obs. well 1 well from the lower-left)	ft	2964.58
Y_{o1} (y-coordinate of obs. well 1 from the lower-left)	ft	550.00
X_{o2} (x-coordinate of obs. well 2 well from the lower-left)	ft	902.35
Y_{o2} (y-coordinate of obs. well 2 from the lower-left)	ft	375.00

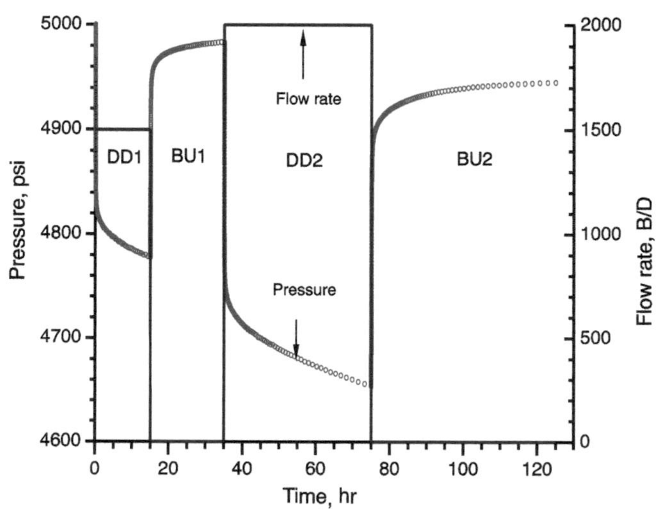

Figure 6.6 Wellbore pressure and flow-rate history at the active well.

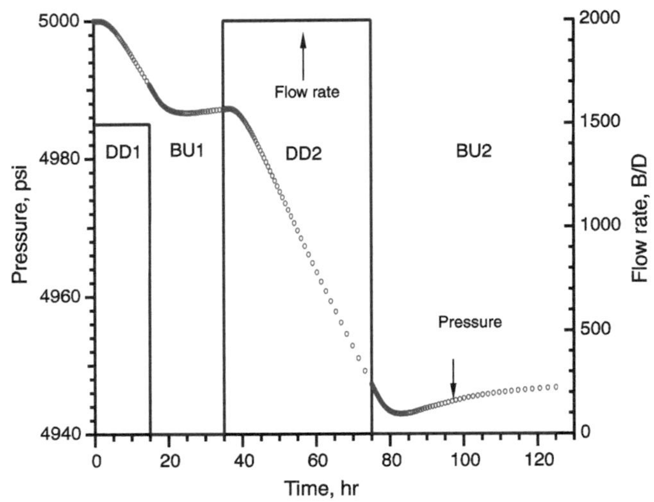

Figure 6.7 Wellbore pressure at observation well 1 and flow-rate history at the active well.

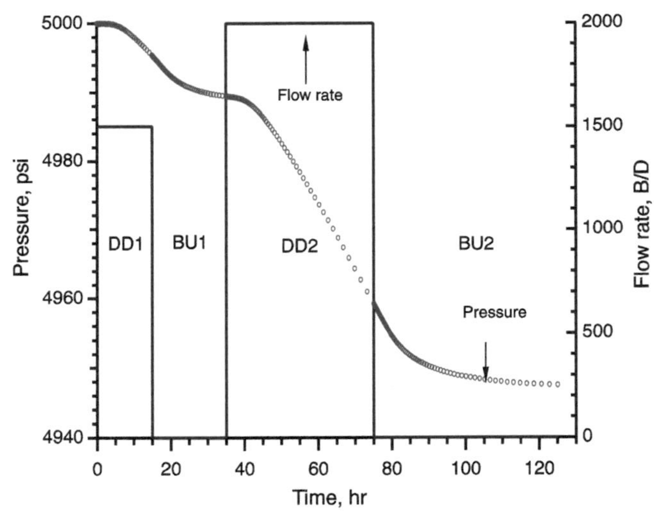

Figure 6.8 Wellbore pressure at observation well 2 and flow-rate history at the active well.

by using the flow rate of 2000 B/D prior to the second buildup period, and are taken with respect to the corresponding elapsed times or superposition times (see Chapters 2 and 3). In general, as also shown in the earlier chapters, all pressure and deconvolved derivatives $\Delta \acute{p}t$ are displayed on log-log plots as a function of elapsed times t, where $\Delta \acute{p} = \frac{d\Delta p}{dt}$, and elapsed times for drawdown tests are t and for buildup tests are Δt. It should be pointed out that the log-log diagnostic plots for the buildup (BU1 and BU2) periods for

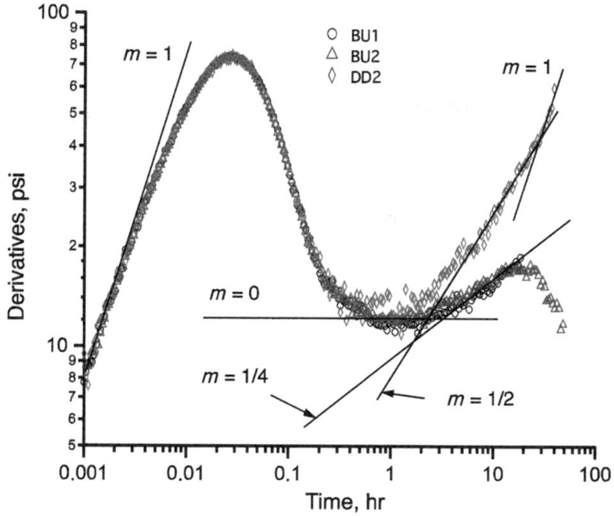

Figure 6.9 Derivatives for DD2, BU1, and BU2 periods at the active well.

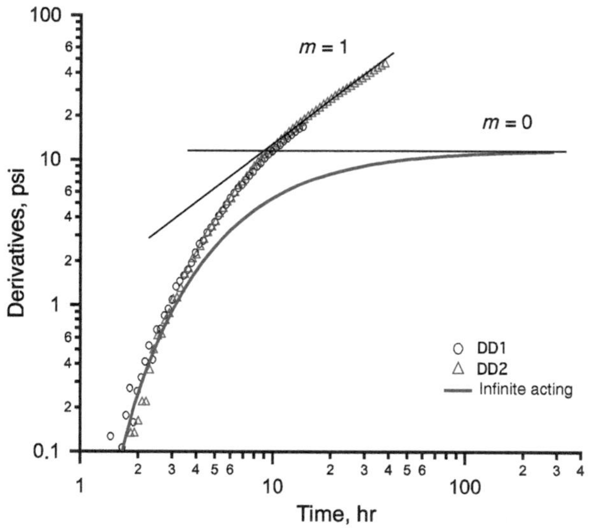

Figure 6.10 Derivatives for DD1 and DD2 periods at observation well 1 and the derivative computed from the line-source solution at the same well location in an infinite system.

both observation wells are not shown in Figures 6.10 and 6.11 because they do not indicate any meaningful signatures.

For comparison purposes, we have also shown the derivatives in Figures 6.10 and 6.11 for both observation wells obtained from the infinite-acting line source solution using the constant rate of 2000 B/D and the

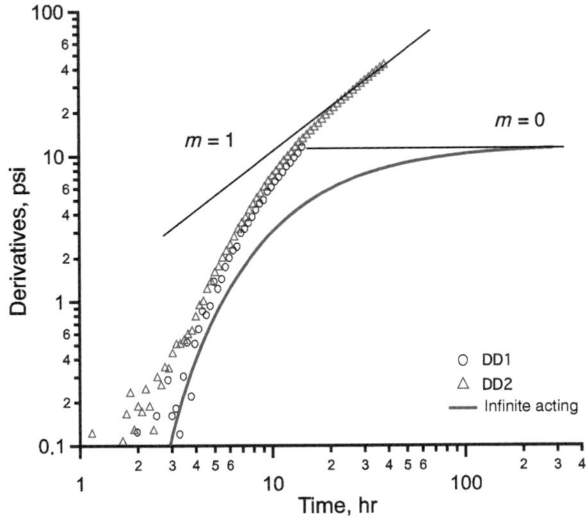

Figure 6.11 Derivatives for DD1 and DD2 periods at observation well 2 and the derivative computed from the line-source solution at the same well location in an infinite system.

values of kh/μ, $\phi c_t h$, inter-well distances, and other relevant parameters given in Table 6.1 and Figure 6.5. As well known, if the inter-well distances between the active and observation wells are relatively large, and the active well wellbore storage coefficient is not very large, then the effects of wellbore storage and skin at the active well are negligible on the observation well responses. As shown in Figures 6.10 and 6.11, a comparison of the derivatives for both observation wells in the closed rectangle reservoir with the infinite-acting derivative clearly reveals the effects of (i) noise, (ii) variable rate at the active well, and (iii) the reservoir boundaries.

As can be observed from Figure 6.10, the derivatives for observation well 1 for both DD1 and DD2 periods match reasonably well with the infinite-acting derivative at very early times (for times less than 4 hr). This means that we can estimate kh/μ and $\phi c_t h$ by performing type-curve matching of the early-time derivative with the infinite-acting derivative. If μ and h are known, then we can estimate k and ϕc_t from the estimated values of kh/μ and $\phi c_t h$ from the type-curve matching.

As can also be observed from Figure 6.11, the derivative data for the DD1 period for observation well 2 for the closed rectangle case can only be reasonably matched well with the corresponding constant-rate drawdown derivative for the infinite acting case at very early times (for times less than 4 hr). As for the well 1 case, we can also estimate kh/μ and $\phi c_t h$ by performing type-curve matching of the early-time derivative with the infinite-acting derivative. It is clear that the derivative for the DD2 period of observation well 2 for the closed rectangle case does not match the infinite-acting derivative. The main reasons are due to noise, boundary and variable rate effects.

The active well derivatives for BU1, DD2, and BU2 periods shown in Figure 6.9 indicate a well defined wellbore storage unit-slope line ($m = 1$ slope) from 0.001 to 0.004 hr, from which one can determine the wellbore storage coefficient (C_w). In addition, the derivatives exhibit a short infinite-acting radial flow regime ($m = 0$ slope) from 0.4 to 2 hr, from which one can determine the transmissivity (kh/μ) and skin factor (S) using the flow regime equations given in Chapter 2. The active well derivatives, as shown in Figure 6.9, indicate a bilinear flow regime ($m = 1/4$) from 2 to 20 hr instead of the linear flow regime observed on DD2. This is quite misleading because the reservoir model does not warrant such a flow regime. As can also be seen in the same plot, the BU2 derivative goes down after 20 hr, which could be interpreted as a pressure support and is also misleading. Therefore, if the system identification depends only on the BU1 and BU2 derivatives, we will end up with an incorrect reservoir model. These derivative behaviors are due to the complicated effects of the boundaries (Larsen, 2005), the active well location, which is not symmetric about the boundaries, and flow-rate superposition.

As shown in Figure 6.9, the derivative of DD2 from 5 to 30 hr indicates a linear flow regime ($m = 1/2$) due to the channel between the north and south no-flow boundaries, as expected given the reservoir geometry and active well location. The analyses of these linear (seen from the DD2 period) and infinite-acting radial flow regimes yields the channel width and the location of the well between the two no-flow boundaries parallel to the E-W direction (see Figure 6.5). Of course, it is well-known that one would not be able to tell whether the parallel no-flow boundaries are lying in the E-W direction from the linear flow regime analysis of these tests from the active well. After the linear flow regime, the derivative of DD2 indicates a very short pseudosteady-state flow regime ($m = 1$ slope) from 30 hr to the end of the test (40 hr). It should be noticed that the pseudosteady-state flow regime shown in Figure 6.9 is not fully developed.

Both observation well derivatives for DD1 and DD2 tests, shown in Figures 6.10 and 6.11, indicate a pseudosteady-state flow regime ($m = 1$ slope) from 30 hr to the end of the tests. This is also confirmed by the derivative of DD2. It is also interesting that the derivatives for both observation wells, shown in Figures 6.10 and 6.11, do not exhibit an infinite-acting radial flow regime. Both derivatives continuously bend as boundary effects are felt at the observation wells. In order to observe an infinite-acting radial flow regime at an observation well, the following inequality (see Equation (2.37) in Chapter 2 and Earlougher (1977)) has to be satisfied:

$$\frac{t_D}{r_D^2} > 100 \quad \text{or} > 10 \text{ with less than 1\% error,} \tag{6.1}$$

where t_D is the dimensionless time which is defined as

$$t_D = \frac{0.0002637kt}{\phi \mu c_t r_w^2}, \tag{6.2}$$

and r_D is the dimensionless radius which is defined as

$$r_D = \frac{d_w}{r_w}, \tag{6.3}$$

and d_w is the distance between the observation well and the active well. If we use the input formation and fluid properties, and the relevant reservoir dimensions given in Table 6.1, $d_w = 900.0$ ft, and the test duration of 40 hr for DD2, we obtain t_D/r_D^2 to be 13, which is much smaller than 100 or almost the same as 10 (just the beginning). This is one of the reasons why we have not observed an infinite-acting radial flow regime for both observation wells. As can be observed from the infinite-acting behaviors of both wells shown in Figures 6.10 and 6.11, we would have observed infinite-acting radial flow regimes after 100 hr for well 1 and 140 hr for well 2. As can also be observed from these figures, the boundary effects are felt by both wells approximately after 4 hr, which is much earlier than the beginning of infinite-acting radial flow regimes, as shown in Figures 6.10 and 6.11.

Based on the interpretation methodology given in this chapter, the next task is to perform pressure-rate and/or pressure-pressure deconvolution to minimize the effects of variable rate history by converting the active well and observation well pressure data to equivalent constant-rate responses. This will enable us to have a much better idea about this fairly simple reservoir/well system. It is a homogeneous, single layer, isotropic, and geometrically bounded reservoir. It is very difficulty to encounter such systems in reality. However, the complexity of this interference test comes from the fact that it is variable rate with a few limited-duration flow periods and for the active well we depend on only buildup test data. As discussed in detail in Chapter 4, the more recent deconvolution algorithm of Pimonov, Ayan, Onur, and Kuchuk (2009a) will be used for this example. Here, for simplicity, we will assume that flow-rate data at the active and initial static pressure are accurately known so that they can be treated as known during the deconvolution process. Of course, in cases where such data are uncertain due to errors, one should consider the general deconvolution methodology given in Chapter 4 to get around to these problems. In addition, we will consider only pressure-rate deconvolution because we assume that flow-rate data at the active well are accurately known.

Figure 6.12 compares the deconvolved derivative for the active well, for which only the BU2 pressures were used for deconvolution, with the DD2 and BU2 derivatives based on the superposition time. Similarly, Figures 6.13 and 6.14 show the deconvolved derivatives for both observation

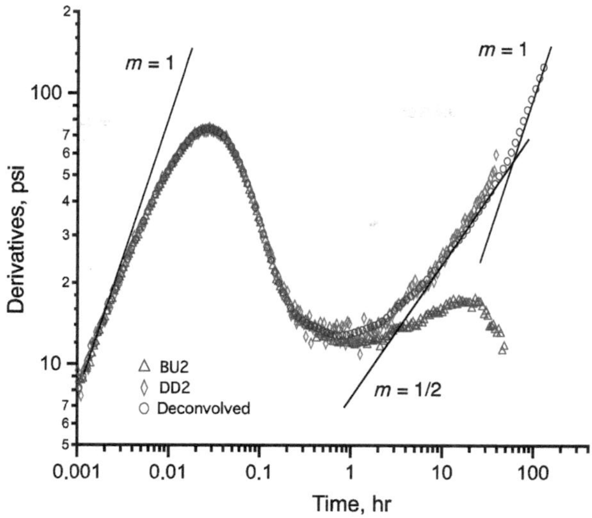

Figure 6.12 Comparison of the deconvolved derivative with the derivatives of the DD2 and BU2 periods for the active well.

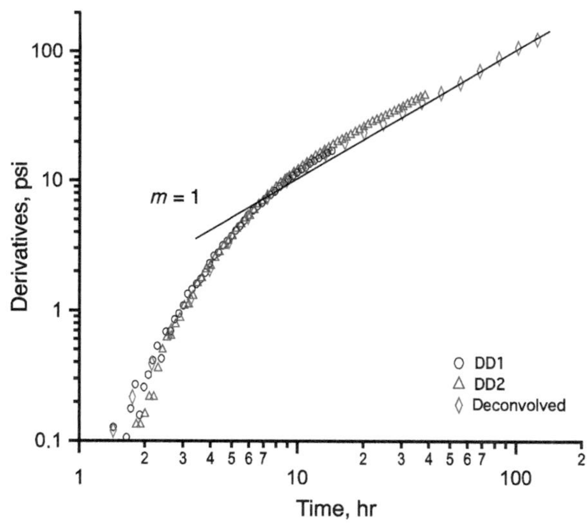

Figure 6.13 Comparison of the deconvolved derivative with the derivatives of the DD1 and DD2 periods for observation well 1.

wells, for which only BU2 pressures were used for deconvolution, in comparison with the DD1 and DD2 derivatives based on the superposition time. All deconvolved derivatives shown in Figures 6.12–6.14 are rate-normalized using the same constant reference rate of 2000 B/D. These figures clearly exhibit the effects of the rate superposition on the drawdown

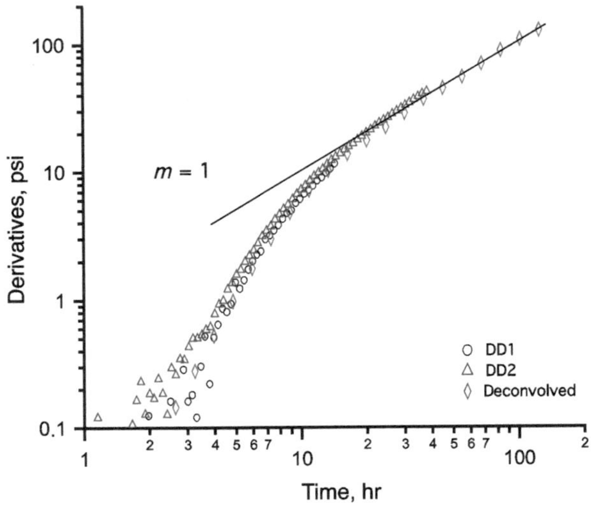

Figure 6.14 Comparison of the deconvolved derivative with the derivatives of the DD1 and DD2 periods for observation well 2.

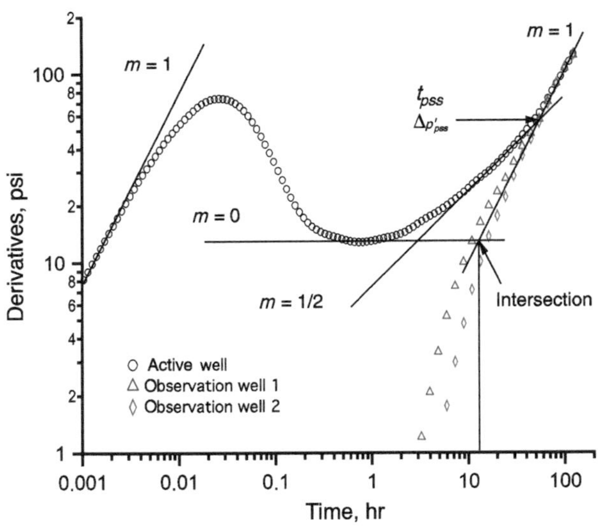

Figure 6.15 Comparison of the deconvolved derivatives for the active well and both observation wells.

and buildup derivatives of these three wells compared to the deconvolved derivative.

Figure 6.15 shows all deconvolved derivatives for the active well and both observation wells. It is clear from Figure 6.15 that deconvolved derivatives for all the three wells merge to the same unit-slope line at late times (for

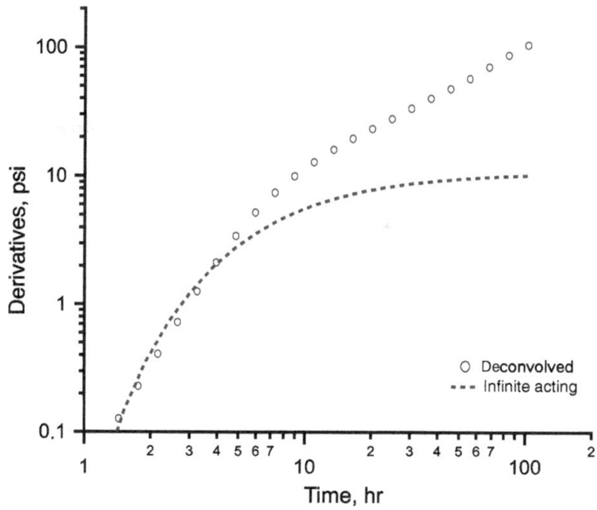

Figure 6.16 Type-curve match of the deconvolved derivative for observation well 1 with the infinite-acting line-source derivative.

times greater than 50 hr), indicating the closed (no-flow boundary) reservoir behavior, from which one can determine the reservoir pore volume. Most importantly, we can estimate most of the well/reservoir parameters from the flow regimes of the deconvolved derivatives shown in Figure 6.15 for this test example. Consequently, it is without doubt that deconvolution provides an excellent tool for identifying the correct well/reservoir model and estimating most of the parameters of interest for this synthetic example.

It should be noted that we can estimate kh/μ and $\phi c_t h$ by type-curve matching the early-time deconvolved derivative data, which are not influenced by reservoir boundaries for both observation wells, with the infinite-acting line-source solution derivative. In the type-curve matching, we did not consider deconvolved derivatives having magnitudes less than 0.1 psi because such data were not reliable due to noise. The best matches obtained are shown in Figures 6.16 and 6.17, and the estimated values of kh/μ and $\phi c_t h$, k and ϕc_t if μ and h are known, and ϕ if c_t is known as shown in Table 6.2. As can be observed from this table, the values of k and ϕc_t estimated by type-curve matching are close to the true values.

From the deconvolved active well derivative, we can estimate the wellbore storage coefficient (C_w) from the unit-slope ($m = 1$ from 0.001 to 0.003 hr), the permeability (k) and skin factor (S) from the infinite-acting radial flow regime ($m = 0$ from 0.4 to 1.5 hr), the channel width (L_y) and the perpendicular distances from the active well to the nearest and farthest no-flow boundaries (L_{y1} and L_{y2}) from the linear flow regime ($m=$ 1/2 from 5 to 30 hr), the total reservoir-pore volume (V_p) from the pseudo-steady state flow regime ($m = 1$ from 50 hr to the end of the test), and the

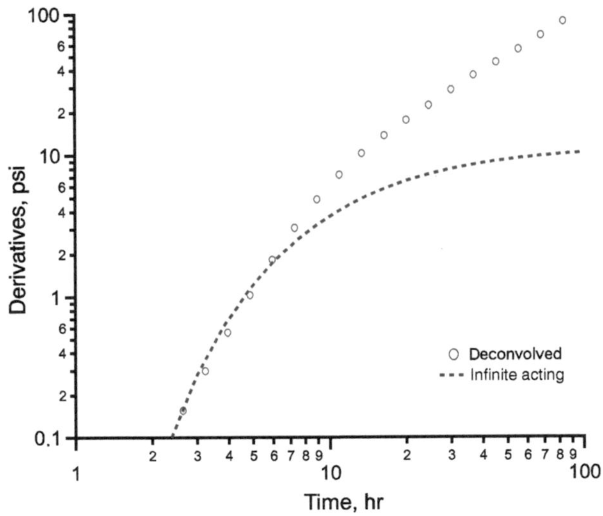

Figure 6.17 Type-curve match of the deconvolved derivative for observation well 2 with the infinite-acting line-source derivative.

Table 6.2 Comparison of the true values with the estimated values from the type-curve matching of the deconvolved observation well derivatives with the derivative computed from the line-source solution at the same well location in an infinite system

Parameters	Unit	Estimated	True
Observation well 1			
kh/μ	md-ft/cp	12800	12000
$\phi c_t h$	ft/psi	1.116×10^{-4}	1.2×10^{-4}
k	md	213	200
ϕc_t	1/psi	1.860×10^{-6}	2.0×10^{-6}
ϕ	fraction	0.186	0.20
Observation well 2			
kh/μ	md-ft/cp	12243	12000
$\phi c_t h$	ft/psi	1.02×10^{-4}	1.2×10^{-4}
k	md	204	200
ϕc_t	1/psi	1.70×10^{-6}	2.0×10^{-6}
ϕ	fraction	0.170	0.20

Table 6.3 Comparison of the parameters estimated from the flow regime analysis of the deconvolved active-well derivative with the true values

Parameters	Unit	Estimated	True
k	md	178.8	200
S (at the active well)	unitless	2.93	4
C_w (at the active well)	B/psi	1.03×10^{-2}	1.0×10^{-2}
L_x (length of the reservoir in the E-W direction)	ft	3995.4	4200
L_y (length of the reservoir in the S-N direction)	ft	939.0	900.00
L_{y1} (perpendicular distance from the active well to the nearest no-flow boundary)	ft	359.2	300
L_{y2} (perpendicular distance from the active well to the farthest no-flow boundary)	ft	579.8	600
V_p (reservoir pore volume)	B	8.02×10^6	8.1×10^6
C_A (Dietz shape factor)		4.145	2.362

side length (L_x) from the values of the channel width estimated from the linear flow regime and the pore volume estimated from the pseudo-steady state flow regime. Table 6.3 compares the values of the parameters estimated from the flow regime analysis of the active-well deconvolved derivative with the true values.

Recently, Whittle and Gringarten (2008) suggested using the start of the pseudosteady-state flow regime or the intersection (Kuchuk, 2009a) shown in Figure 6.15 for calculating the pore volume from derivative and/or deconvolution derivative curves if the system reaches the pseudosteady-state condition. The pore volume of Whittle and Gringarten (2008) can be expressed as

$$V_p = \frac{q t_{pss}}{24 c_t \Delta \acute{p}_{pss}} = \frac{q t_{\text{int}}}{24 c_t \Delta \acute{p}_{\text{int}}}, \tag{6.4}$$

where t_{pss} is the start of the pseudosteady-state flow regime, $\Delta \acute{p}_{pss}$ is the value of the derivative at t_{pss}, t_{int} is the time at which the pseudosteady-state flow unit slope intersects the infinite-acting line (zero slope), and $\Delta \acute{p}_{\text{int}}$ is the value of the derivative of the infinite acting period. These two pore volume expressions given by Equation (6.4) are identical, but in some cases the intersection time (t_{int}) in the second expression is better defined graphically.

The pore volume estimated from Equation (6.4), using the values indicated in Figure 6.15, is 8.02×10^6 barrels, which is the same as the value given in Table 6.3.

In addition to the estimated parameters given in the previous paragraph, we can estimate the Dietz shape factor (C_A) from the radial and pseudo-steady state flow regimes (in combination). The active well location inside the rectangle may be inferred from the Dietz shape factor. However, it should be pointed out that the value of the Dietz shape factor (C_A) computed by this procedure is very sensitive to small errors in slope as well as intercept values determined from the semilog plot of the radial flow regime and the Cartesian plot of the pseudo-steady state flow regime (see Table 6.3).

All these parameters from flow regimes can be accurately estimated if the given input values of porosity, viscosity, thickness, and total compressibility are accurately known *a priori*. An uncertainty in these input parameters will propagate into the estimated parameters. For actual field cases, porosity, viscosity, thickness, and total compressibility values may contain high uncertainties. In many instances, the effective formation thickness could not be obtained reliably from openhole logs if the wellbore production profile is not available. Almost for all exploration DST tests, the production profile is not available. Porosity values from openhole logs are normally reliable and do not have more than a few percent error for most formations. Reliable values of viscosity can be obtained from fluid samples and even in situ by using wireline formation testers. However, the uncertainty in the total system compressibility could be quite large because of uncertainties in reservoir pressure and temperature, remaining (unproduced) hydrocarbon composition, formation rock and connate water compressibilities, subsidence, and thermal compaction or expansion of the formation rocks. In fact, multiwell interference test data may provide a unique opportunity to estimate a reliable ϕc_t product in addition to further verifying the estimated distances to no-flow boundaries and the reservoir pore volume provided that the interference test exhibits both wellbore-storage free early time (before infinite acting period) and the infinite-acting flow periods. It may also be difficult to reliably resolve ϕc_t if these two periods are also affected by the outer reservoir boundaries.

The next interpretation step is to history match (perform nonlinear regression) the active and/or observation well observed pressure data to the reservoir model to estimate a total of 8 well/reservoir parameters, namely k, ϕ, S, C_w, lengths of the reservoir boundaries in the x- and y-direction (L_x and L_y) and the active well coordinates (X_a and Y_a) in the rectangle, by assuming that h and c_t are known. It should be pointed out that in cases where we regress on observation well pressures, the active well coordinates (X_a, Y_a) as well as observation well coordinates (X_{o1}, Y_{o1}, X_{o2}, and Y_{o2}) are unknown. In such cases we will treat only the active well coordinates as unknown in nonlinear regression because the inter-well distance between an

Table 6.4 Initial guesses, and lower and upper bounds of the parameters for the nonlinear regression

Parameters	Unit	Initial guess	Lower limit	Upper limit
k	md	198.6	1	500
ϕ		0.18	0.01	0.4
S (at the active well)		2.93	−2	20
C_w (at the active well)	B/psi	0.0103	0.0	0.10
L_x (length of the reservoir in the E-W direction)	ft	3995.4	100	5000
L_y (length of the reservoir in the S-N direction)	ft	939.0	100	1500
X_a (x-coordinate of the active well measured from the lower-left)	ft	2000	0	[a]
Y_a (y-coordinate of the active well from the lower-left)	ft	359.2	0	[a]

[a] The upper bounds for these parameters are set automatically inside the minimization algorithm to the assumed values of L_x and L_y in each iteration.

active well and observation well is always known, and hence, the coordinates of the active well determines the coordinates of the observation wells.

We will history match pressure data pertaining to only buildup portions for the active well and the entire pressure data sets for both observation wells by using the maximum likelihood estimation (MLE) method discussed in Chapter 5. The reason we use the MLE method is because it better handles multiwell data sets with disparate orders of magnitude or noise (or both) and allows one to treat the data error variances as unknown in history matching if the data error variances are uncertain (see Chapter 5).

Table 6.4 presents the initial guesses along with the lower and upper bounds used for the parameter estimation by the MLE method. The initial guesses for k and ϕ were determined as the arithmetic average of the individual estimates of k and ϕ determined from the type-curve matching of the early-time observation well data and flow regime analyses of the active well data, as given in Tables 6.2 and 6.3. The initial guesses for S, C_w, L_x, L_y, and Y_a were taken as the corresponding estimated values from the flow regime analysis of deconvolved active well data (see Table 6.3). The initial guesses for the active well coordinate in the x-direction were somewhat arbitrarily chosen. When the observation well pressure data are history matched, the active well coordinates are constrained in such a way that the known inter-well distance between the active and observation well remains the same and the observation well coordinates remain inside the rectangle in each iteration during the regression minimization. The lower and upper bounds for the parameters given in Table 6.4 should normally be determined from the prior knowledge of the reservoir and locations of the well. These bounds will be used in the nonlinear regression to be given next.

Six different combinations, which are called options, of the data from all three wells were history matched. The results obtained from six different history matches (nonlinear regressions) based on the various combinations of the active and observation well pressure data sets are summarized in Tables 6.5 and 6.6. The numbers given with ± denote 95% confidence intervals. As can be seen from the last three columns of Table 6.5, the standard deviations of errors in active and observation wells estimated from the MLE are in perfect agreement with input noise levels in each data set. Therefore, the active and observation well pressures as well as pressure derivatives for the drawdown and buildup periods (not shown here) were successfully reproduced for the entire test duration for all options, except Option 3. The nonlinear regression algorithm for Option 3 converged to an unacceptable local minimum, as is evident from the standard deviation of errors estimated from the MLE for the active well and observation pressures, which are considerably higher than the true input values. As discussed in Chapter 5, in fact this could happen sometimes in any gradient-based nonlinear regression algorithm (based on WLS or MLE). Basically, the regression search gets stuck at a local minimum for some sets of initial guesses of the unknown parameters. A critical examination of the matches obtained for pressure data sets, standard deviation of errors, correlation coefficients, and confidence intervals for the unknown parameters usually helps to detect undesirable local minimums. For instance, as can be seen from the history matches of the active well and observation well 2 pressure data, shown in Figures 6.18 and 6.19, and of the active well derivative data for the second buildup (BU2) period and the observation well 2 derivative for the second drawdown (DD2) period, shown in Figures 6.20 and 6.21, we do not have a good match for the active well pressure derivatives from 1 to 10 hr for Option 3. Note also that the confidence intervals for the estimated parameters for this option are significantly larger than those for the other two options. This is an indication that the match obtained for Option 3 is not good and the estimated parameters are doubtful.

As can be observed from Tables 6.5 and 6.6, all history matches essentially yield slightly different values for the reservoir lengths in the x and y directions and the porosity. On the other hand, the estimated reservoir pore volumes, obtained from the estimated L_x, L_y, and ϕ values, are almost identical and equal to the true value given in Table 6.2 for all cases except Option 3. This is not surprising because the pseudo-steady state flow regime has been observed from the derivatives of all three wells and hence the pore volume should be estimated uniquely, although the individual values of L_x, L_y, and ϕ may change slightly. One of the other reasons that we cannot obtain uniquely the individual values of L_x, L_y,

Table 6.5 Summary of the nonlinear regression results

Regression options	k (md)	ϕ	S	C (B/psi)	σ_a[b] (psi)	σ_{o1}[b] (psi)	σ_{o2}[b] (psi)
Option 1[c]	200 ± 0.2	$0.2 \pm 1.5 \times 10^{-4}$	$4.0 \pm 8 \times 10^{-3}$	$0.01 \pm 3 \times 10^{-6}$	0.055	0.01	0.01
Option 2[d]	198 ± 0.2	$0.2 \pm 2 \times 10^{-4}$	$4.0 \pm 8 \times 10^{-3}$	$0.01 \pm 3 \times 10^{-6}$	0.055	0.01	–
Option 3[e]	198 ± 6	$0.195 \pm 8 \times 10^{-3}$	3.9 ± 0.3	$0.01 \pm 1 \times 10^{-4}$	1.8	–	0.05
Option 4[f]	183 ± 1.0	$0.182 \pm 9 \times 10^{-4}$	2.93[a]	0.01[a]	–	0.013	0.012
Option 5[g]	200 ± 0.2	0.18[a]	$3.95 \pm 9 \times 10^{-3}$	$0.01 \pm 4 \times 10^{-6}$	0.056	–	–
Option 6[h]	193 ± 2.5	$0.196 \pm 3.0 \times 10^{-3}$	2.93[a]	0.01[a]	–	0.011	–

[a] Parameter fixed in regression.

[b] Subscripts $_a$, $_{o1}$, and $_{o2}$ denote the active well, and observation wells 1 and 2, respectively.

[c] Option 1 = active well and both observation wells.

[d] Option 2 = active well and observation well 1.

[e] Option 3 = active well and observation well 2.

[f] Option 4 = both observation wells.

[g] Option 5 = active well.

[h] Option 6 = observation well 1.

Table 6.6 Summary of the nonlinear regression results (continuation of Table 6.5)

Regression options	L_x (ft)	L_y (ft)	X_a (ft)	Y_a (ft)	X_{o1}[a] (ft)	Y_{o1}[a] (ft)	X_{o2}[a] (ft)	Y_{o2}[a] (ft)
Option 1[b]	4199 ± 0.4	899.5 ± 0.7	2099.6 ± 0.05	299.4 ± 0.8	2964	549	902	374
Option 2[c]	4201 ± 1.1	899 ± 0.8	2101 ± 0.5	299.6 ± 0.8	2966	550	–	–
Option 3[d]	4188 ± 62	989 ± 33	1928 ± 17	244 ± 24	–	–	730	319
Option 4[e]	4204 ± 0.7	989 ± 5	2104 ± 0.02	717 ± 2.6	2969	642	906	467
Option 5[f]	4472 ± 25	950 ± 0.8	2062 ± 35	315 ± 1.0	–	–	–	–
Option 6[g]	4165 ± 7	919 ± 15	2056 ± 4	407 ± 6	2920	657	–	–

[a] Computed from the estimated values of X_a and Y_a.
[b] Option 1 = active well and both observation wells.
[c] Option 2 = active well and observation well 1.
[d] Option 3 = active well and observation well 2.
[e] Option 4 = both observation wells.
[f] Option 5 = active well.
[g] Option 6 = observation well 1.

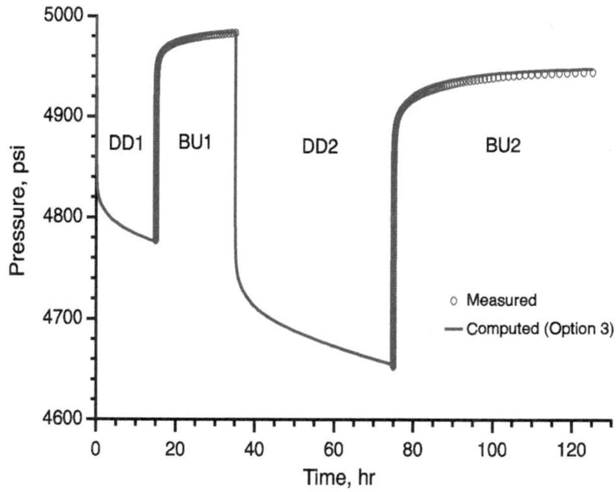

Figure 6.18 Comparison of the measured and computed (based on Option 3) shut-in pressures of the first and second buildup periods for the active well.

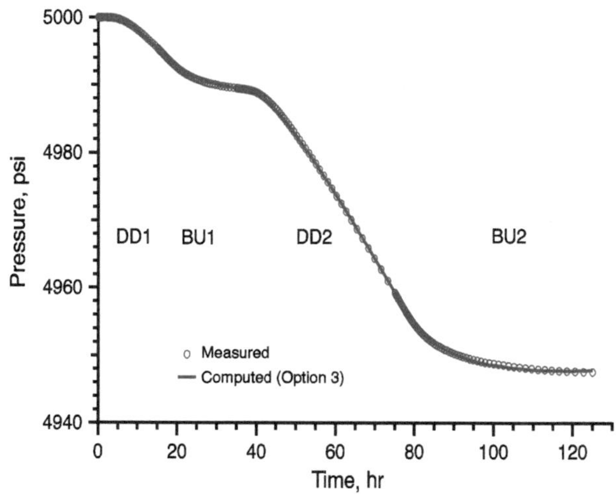

Figure 6.19 Comparison of the measured and computed (based on Option 3) shut-in pressures of the first and second buildup periods for observation well 2.

and ϕ is because they are highly correlated. This correlation may introduce many local minima on the surface of the objective function. Therefore, the matrix of correlation coefficients for the estimated parameter pairs should be scrutinized to understand the correlations among the parameters. For instance, Table 6.7 shows the matrix of correlation coefficients for Option 3. As can be observed from this table, ϕ is strongly correlated (with a correlation coefficient of nearly -1) with L_x and L_y. This means that we can obtain the

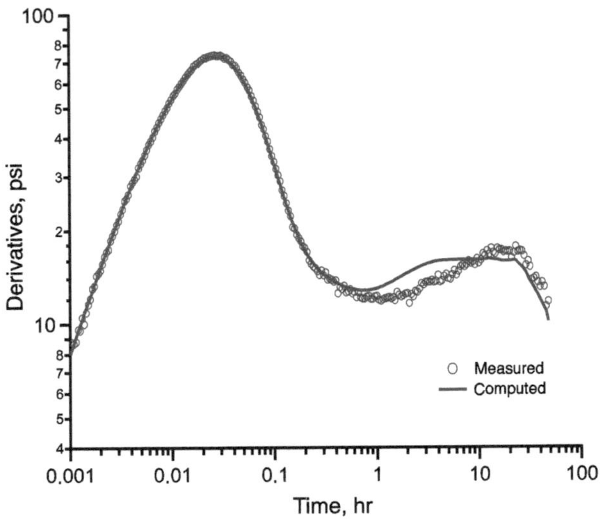

Figure 6.20 Comparison of the measured and computed (based on Option 3) BU2 derivatives for the active well.

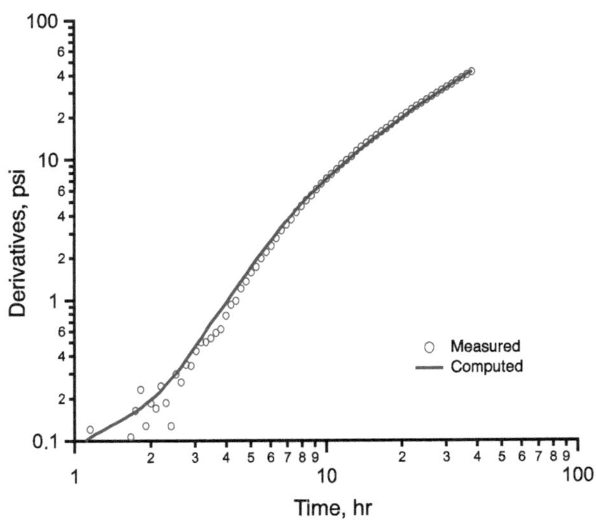

Figure 6.21 Comparison of the measured and computed (based on Option 3) DD2 derivatives for observation well 2.

same observation well pressure if we increase L_x (or L_y) while we decrease ϕ. Similarly, k and ϕ are strongly correlated (with a correlation coefficient of unity) as shown Table 6.7. This is why we should analyze flow regimes prudently. For instance, permeability k obtained from the semilog straight line using derivative or Horner plots is not correlated with porosity ϕ.

Table 6.7 Matrix of correlation coefficients of the estimated parameters from Option 3 in Table 6.5

	L_x (ft)	L_y (ft)	X_a (ft)	Y_a (ft)	k (md)	ϕ	S	C_w (B/psi)
L_x	1.00	0.90	0.83	0.66	−0.66	−0.93	−0.69	−0.43
L_y		1.00	0.78	0.83	−0.83	−0.99	−0.85	−0.51
X_a	–	–	1.00	0.33	−0.30	−0.85	−0.35	−0.20
Y_a	–	–	–	1.00	−0.98	−0.76	−0.97	−0.54
k	–	–	–	–	1.00	0.75	1.00	0.61
ϕ	–	–	–	–	–	1.00	0.78	0.46
S	–	–	–	–	–	–	1.00	0.60
C_w	–	–	–	–	–	–	–	1.00

Finally, it is clear from the regression results given in Tables 6.5 and 6.6, including observation well data in history matching process improves the accuracy of estimated parameters and their confidence intervals. When only an active or observation well pressure data were history matched, some of the parameter(s) were fixed, as shown in Tables 6.5 and 6.6, because the data set has no significant sensitivity to these parameters and/or is exactly correlated with some other unknown parameters. For example, it is well known that skin and porosity cannot be uniquely resolved from the active well pressure data as these two parameters are strongly correlated. Hence, we fixed the porosity value obtained from other sources. If we do treat these two parameters as unknown and attempt to estimate both of them from a single well test pressure data (Option 5 in Tables 6.5 and 6.6), the parameter search may be stuck in a local minima from which it cannot escape. Similarly, if only a single (Option 6) or two observation (Option 4) well pressure data were history matched, we fixed the wellbore storage and skin at the active well at their best values because the observation well pressures from this interference test do not show any significant sensitivity to these two parameters.

6.4. Horizontal Well Test Interpretation of a Field Example

This example given by Kuchuk (1997) consists of two buildup well tests conducted for a horizontal Well X-184 in a carbonate reservoir. The field had been producing for a little more than 10 years when Well X-184, the first horizontal well in the field, was drilled. The drilled length of the well is about 1296 ft with a whole diameter of 7 1/2 inch and an acid-wash

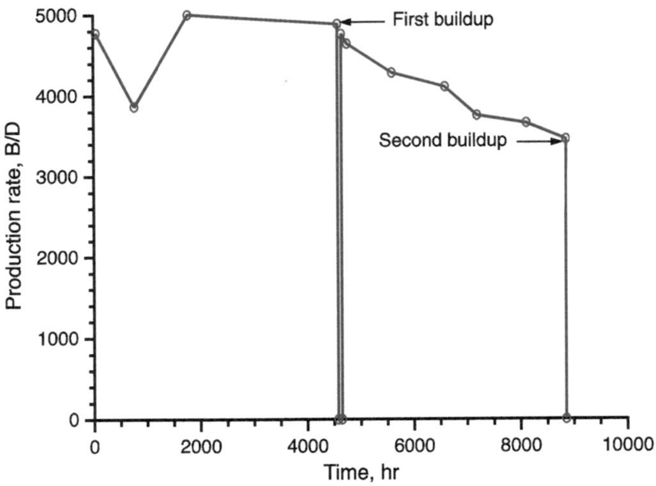

Figure 6.22 Flow-rate history for Well X-184.

barefoot completion. The standoff (z_w) is about 34 ft from the bottom. For the first buildup test, Well X-184 was shut in at the surface for almost 3 days after producing for approximately 190.54 days (4573 hr). The production history is shown in Figure 6.22 (the circles denote the actual surface test-tank rate measurements but converted to reservoir barrel per day). It should be pointed out that no drawdown test was conducted before both buildup tests. Both tests were conducted after long production periods. The stabilized flow rate before the first buildup test (BU1) was 4876 B/D (all rates given here are in reservoir B/D, not STB/D). After producing for another six months (4279 hr), as shown in Figure 6.22 (during which the production rate declined from 4876 to 3454 B/D), the well was shut in for a second buildup test (BU2) at downhole for 4.5 days in order to assess the production decline.

Although the basic openhole logs for resistivity and porosity were run in the pilot-hole section (from the top to bottom of the formation) of the well, with the exception of the LWD Gamma Ray, no openhole log was run in the horizontal section of the well. Therefore, a full characterization of the lateral heterogeneity along the wellbore was not possible. As shown in Figure 6.23, the formation (pay zones) consists of three distinct layers (flow units), and is bounded by the overlain and underlain thick non-permeable anhyritic zones. The openhole-log average porosities of the top and bottom layers (Layer 1 and 3 in Figure 6.23) are about 25.5% and 25%, respectively. The neighboring wells horizontal core permeabilities for these two layers are between 50 md to 300 md, and 20 md to 70 md for the vertical permeabilities. These two layers are predominantly limestone with some dolomitic patches.

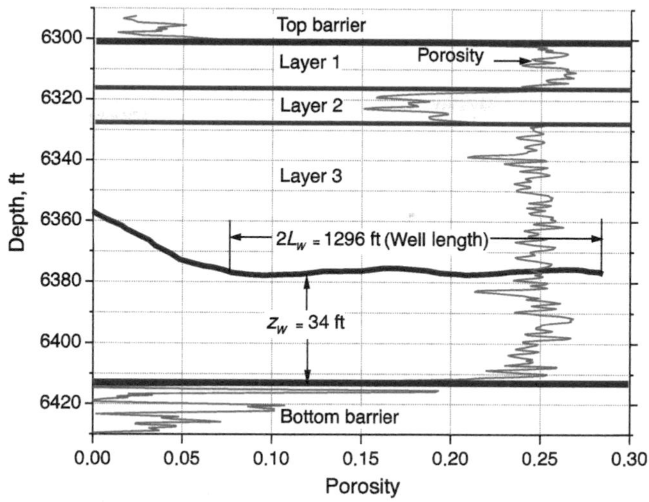

Figure 6.23 Reservoir model and log porosity for Well X-184.

The middle layer, with an average 17% porosity, as shown in Figure 6.23, consists mainly of dolomatic carbonate and is known to be tight and patchy but correlatable throughout the field. Its core horizontal and vertical permeabilities from the nearby wells are between 0.01 md to 1 md. In addition to the field geology, openhole logs from the vertical pilot hole of the horizontal Well X-184 and openhole logs and cores from nearby vertical wells are used to develop a reservoir model for the interpretation of these two buildup tests. However, given the uncertainties in vertical communication among the layers (flow units), we will at least consider two plausible models for interpretation. The first model (Model 1) consists of only the bottom layer with no-flow boundaries at the top and bottom, which implies that the middle tight layer is a barrier. The second model (Model 2) assumes that all three layers communicate with the wellbore, i.e., the vertical permeabilities of all their layers are not zero or very small.

The formation and fluid properties, and the other relevant information for Well X-184 are given in Table 6.8. It should be pointed out that the first layer fluid properties given in the table were the initial values. As will be discussed later, because of a possible presence of a second gas cap in the first layer, its fluid properties could be different and will be treated as unknowns in the history matching. This implies that the gas migrated into the first layer during the production period of this well, because the openhole logs did not show any presence of gas. Figure 6.23 shows the well trajectory, the porosity log from the pilot hole, and the layer definitions, which are determined from open-hole logs with the local geological knowledge and core data.

Figure 6.24 presents pressure data for the first and second buildup tests. As can be seen from this plot, the behavior of these two buildup tests are quite

Table 6.8 Formation and fluid properties, and well geometric dimensions for Well X-184

ϕ_1	fraction	0.255
ϕ_2	fraction	0.17
ϕ_3	fraction	0.25
h_1	ft	16.2
h_2	ft	9.5
h_3	ft	84.7
$c_{t1} = c_{t2} = c_{t3}$	psi^{-1}	5.7×10^{-5}
$\mu_1 = \mu_2 = \mu_3$	cp	0.78
r_w	ft	0.354
z_w	ft	34.0
L_w	ft	648

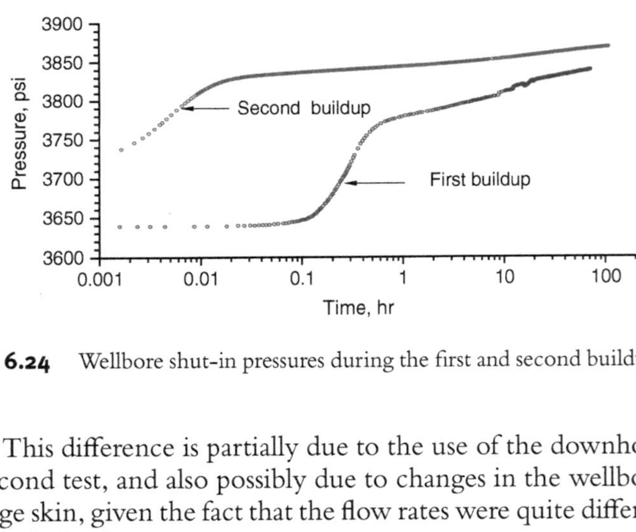

Figure 6.24 Wellbore shut-in pressures during the first and second buildup tests.

different. This difference is partially due to the use of the downhole shut-in for the second test, and also possibly due to changes in the wellbore storage and damage skin, given the fact that the flow rates were quite different before the shut-in for both tests.

6.4.1. First buildup test (BU1) interpretation

Figure 6.25 presents the pressure change and its derivative with respect to the multirate superposition time for the first buildup. The pressure change and derivative exhibit a wellbore storage dominated flow period, during which the slopes of both curves are much greater than the unit slope. This is most likely due to a rapid variation of the compressibility of the wellbore fluid because of the strong presence of gas in the production string. Therefore,

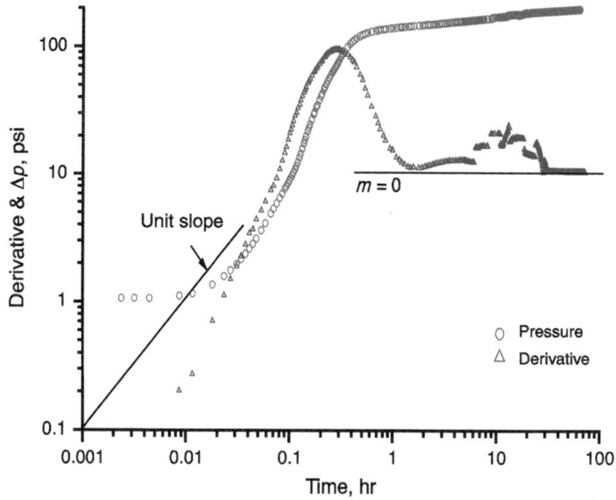

Figure 6.25 The pressure change and its derivative for the first buildup test.

the wellbore-storage-dominated flow period will not fit a constant wellbore-storage reservoir model.

The pressure derivative stabilizes after 1 hr, and then it fluctuates from 7 to 25 hr, as shown in Figure 6.25. This could be due to an additional water and/or gas flow in the opposite directions in the wellbore. In horizontal wells, it is common to have water humps and gas traps along the wellbore during production, and back flow of this water and gas into the formation during buildup tests. As shown in Figure 6.25, the pressure derivative becomes almost flat, indicating a possible radial flow regime at about 25 hr. Note that the value of the derivative is slightly smaller than that of the earlier stabilized period. Because horizontal wells, in general, may exhibit many different flow regimes, without a careful analysis it may be impossible to classify this flat portion of the derivative as an indication of a radial flow regime.

Because Well X-184 was approximately completed in the middle of the bottom layer, we may therefore expect to observe only possible first and third radial flow regimes. We should not expect to observe the intermediate-time linear flow regime because the inequality

$$L_w \geq 10h\sqrt{\left(\frac{k_h}{k_v}\right)^{2/3}}, \tag{6.5}$$

given by Ozkan (2009) has to be satisfied. If we assume that $\frac{k_h}{k_v}$ is 1, which is very small compared to core values, the value of L_w obtained from Equation (6.5) would be 847 ft, which is larger than the drilled L_w value of 648 ft.

We used the third layer thickness (h) of 84.7 ft in Equation (6.5) to obtain L_w. Therefore, we should not expect to observe an intermediate-time linear flow regime for this well because the inequality given by Equation (6.5) is not satisfied.

Let us suppose that the stabilized-derivative shown in Figure 6.25 is the first radial flow regime with a semilog-straight line slope of 23.03 psi/cycle. The first radial flow regime semilog straight line slope can be expressed (Kuchuk, Goode, Wilkinson, & Thambynayagam, 1991) as

$$m_{r1} = \frac{162.6q\mu}{2\sqrt{k_h k_v} L_w}, \tag{6.6}$$

where $\sqrt{k_h k_v}$, k_h, and k_v are the geometric, horizontal, and vertical permeabilities, respectively, L_w is the half well length, and the damage skin as

$$S = 1.151 \left[\frac{\Delta p_{1hr}}{m_{r1}} + 3.2275 + 2 \log \frac{1}{2} \left(\sqrt[4]{\frac{k_h}{k_v}} + \sqrt[4]{\frac{k_v}{k_h}} \right) - \log \left(\frac{\sqrt{k_h k_v}}{\phi \mu c_t r_w^2} \right) \right], \tag{6.7}$$

and $\Delta p_{1hr} = p_{wf}(\Delta t = 1hr) - p_{wf}(\Delta t = 0)$ for buildup tests. Δp_{1hr} should be obtained from the Horner or Horner superposition plot.

The geometric mean permeability $\sqrt{k_h k_v}$ is calculated from Equation (6.6) and the semilog-straight line slope of 23.03 psi/cycle to be 20.73 md. As can be seen from Figure 6.25, the apparent first radial flow regime continues to the end of the test (72 hr), which means that the well pressure behavior has not yet been affected by the top or bottom boundary effect during the entire test period. We can use this information to obtain a maximum vertical permeability (k_v) (Kuchuk et al., 1991) from

$$k_v = \frac{\phi \mu c_t}{0.0002637 \pi t_{snbe}} \min\{z_w^2, (h - z_w)^2\}, \tag{6.8}$$

where t_{snbe} is the time to feel the nearest boundary effect (the time of onset of the deviation of the pressure or pressure derivative from the first radial flow regime). From this equation, the maximum vertical permeability (k_v) is calculated to be 0.21 md. From the geometric mean and vertical permeability, a horizontal permeability (k_h) of 1993 md is obtained. It should be pointed out that a two-Darcy horizontal permeability is too large for this formation, and it is unlikely to have an anisotropy ratio ($\frac{k_h}{k_v}$) of 9254, where core anisotropy ratios are between 2 and 15. Using these permeabilities, and m_{r1} and Δp_{1hr} from Figure 6.25, the damage skin is obtained from Equation (6.7) to be about 5.

Let us suppose the stabilized-derivative shown in Figure 6.25 is not the first radial flow regime, but it is the third radial flow regime from which the semilog slope and skin expressions (Kuchuk et al., 1991) are given as.

$$m_{r3} = \frac{162.6q\mu}{k_h h} \tag{6.9}$$

and the damage skin

$$S = 2.303\sqrt{\frac{k_v}{k_h}\frac{L_w}{h}}\left[\frac{\Delta p_{1hr}}{m_{r3}} - \log\left(\frac{k_h}{\phi\mu c_t L_w^2}\right) + 2.5267\right] - S_z, \tag{6.10}$$

where

$$S_z = -2.303\log\left[\frac{\pi r_w}{h}\left(1+\sqrt{\frac{k_v}{k_h}}\right)\sin\left(\frac{\pi z_w}{h}\right)\right] - \sqrt{\frac{k_h}{k_v}\frac{h}{L_w}}\left(\frac{1}{3} - \frac{z_w}{h} + \frac{z_w^2}{h^2}\right). \tag{6.11}$$

The horizontal permeability (k_h) is calculated from Equation (6.9) using the semilog-straight line slope of 23.03 psi/cycle and the third layer thickness (h) of 84.7 ft to be 317 md. If the stabilized-derivative shown in Figure 6.25 is the third radial flow regime, this implies that the vertical permeability is very high such that the effects of the upper and lower boundaries are felt by the well before 1 hr, at which the wellbore storage period ends. Therefore, the vertical permeability and damage skin cannot be obtained. The start of the third radial flow regime is given by Ozkan (2009) as

$$t \geq \max\left\{\frac{988\phi\mu c_t(2L_w)^2}{k_h}, \frac{2515\phi\mu c_t h^2}{k_v}\right\}. \tag{6.12}$$

Using the horizontal permeability of 317 md, the start of the third radial flow regime is estimated from Equation (6.12) to be 29 hr. The second expression given in Equation (6.12) cannot be used to calculate the start of the third radial flow regime because we cannot obtain the vertical permeability if we assume that the stabilized-derivative shown in Figure 6.25 is the third radial flow regime.

Figure 6.26 presents the shut-in pressure as a function of the Horner time for the first buildup test. The semilog straight-line slope, as shown

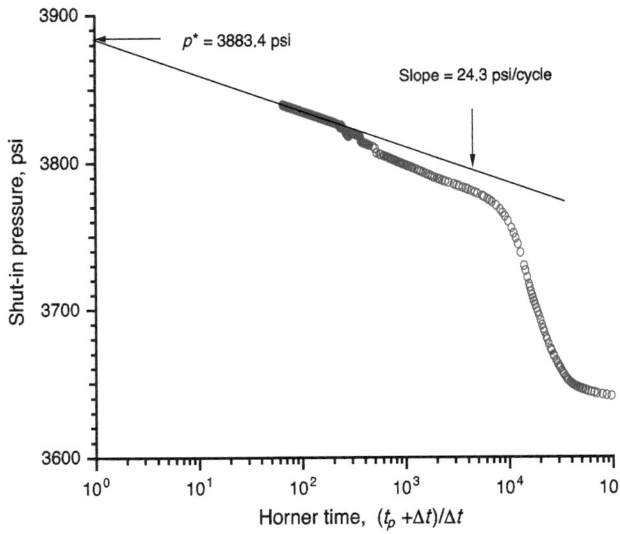

Figure 6.26 Horner plot of the shut-in pressure for the first buildup test.

in Figure 6.26, is almost the same as that from the derivative plot. The extrapolated pressure, $p*$, obtained from the Horner semilog straight line is 3883.4 psi.

Therefore, the first buildup test data are inconsistent because the test derivative behavior is strange, mostly because the surface shut-in and estimated parameters from the flow regime are contradictory. As we have shown in Figure 1.4 (Chapter 1) and Figure 2.10 (Chapter 2), downhole shutting is critical for all wells, and particularly for horizontal wells. Because of these reasons, we have not attempted to perform deconvolution for the first buildup test.

The above identification and flow regime analysis do not lead to a well-defined reservoir model due to a variable wellbore storage dominated flow period and inconsistencies in flow regimes. As a final interpretation step, we perform a history matching (a nonlinear estimation) with the Model 1 (M1) described above. For the history match, the prior production history before the buildup is partially accounted for because it was not measured frequently. Figure 6.27 presents the match of the measured pressure and its derivative with the model response (Model 1); it is an inferior and unsatisfactory match. The estimates from this match are $k_h = 123.5$ md, $k_v = 107.9$ md, $S = 38.9$, $2L_w = 1296$ ft (the wellbore length), and $C = 0.63$ B/psi. Note that the estimated vertical permeability and skin are very high, and that the wellbore length value has not changed. The estimated wellbore storage, C, is too large for this well. As also shown in Figure 6.27, the history match with Model 2 is also very inferior.

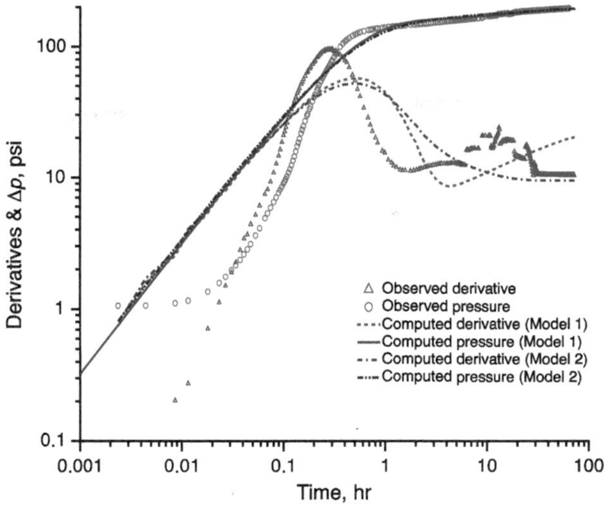

Figure 6.27 Comparison of the measured and computed pressures and their derivatives of the first buildup test.

The estimation of the crucial reservoir parameters, such as the permeabilities, skin, and effective well length, cannot be done satisfactorily from the first buildup, because the test is too short and the early time data is dominated by the variable wellbore storage and phase segregation. Although we could improve the appearance of the match by using a variable storage model, it would not improve the estimates of the reservoir parameters. In the following section, we will present an interpretation for the second buildup.

6.4.2. Second buildup test (BU2) interpretation

As we stated above, the second buildup was conducted after six months (4279 hr from the first test, for which the well was shut in downhole for 4.5 days. It should be stated that the production declined from 4876 to 3454 B/D) between the first test and the second test. Figure 6.28 presents the derivatives with respect to the multirate superposition time for the first and second tests. The derivatives were rate-normalized by a reference flow rate of 3454 B/D, which is the rate prior to the second buildup period. As shown in Figure 6.28, the derivatives of these two tests are quite different, and comparison of these derivative signatures shows very clearly the usefulness of the downhole shut-in for testing horizontal wells. As can be seen from the BU2 derivative plot in Figure 6.28, the wellbore storage effects die out at about 0.06 hr. For times less than 0.0004 hr, the BU2 derivative exhibits a period during which its slope is greater than the unit slope. In the time interval from 0.0004 to 0.003 hr, the BU2 derivative displays a unit-slope ($m = 1$) wellbore-storage dominated flow period. The wellbore storage

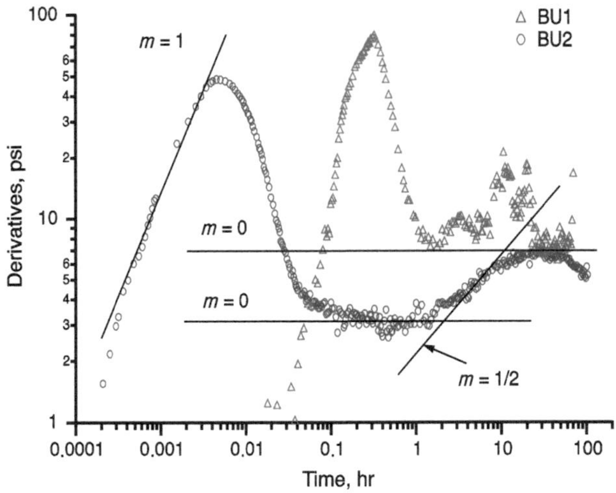

Figure 6.28 The derivatives for the first and second buildup tests.

calculated from this unit slope is 0.01 B/psi, which compares well with 0.008, which is computed from the wellbore volume below the downhole shut-in tool. In the time interval from 0.1 to 1 hr, the derivative is almost flat, indicating an early-radial flow (first radial flow regime, $m = 0$). In the time interval from 3 to 8 hr, the derivative exhibits a not well defined half-slope line ($m = 1/2$), which could be an indication of an intermediate linear flow. As can be seen in Figure 6.28, the BU2 derivative flattens ($m = 0$) in the time interval from 15 to 60 hr before it goes down, probably due to the presence of gas in the first (top) layer, where a secondary gas cap has been observed in some of the neighboring vertical wells. It is also possible that the top layer may have a high permeability due to fractures, which is not unusual in this field. This will be further investigated by performing deconvolution and nonlinear regression in the following paragraphs.

Next, the pressure-rate deconvolution will be performed to minimize the effects of variable rate history by converting the second buildup response to an equivalent constant-rate response. However, there are several issues that should be mentioned before deconvolution of the second BU test data. First, we do not know the initial (static) reservoir pressure. From Chapter 4, we know that uncertainty in initial reservoir pressure significantly affects the late-portion of the deconvolved response (see Chapter 4). Therefore, it will be very difficult to conclusively interpret the late-portion of the reconstructed deconvolved response for the second BU test. Second, the first buildup test is not consistent with the second buildup test due to different wellbore storage effects and skin in both tests, and hence we cannot use a large portion of the 3-day long first (surface shut-in) buildup pressure data, except maybe the buildup pressure data after 30 hr, in deconvolution.

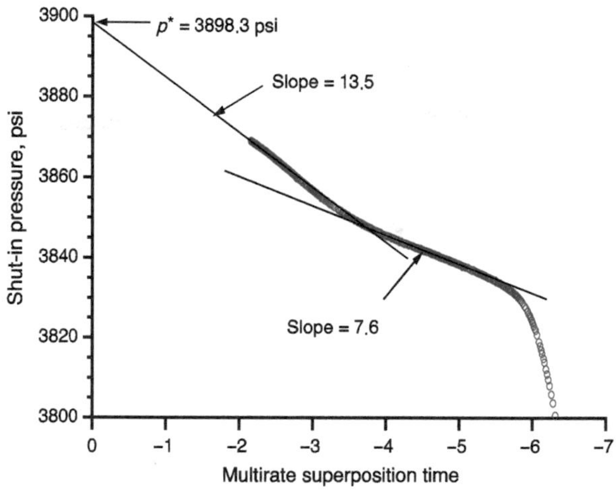

Figure 6.29 Multirate Horner plot for the BU2 pressure data.

Hence, we will be mainly limited to the second buildup pressure data in deconvolution. Unfortunately, the duration of the second buildup test is quite short (although this is considered a long test in the conventional perception) compared to the entire span of the whole flow-rate history over which we wish to reconstruct the constant-rate drawdown response. As discussed in Chapter 4, this will introduce large time gaps in the deconvolved response to be reconstructed. In other words, we have a 4.5-day long buildup data, and using such short buildup pressures we wish to construct a constant-rate drawdown response for about 373 days. Despite all these shortcomings, we will attempt to perform deconvolution and investigate whether we can obtain some meaningful results from the deconvolved (reconstructed) constant-rate response.

In an attempt to have an estimate of initial static pressure to be used in deconvolution, we considered the use of both the multirate Horner plot (Figure 6.29) and the inverse time plot (Figure 6.30) of Kuchuk (1999) for the second buildup test. As can be seen from Figures 6.29 and 6.30, the extrapolated pressures for an infinite shut-in time are 3898.30 psi and 3874.00 psi, respectively, and differ by 24.30 psi. Strictly speaking, both extrapolated pressures are approximations to and may not be representative of the unknown initial pressure because the underlying model for both extrapolations are not confirmed by the BU2 derivative. However, we have no other sources from which to obtain an estimate of the initial pressure to be used in deconvolution, considering the fact that we are limited to only a single meaningful buildup test.

We will use both extrapolated pressure values for deconvolution. The BU2 pressure data earlier than 0.007 hr will not be used for deconvolution because the derivative earlier than this period is not consistent with

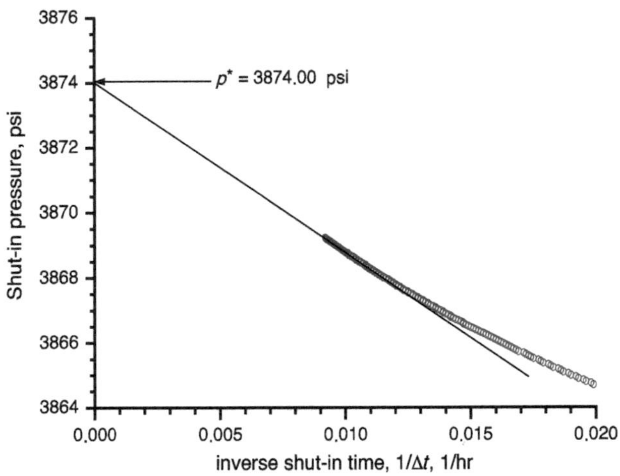

Figure 6.30 Inverse shut-in time plot of Kuchuk (1999) for the BU2 pressure data.

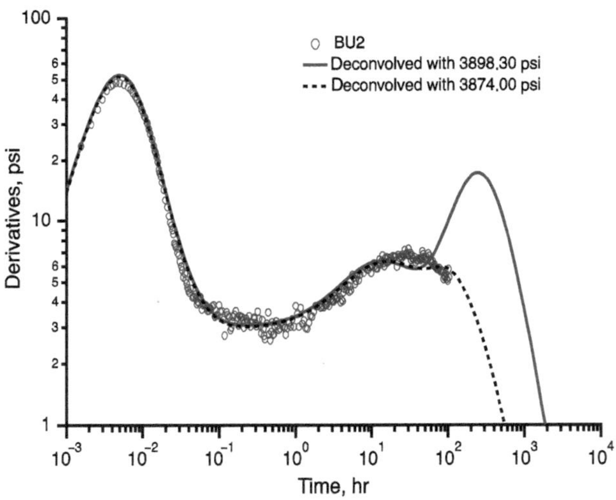

Figure 6.31 Comparison of BU2 and deconvolved derivatives with two different initial (extrapolated) reservoir pressures.

the unit-slope wellbore storage model. The flow-rate history shown in Figure 6.22 is assumed to be accurate and will be treated as known. Figure 6.31 shows the reconstructed deconvolved responses for the two different values of initial (extrapolated) reservoir pressures in comparison with the BU2 derivative. As can be observed in this figure, the late-portion of the deconvolved derivatives are significantly affected by the initial reservoir pressures of 3874.00 psi and 3898.30 psi. Figure 6.32 compares

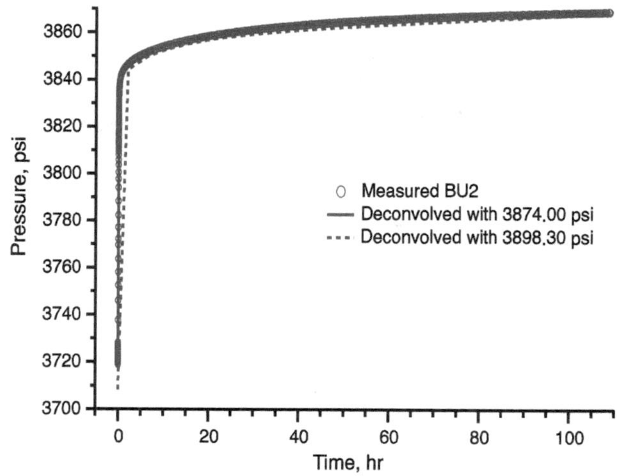

Figure 6.32 Comparison of measured BU2 and reconstructed pressures with two different initial (extrapolated) reservoir pressures.

the BU2 pressure with the reconstructed deconvolved pressures obtained using the two different values of the above initial reservoir pressures. The initial reservoir pressure of 3874.00 psi reconstructs the measured BU2 pressures quite well (with an average RMS value of 0.06 psi) compared to the other initial pressure of 3898.30 psi, which yields an average RMS value of 0.5 psi. We, therefore, think that the deconvolution derivative with the initial reservoir pressure of 3874.00 psi is much more credible than the other case. Furthermore, the fact that a secondary gas cap exists in some of the neighboring vertical wells, the deconvolved derivative with 3874.00 psi, shown in Figure 6.31, is much more representative of the system. It should be stated that unlike the above interference test example, the deconvolved derivative did not improve the BU2 derivative considerably. However, it provides better confidence on the BU2 derivative and further reinforces the presence of the secondary gas cap, although the uncertainty on the initial reservoir pressure remains.

The BU2 and deconvolved derivatives exhibit the same behavior as shown in Figure 6.31 until the end of the BU2 test. Therefore, we use the BU2 derivative for flow regime analysis. As shown in Figure 6.28 and also in Figure 6.31, the BU2 derivative exhibits a first radial flow regime with a semilog straight-line slope of 6.9 psi/cycle because Well X-184 is approximately completed in the middle of the bottom layer, the wellbore storage effect ends about before 0.1 hr, and the effects of the upper and lower boundaries start about 1 hr. Perhaps that is why the BU1 derivative did not show the effects of the upper and lower boundaries because the wellbore storage effect ended before 1 hr for the BU1 test, as observed in Figure 6.28. The first radial flow regime is also observed well in the multirate

Horner plot shown in Figure 6.29. Using the first radial flow regime semilog straight slope from Figure 6.31 and the horizontal well length in Equation (6.6), $\sqrt{k_h k_v}$ is obtained to be 48.93 md. The vertical permeability (k_v) is obtained from Equation (6.7) using $t_{snbe} = 1$ hr and $z_w = 34$ ft to be 15.50 md. From the geometric mean and vertical permeability, a horizontal permeability (k_h) of 154 md is obtained. It should be pointed out that there are large uncertainties in $\sqrt{k_h k_v}$ and k_h estimates because the effective well length is not known accurately, except the drilled length which we use for calculation of these parameters. From Equation (6.7), we obtain a damage skin of 15.86 using $k_h = 154$ md and $k_v = 15.50$ md.

The end of the first radial flow regime is given by Ozkan (2009) as

$$t \leq \frac{\phi \mu c_t}{0.0002637} \min \left\{ \frac{L_w^2}{k_h}, \frac{z_w^2}{5k_v} \frac{(h - z_w)^2}{5k_v} \right\}. \tag{6.13}$$

Using the above horizontal and vertical permeabilities and other parameters given in Table 6.8, the end of the first radial flow regime is estimated from Equation (6.13) to be 0.7 hr, which is comparable to t_{snbe} (the beginning of the upper or lower boundary effect) = 1 hr. As also shown in Figure 6.28, the BU2 derivative does not exhibit a linear flow regime. This observation is consistent with the condition given in Equation (6.5) for the observability of a linear flow regime.

As shown in Figure 6.28, the derivative becomes flatter after 20 hr and then at 50 hr goes downward. Therefore it is not a fully developed third radial flow regime before the effects of the upper layers, although the multirate Horner plot, shown in Figure 6.29, seems to exhibit a third radial flow regime for BU2. The horizontal permeability is estimated from Equation (6.9) to be 323 md using the third radial flow regime semilog straight-line slope of 16 psi/cycle. Using the horizontal permeability of 323 md, and geometric mean permeability of 48.93 md from the first radial flow, we obtain k_v of 7.4 md. We can also estimate a vertical permeability k_v of 15.50 md from the start of the bottom (nearest) boundary effect, which is clearly identified on the derivative plot, as shown in Figure 6.28, and Equation (6.8), which is independent of the horizontal well length. Using Equation (6.10), we obtain a damage skin of 13.5 using $k_h = 323$ md and $k_v = 15.50$ md (more reliable $\sqrt{k_h k_v}$ values). Table 6.9 compares the estimated parameters from the flow regime analysis of BU1 and BU2.

Next, we perform nonlinear regression analysis of the data. As indicated above, the derivatives of BU2 and deconvolution do not exhibit a single-layer model behavior. For the nonlinear regression, we, therefore, will use a three-layer crossflow model (Model 2) identified from openhole logs and partially from the derivatives of BU2 and deconvolution. Each layer is assumed to extend to infinity in the lateral directions. We will regress on the total of twelve parameters, namely, horizontal and vertical permeability (k_h and k_v)

Table 6.9 Comparison of the estimated parameters from BU1 and BU2 flow regime analyses

Parameters	Unit	BU1	BU2
$\sqrt{k_h k_v}$ (FRFR[a])	md	20.73	48.93
k_h (FRFR)	md	1993	154
(TRFR[b])		317	323[c]
k_v (FRFR)	md	0.21	15.50[c]
(TRFR)			7.4
S (FRFR)		5	15.86
(TRFR)			13.50[c]
C	B/psi	0.63	0.01[c]

[a] FRFR = obtained from the first radial flow regime.
[b] TRFR = obtained from the third radial flow regime.
[c] More reliable values because the flow regimes were well defined.

Table 6.10 Initial guesses, lower and upper constraints used for the parameters to be estimated by nonlinear regression

Parameters	Unit	Initial guess	Lower limit	Upper limit
k_{h1}	md	418	4	600
k_{v1}	md	35	3	100
μ_1	cp	0.019	0.01	0.8
c_{t1}	B/psi	2.6×10^{-4}	6.0×10^{-5}	2.6×10^{-3}
k_{h2}	md	0.39	0.05	10
k_{v2}	md	0.14	0.05	10
k_{h3}	md	323	100	600
k_{v3}	md	15.50	5	100
S		13.50	−2	30
C	B/psi	0.01	0.0	0.10
L_w	ft	486	200	650
p_o	psi	3874	3868	4000

for each layer, viscosity of the first layer (μ_1), total compressibility of the first layer (c_{t1}), horizontal half well-length L_w, wellbore storage coefficient (C), the mechanical skin (S), and the initial pressure (p_o). The other parameters of the fluid, formation, and the location of the well are known and given in Table 6.8. The more reliable parameter values in Table 6.9 will be used as initial guesses. Table 6.10 gives all the initial guesses and the lower and upper bounds of the parameters to be used in the nonlinear regression estimation.

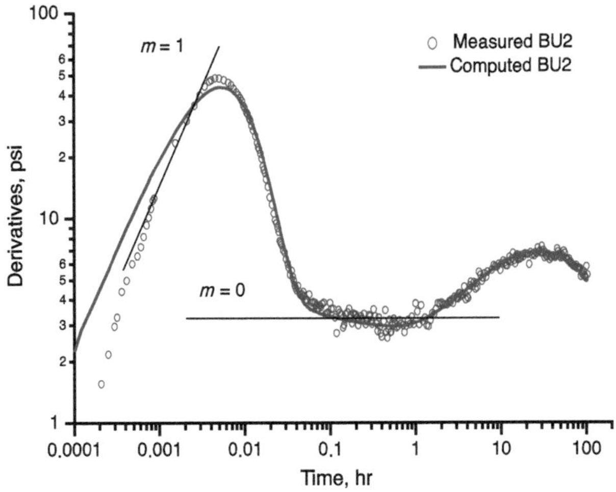

Figure 6.33 History match of the measured and computed derivatives of BU2, where the computed response is based on the Option 1 nonlinear regression (see Table 6.11).

We will history match only the BU2 pressure data for times greater than 0.007 hr. Furthermore, because of a possible presence of gas in the first (top) layer, we, therefore, will treat its viscosity and total compressibility as unknown parameters.

The results obtained from nonlinear regression (based on the maximum likelihood estimation method) for three different options (Options 1, 2, and 3) are summarized in Table 6.11. The numbers given as (±) represent the 95-percent confidence intervals for the estimated parameters. In Option 1, we regress on a total of 12 unknown parameters. For this option, as can be seen from Table 6.11, based on the values of confidence intervals, we may conclude that only the third layer horizontal and vertical permeabilities, the second layer vertical permeability, wellbore storage coefficient, and the damage skin factor were estimated reliably. We are also confident that the initial reservoir pressure of 3874 psi is correct, as explained above from the BU2 and deconvolved derivatives, even through its confidence interval is high, as shown in Table 6.11. The other parameters cannot be determined reliably because the BU2 pressure behavior may not be sensitive to these parameters, and/or the estimated parameters are highly correlated with the other parameters or each other. Figure 6.33 compares the measured BU2 derivative with the three-layer reservoir model derivative response computed by using Option 1 parameters given in Table 6.11. As can be seen form this figure, the match is very good indeed, and the standard deviation of errors in pressure estimated by the MLE is about 0.44 psi. As can be seen from this figure, the entire derivative behavior of BU2 is reproduced, except for a few minutes during the early time, where the measured derivative behavior deviates from the unit slope line. To further check whether the Option 1

Table 6.11 Summary of nonlinear regression analysis results

Regression options	k_{h1} (md)	k_{h2} (md)	k_{h3} (md)	k_{v1} (md)	k_{v2} (md)	k_{v3} (md)	σ (psi)
Option 1[b]	522 ± 230000	5 ± 10000	224 ± 117	34 ± 23000	0.24 ± 0.95	24 ± 10	0.44
Option 2[c]	582 ± 200000	5 ± 200	224[a]	95 ± 23000	0.24[a]	24[a]	0.44
Option 3[d]	519 ± 400000	5 ± 4000	214 ± 499	91 ± 96000	0.27 ± 2	26 ± 55	0.79
	μ_1 (cp)	c_{t1} $(\text{psi})^{-1}$	C (B/psi)	S	L_w (ft)	p_o (psi)	
Option 1[b]	0.014 ± 4.5	0.0021 ± 0.96	0.006 ± 0.00006	18.59 ± 4.5	445 ± 213	3874.6 ± 436	
Option 2[c]	0.012 ± 4.5	0.0004 ± 0.006	0.006[a]	18.59[a]	442 ± 2	3874.6[a]	
Option 3[d]	0.015 ± 11	0.00043 ± 0.01	0.0066 ± 0.0001	17.4 ± 10	455 ± 1000	3874.6[a]	

[a] Parameter fixed in regression.
[b] Option 1 = Total of 12 unknown parameters.
[c] Option 2 = Total of 6 unknown parameters.
[d] Option 3 = Total of 11 unknown parameters.

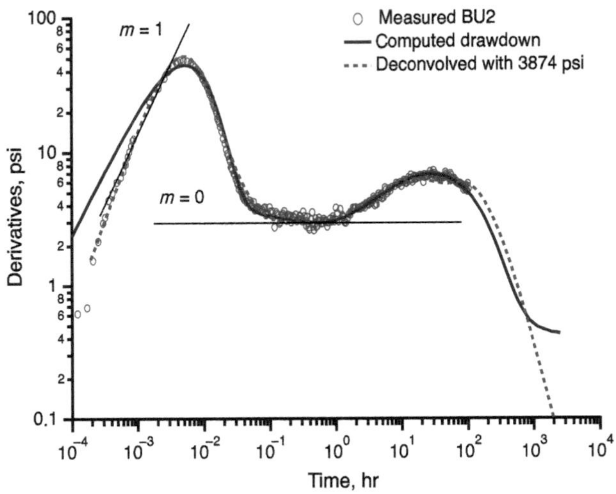

Figure 6.34 Comparison of measured BU2 and deconvolved responses, and drawdown response based on the model parameters estimated by nonlinear regression Option 1.

model and its parameters reproduces the deconvolved response obtained for $p_O = 3874$ psi previously (see Figure 6.31), we simulated a constant rate drawdown (based on a reference rate of 3454 B/D) test for 8960 hr using the Option 1 model and its parameters. Figure 6.34 compares the pressure change and derivative responses generated from this drawdown test with the corresponding deconvolved response. Clearly, the agreement is quite good until 700 hr, at which point the model derivative curve starts to flatten out due to finite thickness of the gas cap at the top layer. It is most likely that the deconvolved derivative is inaccurate after 700 hr, because there is a very large gap in the deconvolved response, because the BU2 test duration is not sufficiently long enough compared to the entire production period.

As a final interpretation and validation step, we further investigate the sensitivity of the BU2 pressure data to the first (top) and second layer parameters, and performed a second nonlinear regression as shown by Option 2 in Table 6.11, for which we reduced the total of unknown parameters from 12 (Option 1) to 6. As shown in Table 6.11, those six parameters determined reliably from the Option 1 nonlinear regression were fixed for the Option 2 nonlinear regression. The match obtained between model and measured buildup pressures for Option 2 is as good as the match obtained from Option 1 because the MLE for Option 2 yields the same standard deviation of errors equal to 0.44 psi. As shown in Table 6.11, confidence intervals for some of the parameters have been improved, particularly for the half horizontal well length, and the total compressibility of the top layer. However, there are still large uncertainties in the top layer horizontal and vertical permeabilities and the second layer

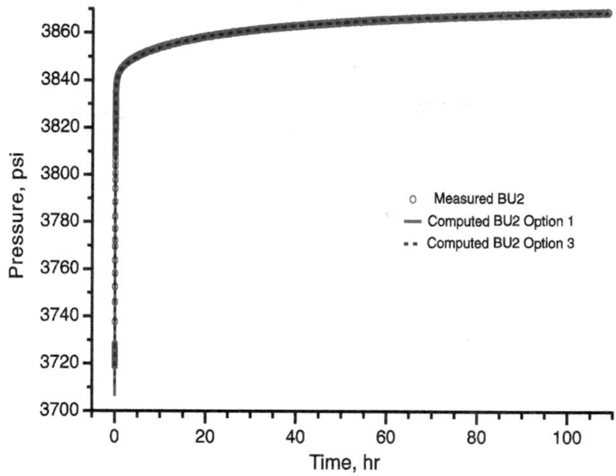

Figure 6.35 Comparison of the measured BU2 pressure and computed pressures from Options 1 and 3 nonlinear regression.

horizontal permeability. The above results indicate that these parameters cannot reliably be determined from the BU2 pressure data.

As a final regression, we also regressed using the entire BU2 pressure data, including the pressures earlier than 0.007 hr. This regression option is called Option 3, and its estimated parameters are given in Table 6.11. As can be observed from this table, the estimated parameters for this option did not change significantly compared to the estimated parameters for Option 1, for which the pressure data earlier than 0.007 hr were not used. However, the standard deviation of errors for shut-in pressure mismatch increased about twofold (0.79 psi), compared to that of 0.44 psi obtained from Option 1. Figure 6.35 compares the measured and computed BU2 pressures from the three-layer reservoir model using Option 1 and 3 parameters given in Table 6.11. As can be seen from this figure, the agreement between the measured and computed pressures is excellent, and the standard deviation of errors is 0.44 psi for Option 1, whereas it is 0.79 psi for Option 3. Figure 6.36 shows the comparison of measured and computed pressure changes ($\Delta p = p_{ws} - p_{wf}(\Delta t = 0)$, where $p_{wf}(\Delta t = 0)$ is the flowing bottomhole pressure at the moment of shut-in) for the same data given in Figure 6.35. As can be observed from Figure 6.36, the agreement among the measured and computed Δp is not as good as the agreement obtained among the measured and computed pressures. The difference between the measured and computed pressure changes for Option 1 is about 12.7 psi and is mainly due to the difference in the measured and computed values of $p_{wf}(\Delta t = 0)$. The measured $p_{wf}(\Delta t = 0)$ is equal to 3718.98 psi, whereas the computed one is 3706.29 psi, and the difference is 12.69 psi. This difference is expected because the computed buildup pressures and buildup pressure changes for

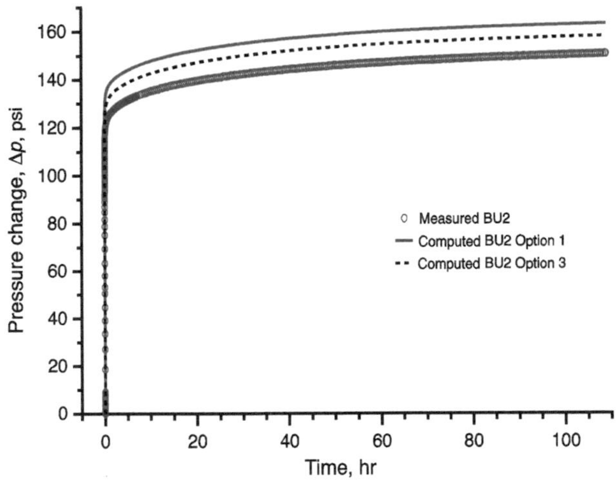

Figure 6.36 Comparison of the measured BU2 pressure change and computed pressure changes from Options 1 and 3 nonlinear regression.

Option 1 were obtained only by regressing on the pressure data for times greater than 0.007 hr. Zero weights were assigned for pressure data earlier than 0.007 hr simply because the earlier pressure data deviated from the unit-slope storage behavior.

The computed shut-in pressures and shut-in pressure changes for Option 3 are also shown in Figures 6.35 and 6.36, respectively. Although it is difficult to observe the difference between Option 1 and 3 pressure matches from Figure 6.35, the pressure match for Option 1 is better than that from Option 3 because the standard deviation of errors are 0.44 psi and 0.79 psi, respectively. On the other hand, the difference between Option 1 and 3 pressure change matches can clearly be observed in Figure 6.35, and the match for Option 3 is better than the Option 1 match. The reason for this disagreement between the two matches is that the Option 3 computed $p_{wf}(\Delta t = 0)$ of 3711.42 psi is much closer to the measured $p_{wf}(\Delta t = 0)$ of 3718.98 psi compared to 3706.29 psi from Option 1.

Figure 6.37 shows the match obtained for the measured BU2 derivative and the computed BU2 derivatives from Options 1 and 3 nonlinear regression. As can be seen from this figure, there is no observable difference among these three derivatives, except for times smaller than 0.007 hr.

Table 6.12 presents the matrix of correlation coefficients for Option 1 (12 unknown parameters as shown in Table 6.11). As can be observed from this table, the following parameter pairs are strong correlated:

- k_{h3} and L_w with a correlation coefficient of -0.99,
- k_{v3} and L_w with a correlation coefficient of -0.97,
- p_o and c_{t1} with a correlation coefficient of $+0.97$,

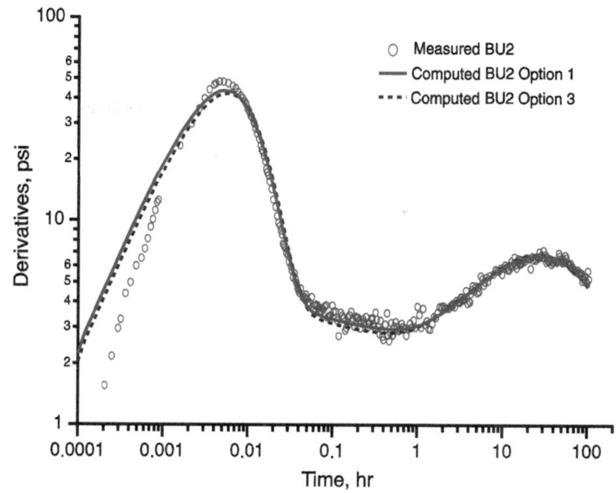

Figure 6.37 History matches of the measured BU2 derivative and computed BU2 derivatives from Options 1 and 3 nonlinear regression.

- k_{v2} and k_{v3} with a correlation coefficient of -0.91, and
- k_{v2} and L_w with a correlation coefficient of -0.89.

These results of the matrix of correlation coefficients indicate that the horizontal well length is highly correlated with the horizontal and vertical permeabilities of the third layer, where the horizontal well is located, and the vertical permeability of the second layer. And also the compressibility of the top layer is highly correlated with the initial pressure, indicating that increasing compressibility of the top layer while increasing initial pressure we would obtain the same pressure response. Because of these strong correlations among the estimated parameters the confidence intervals for some of these correlated parameters are very high, as can be seen from Table 6.11.

6.4.3. Interpretation summary of two buildup tests from well X-184

The reservoir model and its parameters are reasonably well defined from the second buildup test for Well X-184. As shown in Table 6.11, the third layer horizontal and vertical permeabilities, the second layer vertical permeability, wellbore storage coefficient, the damage skin factor and the well length were estimated reliably. It is most likely that the initial reservoir pressure of 3874.00 psi is correct because its deconvolved derivative captured the observed BU2 derivative behavior better than that from 3898.30 psi as shown in Figure 6.31 even though its confidence interval was high, as shown in Table 6.11. The other parameters cannot be determined reliably because the

Table 6.12 Matrix of correlation coefficients of the estimated parameters from Option 1

	k_{h1} (md)	k_{h2} (md)	k_{h3} (md)	k_{v1} (md)	k_{v2} (md)	k_{v3} (md)	μ_1 (cp)	c_{t1} (1/psi)	L_w (ft)	S	p_o (psi)	C (B/psi)
k_{h1}	1.00	0.36	−0.29	−0.2	0.51	−0.33	0.56	−0.70	0.31	0.05	−0.71	0.10
k_{h2}	–	1.00	−0.33	0.47	0.73	−0.6	−0.05	−0.60	0.45	0.62	−0.44	0.01
k_{h3}	–	–	1.00	−0.22	−0.84	0.93	−0.01	0.49	−0.99	0.42	0.30	−0.37
k_{v1}	–	–	–	1.00	0.4	−0.33	−0.55	−0.32	0.27	0.24	−0.21	0.05
k_{v2}	–	–	–	–	1.00	−0.91	−0.06	−0.80	0.89	−0.02	−0.62	0.27
k_{v3}	–	–	–	–	–	1.00	0.01	0.56	−0.97	0.05	0.35	−0.24
μ_1	–	–	–	–	–	–	1.00	0.18	0.00	−0.03	0.19	0.01
c_{t1}	–	–	–	–	–	–	–	1.00	−0.53	−0.07	0.96	−0.16
L_w	–	–	–	–	–	–	–	–	1.00	−0.28	−0.33	0.34
S	–	–	–	–	–	–	–	–	–	1.00	−0.08	−0.35
p_o	–	–	–	–	–	–	–	–	–	–	1.00	−0.10
C	–	–	–	–	–	–	–	–	–	–	–	1.00

BU2 pressure behavior is not sensitive to these parameters. On the other hand, it is still difficult to explain the behavior of the first test. The final estimates and model have the following consequences:

1. The top layer most likely has a secondary gas cap,
2. The top layer most likely is not naturally fractured because its estimated vertical permeability is not very high,
3. The results from derivatives and regression indicate that the second layer definitely has a low vertical permeability, therefore it may act as a barrier for any oil and water crossflow between the first (top) and the third (bottom) layers, and
4. The wellbore is highly damaged (a skin of 18.59) after the first BU test.

Unlike the first buildup test, the wellbore storage does not hinder the interpretation of the second BU test, and all expected flow regimes are observed very well because of the downhole shut-in. The BU2 test could have been run longer to see the third radial flow regime for the whole three-layer system. However, this would have required more than a few hundred-hour tests, which would have been difficult operationally. However, if the downhole drawdown pressure were recorded about 50 to 100 hr during the production before the BU2 test, it would have provided significant improvement in deconvolution and nonlinear parameter estimation without any production loss. As discussed earlier in this chapter, the difficulties of interpretation of these two buildup tests would have been considerably minimized if these test were conducted with a good test design and real-time interpretation.

REFERENCES

Aanonsen, S., Aavatsmark, I., Barkve, T., Cominelli, A., Gonard, R., & Gosselin, O. et al. (2003). Effect of scale dependent data correlations in an integrated history matching loop combining production and 4D seismic data. In *Paper SPE 79665, SPE reservoir simulation symposium.*

Aanonsen, S., Nvda, G., Oliver, D., & Reynolds, A. (2009). The Ensemble Kalman Filter in reservoir engineering–a review. *SPE Journal, 14*(3), 393–412.

Abbaszadeh, M., & Hegeman, P. (1990). Pressure-transient analysis for a slanted well in a reservoir with vertical pressure support. *SPE Formation Evaluation, 5*(3), 277–284.

Abbaszadeh, M., & Kamal, M. (1988). Automatic type-curve matching for well test analysis. *SPE Formation Evaluation, 3*(3), 567–577.

Abel, N. H. (1823). Oplsning af et par opgaver ved hjelp af bestemt integraler. *Magazine for naturvidenskaberne, 1*, 55–68.

Abramowitz, M., & Stegun, I. A. (1972). *Handbook of mathematical functions.* New York: Dover.

Agarwal, G. R., Al-Hussainy, R., & Ramey, H. J. (1970). An investigation of wellbore storage and skin effect in unsteady liquid flow: I. analytical treatment. *SPE Journal, 10*, 279–290.

Akkurt, R., Bowcock, M., Davies, J., Del Campo, C., Hill, B., Joshi, S., et al. (2007). Focusing on downhole fluid sampling and analysis. *Oilfield Review, 18*(4), 4–19.

Al-Hussainy, R., Ramey, H., & Crawford, P. (1966). The flow of real gases through porous media. *Journal of Petroleum Technology, 18*(5), 624–636.

Alabert, F. (1987). The practice of fast conditional simulations through the Lu decomposition of the covariance matrix. *Mathematical Geology, 19*, 369–375.

Anraku, T., & Horne, R. (1995). Discrimination between reservoir models in well-test analysis. *SPE Formation Evaluation, 10*(2), 114–121.

Anterion, F., Karcher, B., & Eymard, R. (1989). Use of parameter gradients for reservoir history matching. In *Paper SPE 18433, SPE reservoir simulation symposium.*

Athichanagorn, S., Horne, R., & Kikani, J. (2002). Processing and interpretation of long-term data acquired from permanent pressure gauges. *SPE Reservoir Evaluation and Engineering, 5*(5), 384–391.

Ayestaran, L. (1987). Testing moves downhole. *Schlumberger Middle East Well Evaluation Review, 2*, 34–47.

Baker, A., Gaskell, J., Jeffery, J., Thomas, A., Veneruso, A., & Unneland, T. (1995). Permanent monitoring - looking at lifetime reservoir dynamics. *Oilfield Review*, 7(4), 32–46.

Ballin, P., Journel, A. G., & Aziz, K. (1992). Prediction of uncertainty in reservoir performance forecast. *Journal of Canadian Petroleum Technology, 31*(4), 52–62.

Ballin, P., Aziz, K., Journel, A. G., & Zuccolo, L. (1993). Quantifying the impact of geological uncertainty on reservoir forecasts. In *Paper SPE 25238, 12th SPE symposium on reservoir simulation.*

Bannies, H. (1957). A new intermitting device which continuously measures bottom-hole pressure and regulates pump submergence. In *Paper SPE 954-G, Fall meeting of the Southern California petroleum section of AIME.*

Bard, Y. (1974). *Nonlinear parameter estimation.* San Diego: Academic Press.

Developments in Petroleum Science, Volume 57
ISSN 0376-7361, DOI: 10.1016/S0376-7361(10)05713-4

Barlow, R. (1989). *Statistics—a guide to the use of statistical methods in the physical sciences.* Chichester: John Wiley and Sons.

Barua, J., Horne, R., Greenstadt, J., & Lopez, L. (1988). Improved estimation algorithms for automated type curve analysis of well tests. *SPE Formation Evaluation, 3*(1), 186–196.

Barua, J., Kucuk, F., & Gomez-Angulo, J. (1985). Application of computers in the analysis of well tests from fractured reservoirs. In *SPE 13662, SPE California regional meeting.*

Bates, D. M., & Watts, D. G. (1988). *Nonlinear regression analysis and its applications.* New York: Wiley.

Baygun, B., Kuchuk, F., & Arikan, O. (1997). Deconvolution under normalized autocorrelation constraints. *SPE Journal, 2*(3), 246–253.

Bezerra, M., Da Silva, S., & Theuveny, B. (1992). Permanent downhole gauges: A key to optimize deepsea production. In *Paper OTC 6991, 24th Annual offshore technology conference.*

Blasingame, T., Johnston, J., Lee, W., & Raghavan, R. (1989). The analysis of gas well test data distorted by wellbore storage using an explicit deconvolution method. In *Paper SPE 19099, SPE gas technology symposium.*

Bostic, J., Agarwal, R., & Carter, R. (1980). Combined analysis of postfracturing performance and pressure buildup data for evaluating an MHF gas well. *Journal of Petroleum Technology, 32*(10), 1711–1719.

Bourdet, D. (2002). *Well test analysis: the use of advanced interpretation models.* Boston: Elsevier.

Bourdet, D., Ayoub, J., & Pirard, Y. M. (1989). Use of pressure derivative in well-test interpretation. *SPE Formation Evaluation, 4*(2), 293–302.

Bourdet, D., Whittle, T., & Douglas, A. (1983). A new set of type curves simplifies well test analysis. *World Oil, 6*, 95–106.

Bourgeois, M., & Horne, R. (1993). Well-test-model recognition with Laplace space. *SPE Formation Evaluation, 8*(1), 17–25.

Bracewell, R. (1973). *The Fourier transform and its applications.* New York City: McGraw-Hill.

Brons, F., & Marting, V. E. (1961). The effect of restricted fluid entry on well productivity. *Journal of Petroleum Technology, 13*(2), 172–174.

Burns, W. (1969). New single well test for determining vertical permeability. *Journal of Petroleum Technology, 21*(6), 743–752.

Caers, J. (2005). *SPE Interdisciplinary primer series, Petroleum geostatistics* (1st ed.). Dallas: Society of Petroleum Engineers.

Cannon, J. R. (1984). *The one-dimensional heat equation.* Menlo Park, CA: Addison-Wesley.

Carroll, R., & Ruppert, D. (1988). *Nonlinear regression analysis and its applications.* New York: Chapman & Hall.

Carslaw, H. C., & Jaeger, J. C. (1959). *Conduction of heat in solids.* Oxford: Clarendon Press.

Carter, P., & Morel, E. (1990). Reservoir monitoring in the development of marginal fields: Ivanhoe, Rob Roy, and Hamish. In *European petroleum conference.*

Carter, R., Kemp, L., Pierce, A., & Williams, D. (1974). Performance matching with constraints. *SPE Journal, 14*(2), 187–196.

Carter, R., & Kuchuk, F. (1990). The numerical inversion of a class of convolution integrals. Schlumberger-Doll Reseach (unpublished report).

Carvalho, R., Redner, R., Thompson, L., & Reynolds, A. (1992). Robust procedures for parameter estimation by automated type-curve matching. In *Paper SPE 24732, 67th SPE annual technical conference and exhibition.*

Carvalho, R., Thompson, L., Redner, R., & Reynolds, A. (1996). Simple procedures for imposing constraints for nonlinear least squares optimization. *SPE Journal, 1*(4), 395–402.

Chavent, G., Dupuy, M., & Lemmonnier, P. (1975). History matching by use of optimal control theory. *SPE Journal, 15*(1), 74–86.

Chen, W., Gavalas, G., Seinfeld, J., & Wasserman, M. (1974). A new algorithm for automatic history matching. *SPE Journal, 14*(6), 593–608.

Chu, L., Reynolds, A., & Oliver, D. (1995). Robust procedures for parameter estimation by automated type-curve matching. In *Paper SPE 29999, International meeting on petroleum engineering*.

Coats, K., Rapoport, L., McCord, J., & Drews, W. (1964). Determination of aquifer influence functions from field data. *Journal of Petroleum Technology, 16*(12), 1417–1424.

Colton, D., Ewing, R. E., & Rundell, W. (1990). *Inverse problems in partial differential equations* (1st ed.). Philadelphia: Society for Industrial and Applied Mathematics.

Corbett, P., Couples, G., & Gardiner, A. (2004). *Geosciences manual* (1st ed.). Heriot–Watt University.

Corbett, P., Ellabad, Y., & Mohammed, K. (2001). The recognition, modeling and validation of hydraulic units in reservoir rock. In *3rd Institute of mathematics and its applications conference on modelling permeable rocks*.

Corbett, P., Zheng, S., Pinisetti, M., Mesmari, A., & Stewart, G. (1998). Geology and well testing for improved fluvial reservoir characterisation. In *Paper SPE 48880, SPE international oil and gas conference and exhibition in China*.

Corbett, P. W. M. (2009). *Distinguished instructor series: No. 12. Petroleum geoengineering: Integration of static and dynamic models* (1st ed.) (pp. 4–5). USA: SEG/EAGE.

Corbett, P. W. M., Geiger, S., Borges, L., Garayev, M., Gonzales, J., & Camilo, R. (2010). Limitation in numerical well test modeling of fractured carbonate rocks. In *Paper SPE 130252, SPE Europec/EAGE annual conference*.

Corduneanu, C. (1991). *Volterra equations and applications*. Cambridge, UK: Cambridge University Press.

Culham, W. (1974). Pressure buildup equations for spherical flow regime problems. *SPE Journal, 14*(6), 545–555.

Cullender, M. (1955). The isochronal performance method of determining the flow characteristics of gas wells. *Transactions, AIME, 204*, 137–142.

De Moor, B. (1993). Structured total least squares and L2 approximation problems. *Linear Algebra and its Applications, 188–189*, 163–207.

Dixon, T., Seinfeld, J., Startzman, R., & Chen, W. (1973). Robust procedures for parameter estimation by automated type-curve matching. In *Paper SPE 4282, 3rd SPE numerical reservoir simulation of reservoir performance symposium*.

Dogru, A., Dixon, T., & Edgar, T. (1977). Confidence limits on the parameters and predictions of slightly compressible. *SPE Journal, 17*(1), 42–56.

Draper, N., & Smith, H. (1966). *Applied regression analysis*. New York: Wiley.

Dubost, F. X., Zheng, S. Y., & Corbett, P. W. M. (2004). Analysis and numerical modelling of wireline pressure tests in thin-bedded turbidites. *Journal of Petroleum Science and Engineering, 45*(3–4), 247–261.

Duhamel, J. (1833). Memoire sur la methode generale relative au mouvement de la chaleur dans les corps solides polonge dans les milieux dont la temprature varie avec le temps. *Journal de L'École Polytechnique, 14*, 20–77.

Earlougher, R. (1977). *Monograph series 5, Advances in well test analysis* (1st ed.). Dallas, Texas: Society of Petroleum Engineers of AIME.

Earlougher, R., & Kersch, K. (1972). Examples of automatic transient test analysis. *Journal of Petroleum Technology, 24*(10), 1271–1277.

Ehlig-Economides, C., Joseph, J., Ambrose, R., & Norwood, C. (1990). A modern approach to reservoir testing. *Journal of Petroleum Technology, 42*(12), 1554–1563.

Ehlig-Economides, C., Joseph, J., Erba, M., & Vik, S. (1986). Evaluation of single-layer transients in a multilayered system. In *Paper SPE 15860, SPE European petroleum conference*.

Engel, R. (1963). Remote reading bottom-hole pressure gauges- an evaluation of installation techniques and practical applications. *Journal of Petroleum Technology, 15*(12), 1303–1313.

Ewing, R., Pilant, M., Wade, G., & Watson, A. (1994). Estimating parameters in scientific computation. *IEEE Computational Science and Engineering, 1*(3), 19–31.

Fair, W. J. (1981). Pressure buildup analysis with wellbore phase re-distribution. *SPE Journal, 21*(2), 259–270.

Fair, W. J. (1996). Generalization of wellbore effects in pressure-transient analysis. *SPE Formation Evaluation, 11*(2), 114–119.

Farmer, C. (1988). The generation of stochastic fields of reservoir parameters with specified geostatistical distributions. In *Mathematics in oil production* (pp. 235–252). Oxford: Oxford Science Publications, Clarendon Press.

Ferris, J., & Knowles, D. (1954). The slug test for estimating transmissibility. *U.S. Geological Survey of Ground Water Note, 26*, 1–7.

Fetkovich, M., & Vienot, M. (1984). Rate normalization of buildup pressure using afterflow data. *Journal of Petroleum Technology, 36*(12), 2211–2224.

Fisher, R. (1950). *Contributions to mathematical statistics*. New York: Wiley.

Fletcher, R. (1986). *Practical methods of optimization*. Chichester: John Wiley and Sons.

Gajdica, R., Wattenbarger, R., & Startzman, R. (1988). A new method of matching aquifer performance and determining original gas in place. *SPE Formation Evaluation, 3*(3), 985–994.

Gao, G., & Reynolds, A. (2006). An improved implementation of the LBFGS algorithm for automatic history matching. *SPE Journal, 1*(3), 5–17.

Gerard, M., & Horne, R. (1985). Effects of external boundaries on the recognition of reservoir pinchout boundaries by pressure transient analysis. *SPE Journal, 25*(3), 427–436.

Gill, P. E., Murray, W., & Wright, M. (1981). *Practical optimization*. London: Academic Press.

Gladfelter, R., Tracy, G., & Wilsey, L. (1955). Selecting wells which will respond to production-stimulation treatment. *Drilling and Production Practice, API*, 117–129.

Gok, I., Onur, M., Hegeman, P., & Kuchuk, F. (2006). Effect of an invaded zone on pressure-transient data from multiprobe and packer-probe wireline formation testers. *SPE Reservoir Evaluation and Engineering, 9*(1), 39–49.

Gok, I., Onur, M., & Kuchuk, F. (2005). Estimating formation properties in heterogeneous reservoirs using 3D interval pressure transient and geostatistical data. In *Paper SPE 93672, 14th SPE Middle East oil & gas show and conference*.

Golub, G., & Van Loan, C. (1993). An analysis of the total least squares problem. *SIAM Journal on Numerical Analysis, 17*, 883–893.

Goode, P., Pop, J., & Murphy, W. (1991). Multiprobe formation testing and vertical reservoir continuity. In *Paper SPE 22738, SPE annual technical conference and exhibition*.

Goode, P., & Thambynayagam, R. (1992). Permeability determination with a multiprobe formation tester. *SPE Formation Evaluation*, 7(4), 297–303.

Gringarten, A., & Ramey, H. (1973). The use of the point source and Green's functions in solving. *SPE Journal, 13*(5), 285–296.

Gringarten, A. C. (2008). From straight lines to deconvolution: The evolution of the state of the art in well test analysis. In *Paper SPE 102079, SPE annual technical conference and exhibition*.

Groetsch, C. (1984). *The theory of Tikhonov regularization for fredholm equations of the first kind*. Boston: Pitman Advanced Publishing.

Guttman, I., Wilks, S., & Hunter, J. (1982). *Introductory engineering statistics* (3rd ed.). New York: John Wiley & Sons, Inc.

Hadamard, J. (1953). *Lectures on Cauchy's problem in linear partial differential equations*. New York: Dover.

Hanson, J. (1986). Nonlinear inversion of pressure-transient data. *SPE Formation Evaluation, 1*(4), 353–362.

Hantush, M. S. (1957). Non-steady flow to a well partially penetrating an infinite leaky aquifer. *Proceedings of the Iraqi Scientific Societies, 1*, 10–19.

Hasan, A., & Kabir, C. (1995). Generalization of wellbore effects in pressure-transient analysis. *SPE Formation Evaluation, 2*(19), 155–161.

Hawkins, M. (1956). A note on the skin effect. *Journal of Petroleum Technology, 8*(12), 65–66.

He, N., Oliver, D., & Reynolds, A. (2000). Conditioning stochastic reservoir models to well-test data. *SPE Reservoir Evaluation and Engineering, 3*(1), 74–79.

Hegeman, P. (2009). Formation testing. In M. Kamal (Ed.), *Monograph series: Vol. 23. Transient well testing* (1st ed.) (pp. 729–756). Dallas: Society of Petroleum Engineers (chapter 21).

Hirasaki, G. J. (1974). Pulse tests and other early transient pressure analysies for in-situ estimation of vertical permeability. *SPE Journal, 14*(1), 75–90.

Horne, R. (1994). Advances in computer-aided well-test interpretation. *Journal of Petroleum Technology, 46*(7), 599–606.

Horne, R. (2007). Listening to the reservoir–interpreting data from permanent downhole gauges. *Journal of Petroleum Technology, 59*(12), 78–86.

Horne, R. N. (1995). *Modern well test analysis* (2nd ed.). Palo Alto: Petroway Inc.

Horner, D. (1951). Pressure buildup in wells. In Brill, E. J. (Ed.) *Proceedings, Leiden. Third World Petroleum Congress.*

Houze, O. P., Kikani, J., & Horne, R. N. (2009). Permanent gauges and production analysis. In M. Kamal (Ed.), *Monograph series: Vol. 23. Transient well testing* (1st ed.) (pp. 529–552). Dallas: Society of Petroleum Engineers (chapter 17).

Hurst, W. (1934). Unsteady flow of fluids in oil reservoirs. *Journal of Applied Physics, 1*(5), 20–30.

Hurst, W. (1953). Establishment of the skin effect and its impediment to fluid flow into a well bore. *Petroleum Engineers, 25*, B6–B16.

Hutchinson, T., & Sikora, V. (1959). A generalized water-drive analysis. *Transactions, AIME, 216*, 169–177.

Jargon, J. (1976). Effect of wellbore storage and wellbore damage at the active well on interference test analysis. *Journal of Petroleum Technology, 28*(8), 851–858.

Jargon, J., & van Poollen, H. (1965). Unit response function from varying-rate data. *Journal of Petroleum Technology, 17*(8), 965–969.

Jeffreys, H. (1961). *Theory of probability* (3rd ed.). London and New York: Oxford University Press.

Journel, A. G., & Huijbregts, C. (1978). *Mining geostatistics.* London: Academic Press.

Kabir, S. (2009). Wellbore effects. In M. Kamal (Ed.), *Monograph series: Vol. 23. Transient well testing* (1st ed.) (pp. 91–127). Dallas: Society of Petroleum Engineers (chapter 6).

Kabir, S., & Kuchuk, F. (1985). Well testing of low transmissivity reservoirs. In *Paper SPE 13666, SPE California regional meeting.*

Kamal, M., Abbaszadeh, M., Cinco-Ley, H., Hegeman, P., Horne, R., Houze, O., et al. (2009). In M. Kamal (Ed.), *Monograph series: Vol. 23. Transient well testing* (1st ed.). Dallas: Society of Petroleum Engineers.

Kamal, M., & Brigham, W. (1976). Design and analysis of pulse tests with unequal pulse and shut-in periods. *Journal of Petroleum Technology, 28*(2), 205–212.

Kamal, M. M., Fryder, D. G., & Murray, M. A. (1995). Use of transient testing in reservoir management. *Journal of Petroleum Technology, 47*(11), 9929–9999.

Katz, D., Tek, M., & Jones, S. (1962). A generalized model for predicting the performance of gas reservoirs subject to water drive. In *Paper SPE 428, SPE annual meeting.*

Kikani, J. (2009). Well testing measurements. In M. Kamal (Ed.), *Monograph series: Vol. 23. Transient well testing* (1st ed.) (pp. 27–52). Dallas: Society of Petroleum Engineers (chapter 3).

Kitanidis, P. K. (1986). Parameter uncertainty in estimation of spatial functions: Bayesian estimation. *Water Resource Research, 22*(4), 499–507.

Kolb, R. (1960). Two bottom-hole pressure instruments providing automatic surface recording. *Journal of Petroleum Technology, 12*(2), 79–82.

Kress, R. (1989). *Linear integral equations.* New York: Springer-Verlag.

Kuchuk, F. (1990a). Application of convolution and deconvolution to transient well tests. *SPE Formation Evaluation, 5*(4), 375–384.

Kuchuk, F. (1990b). Gladfelter deconvolution. *SPE Formation Evaluation, 5*(3), 285–292.

Kuchuk, F. (1994). Pressure behavior of MDT packer module and DST in crossflow-multilayer reservoirs. *Journal of Petroleum Science and Engineering, 11*(2), 123–135.

Kuchuk, F. (1997). Pressure transient testing and interpretation for horizontal wells: field examples. In *Paper SPE 37796, SPE Middle East oil show and conference.*

Kuchuk, F. (1999). A new method for determination of reservoir pressure. In *Paper SPE 56418, SPE annual technical conference and exhibition.*

Kuchuk, F. (2009a). Radius of investigation for reserve estimation from pressure transient well tests. In *Paper SPE 20515, SPE Middle East oil and gas show and conference.*

Kuchuk, F. (2009b). Interpretation of simultaneously measured transient flow-rate and pressure data: convolution and deconvolution. In M. Kamal (Ed.), *Monograph series: Vol. 23. Transient well testing* (1st ed.) (pp. 477–527). Dallas: Society of Petroleum Engineers (chapter 16).

Kuchuk, F., & Ayestaran, L. (1983). Analysis of simultaneously measured pressure and sandface flow rate in transient testing. In *Paper SPE 12177, SPE annual technical conference and exhibition.*

Kuchuk, F., & Ayestaran, L. (1985). Analysis of simultaneously measured pressure and sandface flow rate in transient testing. *Journal of Petroleum Technology, 37*(2), 23–334.

Kuchuk, F., Carter, R., & Ayestaran, L. (1990a). Deconvolution of wellbore pressure and flow rate. *SPE Formation Evaluation, 5*(1), 53–59.

Kuchuk, F., Goode, P. A., Wilkinson, D. J., & Thambynayagam, R. K. M. (1991). Pressure-transient behavior of horizontal wells with and without gas cap or aquifer. *SPE Formation Evaluation, 6*(1), 86–94.

Kuchuk, F., Hollaender, F., Gok, I., & Onur, M. (2005). Decline curves from deconvolution of pressure and flow-rate measurements for production optimization and prediction. In *Paper SPE 96002, SPE annual technical conference and exhibition.*

Kuchuk, F., Ramakrishnan, T., & Dave, Y. (1994). Interpretation of wireline formation tester packer and probe pressures. In *Paper SPE 28404, Annual technical conference and exhibition.*

Kuchuk, F. J. (1996). Multiprobe wireline formation tester pressure behavior in crossflow-layered reservoirs. *In Situ, 20*(1), 1–40.

Kuchuk, F. J., Goode, P. A., Brice, B. W., Sherrard, D. W., & Thambynayagam, R. M. (1990b). Pressure transient analysis for horizontal wells. *Journal of Petroleum Technology, 42*(8), 974–979.

Kuchuk, F. J., & Onur, M. (2003). Estimating permeability distribution from 3D interval pressure transient tests. *Journal of Petroleum Science and Engineering, 39*, 5–27.

Kuchuk, F. J., & Wilkinson, D. (1991). Pressure behavior of commingled reservoirs. *SPE Formation Evaluation, 6*(1), 111–120.

Landa, J. (2009). Integration of well testing into reservoir characterization. In M. Kamal (Ed.), *Monograph series: Vol. 23. Transient well testing* (1st ed.) (pp. 787–822). Dallas: Society of Petroleum Engineers (chapter 23).

Landa, J., Horne, R., Kamal, M., & Jenkins, C. (2000). Reservoir characterization constrained to well-test data: a field example. *SPE Reservoir Evaluation and Engineering, 3*(4), 325–334.

Landa, J. L., & Horne, R. N. (1997). A procedure to integrate well test data, reservoir performance history and 4-d seismic data into a reservoir description. In *Paper SPE 38653, Annual technical conference and exhibition.*

Larsen, L. (2005). Uncertainties in standard analyses of boundary effects in buildup data. *SPE Reservoir Evaluation & Engineering, 8*(5), 437–444.

Lebourg, M., Fields, R., & Doh, C. (1957). A method of formation testing on logging cable. *Transactions, AIME, 210*, 260–267.

Lee, W., Rollins, J., & Spivey, J. (2003). *SPE textbook series: Vol. 9. Pressure transient testing.* Society of Petroleum Engineers.

Lenn, C., Kuchuk, F., Rounce, J., & Hook, P. (1990). Horizontal well performance evaluation and fluid entry mechanisms. In *Paper SPE 49089, SPE annual technical conference and exhibition.*

Levenberg, K. (1944). A method for the solution of certain nonlinear problems in least-squares. *Quarterly of Applied Mathematics, 2*, 164–168.

Levitan, M. M. (2005). Practical application of pressure/rate deconvolution to analysis of real well tests. *SPE Reservoir Evaluation and Engineering, 8*(2), 113–121.

Levitan, M. M., Crawford, G. E., & Hardwick, (2006). Practical considerations for pressure-rate deconvolution of well-test data. *SPE Journal, 11*(1), 35–47.

Li, G., Han, M., Banerjee, R., & Reynolds, A. (2009). Integration of well test pressure data into heterogeneous geological reservoir models. In *Paper SPE 124055, SPE annual technical conference and exhibition.*

Li, G., & Reynolds, A. (2009). Ensemble Kalman Filters for data assimilation. *SPE Journal, 14*(3), 496–505.

Li, R., Reynolds, A., & Oliver, D. (2003). History matching of three-phase flow production data. *SPE Journal, 8*(4), 328–340.

Linz, P. (1985). Analytical and numerical methods for volterra equations. *SIAM Studies in Applied Mathematics.*

Ljung, L. (1987). *System identification: theory for the users*. Englewood Cliffs, NJ: Prentice Hall.

Lozano, G., & Harthorn, W. (1959). Field test confirms accuracy of new bottom-hole pressure gauge. *Journal of Petroleum Technology, 11*(2), 26–30.

Luenberger, D. G. (1969). *Optimization by vector space methods*. New York: Wiley.

Marquardt, D. (1963). An algorithm for least squares estimation of nonlinear parameters. *SIAM Journal on Applied Mathematics, 11*, 431–441.

Mathematica (2009). *Mathematica (software), Ver. 7.0.1.* Wolfram Research, Champaign-Urbana, IL.

Matthews, C., Brons, F., & Hazebroek, P. (1954). Determination of average pressure in a bounded reservoi. *Transactions, AIME, 201*, 182–191.

Matthews, C., & Russell, D. (1967). *Monograph series: Vol. 1. Pressure buildup and flow tests in wells* (1st ed.). Dallas, Texas: Society of Petroleum Engineers of AIME.

McEdwards, D. (1981). Multiwell variable-rate well test analysis. *SPE Journal, 21*(4), 444–446.

McKinley, P. M., Vela, S., & Carlton, L. A. (1968). A field application of pulse-testing for detailed reservoir description. *Journal of Petroleum Technology, 20*(3), 313–321.

McKinley, R. M. (1971). Wellbore transmissibility from afterflow-dominated pressure buildup data. *Journal of Petroleum Technology, 23*(7), 863–872.

Mendes, L., Tygel, M., & Correa, A. (1989). A deconvolution algorithm for analysis of variable–rate well test pressure data. In *Paper SPE 19815, SPE annual technical conference and exhibition.*

Mengen, A., & Onur, M. (2004). Application of different optimization techniques to parameter estimation from interference well test data. *Turkish Journal of Oil and Gas, 10*(2), 3–16.

Meunier, D., Wittmann, M., & Stewart, G. (1985). Interpretation of pressure buildup test using in-situ measurement of afterflow. *Journal of Petroleum Technology, 37*(1), 143–152.

Miller, C., Dyes, A., & Hutchinson, C. (1950). The estimation of permeability and reservoir pressure from bottom-hole pressure buildup characteristics. *Transactions, AIME, 189*, 91–104.

Miller, G., Seeds, R., & Shira, H. (1972). A new, surface recording, down-hole pressure gauge. In *Paper SPE 4125, Fall meeting of the society of petroleum engineers of AIME.*

Millikan, C. V., & Sidwell, C. V. (1931). Bottom-hole pressures in oil wells. In *Paper SPE 931194-G.*

Moore, T., Schilthuis, R., & Hurst, W. (1933). The determination of permeability from field data. *Amer. Petrol. Inst., Prod. Bull., 211*, 4–13.

Moran, J., & Finklea, E. (1962). Theoretical analysis of pressure phenomena associated with the wireline formation tester. *Journal of Petroleum Technology, 14*(8), 899–908.

Morozov, V. A. (1984). *Methods for solving incorrectly posed problems*. Berlin: Springer-Verlag.

Murray, M. (2009). Drillstem testing. In M. Kamal (Ed.), *Monograph series: Vol. 23. Transient well testing* (1st ed.) (pp. 757–785). Dallas: Society of Petroleum Engineers (chapter 22).

Muskat, M. (1934). The flow of compressible fluids through porous media and some problems in heat conduction. *Physics*, *5*(71), 71–94.
Muskat, M. (1937a). *The flow of homogeneous fluids through porous media*. Ann Arbor, Michigan: J. W. Edwards, Inc.
Muskat, M. (1937b). Use of data on the buildup of bottom hole pressures. *Transactions, AIME*, *123*, 44.
Nanba, T., & Horne, R. (1992). An improved algorithm for automated well test analysis. *SPE Formation Evaluation*, 7(1), 61–91.
Nisle, R. G. (1958). The effect of partial penetration on pressure build-up in oil wells. *Transactions, AIME*, *213*, 85–90.
Odeh, A., & Jones, L. (1965). Pressure drawdown analysis, variable-rate case. *Journal of Petroleum Technology*, *17*(8), 960–964.
Odeh, A., & Selig, F. (1963). Pressure build-up analysis, variable-rate case. *Journal of Petroleum Technology*, *15*(7), 789–795.
Ogbe, D. O., & Brigham, W. (1984). A model for interference testing with wellbore storage and skin effects at both wells. In *Paper SPE 13253, SPE annual technical conference and exhibition*.
Oliver, D. (1994). Incorporation of transient pressure data into reservoir characterization. *In Situ*, 18 (3) 243–275.
Oliver, D., He, N., & Reynolds, A. (1996). Conditioning permeability fields to pressure data. In *5th European conference on the mathematics of oil recovery*.
Oliver, D., Reynolds, A., & Liu, N. (2008). *Inverse theory for petroleum reservoir characterization and history matching*. Cambridge: Cambridge University Press.
O'Neill, F. E. (1934). Formation testers. In *Paper SPE 934053-G*.
Onur, M., Ayan, C., & Kuchuk, F. J. (2009a). Pressure-pressure deconvolution analysis of multiwell interference and interval pressure transient tests. In *Paper IPTC 13394, International petroleum technology conference held in Doha*.
Onur, M., Cinar, M., Ilk, D., Valko, P., Blasingame, T. S., & Hegeman, P. (2008). An investigation of recent deconvolution methods for well-test data analysis. *SPE Journal*, *13*(2), 226–247.
Onur, M., Gok, I., & Yamanlar, S. (2001). Effect of flow rate variations on the interpretation of pressure transient test data. *Turkish Journal of Oil and Gas*, 7(2), 3–13.
Onur, M., Hegeman, P., Gok, I., & Kuchuk, F. (2009b). A novel analysis procedure for estimating thickness-independent horizontal and vertical permeabilities from pressure data at an observation probe acquired by packer-probe wireline formation testers. In *Paper IPTC 13912, The international petroleum technology conference held in Doha*.
Onur, M., Hegeman, P., & Kuchuk, F. (2004a). Pressure-transient analysis of dual packer-probe wireline formation testers in slanted wells. In *Paper SPE 90250, SPE annual technical conference and exhibition*.
Onur, M., Hegeman, P., & Kuchuk, F. J. (2004b). Pressure/pressure convolution analysis of multiprobe and packer-probe wireline formation tester data. *SPE Reservoir Evaluation and Engineering*, 7(5), 351–364.
Onur, M., & Kuchuk, F. J. (1999). Integrated nonlinear regression analysis of multiprobe wireline formation tester packer and probe pressures and flow rate measurements. In *Paper SPE 56616, SPE annual technical conference and exhibition*.
Onur, M., & Kuchuk, F. J. (2000). Nonlinear regression analysis of well-test pressure data with uncertain variance. In *Paper SPE 62918, SPE annual technical conference and exhibition*.
Onur, M., & Reynolds, A. (2002). Nonlinear regression: the information content of pressure and pressure-derivative data. *SPE Journal*, 7(3), 243–249.
Onur, M., & Reynolds, A. C. (1998). Numerical Laplace transformation of sampled-data for well test analysis. *SPE Reservoir Evaluation and Engineering*, *1*(3), 268–277.
Ozkan, J. (2009). Horizontal wells. In M. Kamal (Ed.), *Monograph series*: *Vol. 23*. *Transient well testing* (1st ed.) (pp. 408–445). Dallas: Society of Petroleum Engineers (chapter 14).

Padmanabhan, L., & Woo, P. T. (1976). A new approach to parameter estimation in well testing. In *Paper SPE 5741, 4th SPE numerical reservoir simulation of reservoir performance symposium*.

Papatzacos, P. (1987). Approximate partial-penetration pseudoskin for infinite-conductivity wells. *SPE Reservoir Engineering*, *2*(2), 227–234.

Pascal, H. (1981). Advances in evaluating gas deliverability using variable ratetests under non-darcy flow. In *Paper SPE 9841, SPE/DOE low permeability sympasium*.

Perrine, R. L. (1956). Analysis of pressure buildup curves. *Drilling and Production Practice, API*, 482–509.

Pimonov, E., Ayan, C., Onur, M., & Kuchuk, F. J. (2009a). A new pressure/rate deconvolution algorithm to analyze wireline formation tester and well-test data. In *SPE 123982, SPE annual technical conference and exhibition*.

Pimonov, E., Onur, M., & Kuchuk, F. (2009b). A new robust algorithm for solution of pressure/rate deconvolution problem. *Journal of Inverse and Ill-posed Problems*, *17*(6), 609–625.

Pop, J., Badry, R., Morris, C., Wilkinson, D., Tottrup, P., & Jonas, J. (1993). Vertical interference testing with a wireline-conveyed straddle-packer tool. In *Paper SPE 26481, SPE annual technical conference and exhibition*.

Prats, M. (1970). A method for determining the net vertical permeability near a well from in-situ. *Journal of Petroleum Technology*, *22*(5), 637–643.

Prats, M., & Scott, J. (1975). Effects of wellbore storage on pulse-teat pressure response. *Journal of Petroleum Technology*, *27*(6), 707–709.

Press, W., Teukolsyk, S., Vetterling, W., & Flannery, B. (1986). *Numerical recipes: the art of scientific computing* (3rd ed.). Cambridge: Cambridge University Press.

Raghavan, R. (1993). *Well test analysis*. Boston: Prentice Hall.

Raghavan, R., & Clark, K. (1975). Vertical permeability from limited entry flow tests in thick formations. *SPE Journal*, *15*(1), 65–73.

Ramey, H. J. J. (1970). Short-time well test data interpretation of the presence of skin effect and wellbore storage. *Journal of Petroleum Technology*, *22*(1), 97–104.

Ramey, H. J. J (1976a). Verification of the Gladfelter, Tracey, and Wilsey concept for wellbore storage-dominated transient pressures during production. *Journal of Canadian Petroleum Technology*, *15*(2), 84–85.

Ramey, H. J. J. (1976b). Practical use of modern well test analysis. In *Paper SPE 5878, SPE California regional meeting*.

Reynolds, A. C., He, N., Chu, L., & Oliver, D. S. (1996). Reparameterization techniques for generating reservoir descriptions conditioned to variograms and well-test pressure data. *SPE Journal*, *1*(4), 413–426.

Riordan, M. (1951). Surface indicating pressure, temperature and flow equipment. *Transactions, AIME*, *192*, 257–262.

Rosa, A., & Horne, R. (1983). Automated type curve matching in well test analysis using Laplace space determination of parameter gradients. In *Paper SPE 12131, SPE annual technical conference*.

Rosa, A., & Horne, R. (1995). Automated well test analysis using robust nonlinear parameter estimation. *SPE Advanced Technology Series*, *3*(1), 95–102.

Rosenzweig, J. J., Korpics, D. C., & Crawford, G. E. (1990). Pressure transient analysis of the JX-2 horizontal well, Prudhoe Bay, Alaska. In *Paper SPE 20610, SPE annual technical conference and exhibition*.

Roumboutsos, A., & Stewart, G. (1988). A direct deconvolution or convolution algorithm for well test analysis. In *Paper SPE 18157, SPE annual technical conference and exhibition*.

Russell, D. (1963). Determination of formation characteristics from two-rate flow tests. *Transactions, AIME*, *228*, 1347–1355.

Samson, P., Fligelman, H., & Braester, C. (1985). New pressure analysis technique using repeat formation tester data. *The APEA Journal*, *25*(1), 275–280.

Schlumberger, (1981). *RFT essentials of pressure test interpretation*. Houston, Texas: Schlumberger.

Schlumberger, (1996). *Schlumberger wireline formation testing and sampling*. Houston, Texas: Schlumberger.

Schlumberger, (2006). *Fundamentals of formation testing*. Houston, Texas: Schlumberger.

Schultz, A., Bell, W., & Urbanosky, H. (1975). Advancements in uncased-hole, wireline formation-tester techniques. *Journal of Petroleum Technology*, *27*(11), 1331–1336.

Seber, G., & Wild, C. (1989). *Nonlinear regression*. New York: Wiley.

Shepherd, C., Neve, P., & Wilson, D. (1991). Use and application of permanent downhole pressure gauges in the Balmoral field and satellite structures. *SPE Production and Facilities*, *6*(3), 271–276.

Simmons, J. (1990). Convolution analysis of surge pressure data. *Journal of Petroleum Technology*, *42*(1), 74–83.

Stehfest, H. (1970). Algorithm 368: Numerical inversion of the Laplace transforms (D5). *Communication of the ACM*, *1*(13), 47–49.

Stewart, G., Meunier, D., & Wittmann, M. (1983). Afterflow measurement and deconvolution in well test analysis. In *Paper SPE 12174, SPE annual technical conference and exhibition*.

Streltsova, T. (1988). *Well testing in heterogeneous formations* (1st ed.). New York: John Wiley & Sons.

Suzuki, S., Caumon, G., & Caers, J. (2008). Dynamic data integration into structural modeling: Model screening approach using a distance-based model parameterization. *Computational Geosciences*, *12*(1), 105–119.

Tarantola, A. (2005). *Inverse problem theory and methods for model parameter estimation*. Philadelphia: SIAM.

Theis, C. V. (1937). The relation between the lowering of the piezometric surface and the rate and duration of discharge of well using ground-water storage. *Transactions, AGU*, *16*, 519–524.

Thompson, L., & Reynolds, A. (1986). Analysis of variable-rate well-test pressure data using Duhamels principle. *SPE Formation Evaluation*, *1*(5), 453–469.

Tiab, D., & Crichlow, H. B. (1979). Pressure analysis of multiple-sealing-fault systems and bounded reservoirs by type-curve matching. *SPE Journal*, *19*(6), 378–392.

Tiab, D., & Kumar, A. (1980). Application of the p'_D function to interference analysis. *Journal of Petroleum Technology*, *32*(8), 1465–1470.

Tikhonov, A. (1963). Resolution of ill-posed problems and the regularization method. *Doklady Akademii Nauk SSSR*, *151*, 501–504 (in Russian).

Tongpenyai, Y., & Raghavan, R. (1981). The effect of wellbore storage and skin on interference test data. *Journal of Petroleum Technology*, *33*(1), 151–160.

Unneland, T., Manin, Y., & Kuchuk, F. (1998). Application of permanent pressure and rate measurement systems in well monitoring and reservoir description: Field cases. *SPE Formation Evaluation*, *1*(3), 224–230.

van Everdingen, A. (1953). The skin effect and its influence on the productive capacity of a well. *Transactions, AIME*, *186*, 171–176.

van Everdingen, A., & Hurst, W. (1949). The application of the Laplace transformation to flow problems in reservoirs. *Transactions, AIME*, *186*, 305–324.

Van Huffel, S. (2003). Total least squares and errors-in-variables modeling: Problem formulation, algorithms, and applications PART II. In *SISTA seminar*.

Vella, M., Veneruso, T., Lefoll, P., McEvoy, T., & Reiss, A. (1992). The nuts and bolts of well testing. *Oilfield Review*, *4*(2), 14–27.

von Schroeter, T., & Gringarten, A. C. (2009). Superposition principle and reciprocity for pressure transient analysis of data from interfering wells. *SPE Journal*, *14*(3), 488–495.

von Schroeter, T., Hollaender, F., & Gringarten, A. (2004). Deconvolution of well-test data as a nonlinear total least-squares problem. *SPE Journal*, *9*(4), 375–390.

Wahba, G. (1977). Practical approximate solutions to linear operator equations when the data are noisy. *SIAM Journal on Numerical Analysis, 29*(4), 491–504.

Weber, K. (1986). How heterogeneity affects oil recovery. Reservoir characterization. In L. W. Lake, & H. B. Carroll (Eds.), *Reservoir characterization* (1st ed.) (pp. 91–127). New York, NY: Academic Press.

Whittle, T., & Gringarten, A. (2008). The determination of minimum tested volume from the deconvolution of well test pressure transients. In *SPE 116575, SPE annual technical conference and exhibition.*

Winestock, A., & Colpitts, G. (1965). Advances in estimating gas well deliverability. *Journal of Canadian Petroleum Technology, 4*(3), 111–119.

Wu, J., Georgi, D., & Ozkan, E. (2009). Deconvolution of wireline formation test data. In *Paper SPE 124220, SPE annual technical conference and exhibition.*

Wu, Z., Reynolds, A., & Oliver, D. (1999). Conditioning geostatistical models to two-phase production. *SPE Journal, 4*(2), 142–155.

Xian, C., Carnegie, A., Al Raisi, M., Petricola, M., & Chen, J. (2004). An integrated efficient approach to perform IPTT interpretation. SPE 88561, SPE Asia Pacific oil & gas conference and exhibition.

Zajic, T., & Kuchuk, F. (1991). Regularized deconvolution. *Schlumberger-Doll Reseach* (unpublished).

Zao, Y., Li, G., & Reynolds, A. (2007). Characterization of the measurement error in time-lapse seismic data and production data with an EM algorithm. *Oil & Gas Science and Technology-Reviews IFP, 62*(2), 181–193.

Zheng, S. Y., & Corbett, P. W. M. (2009). Title well testing best practice. In *Paper SPE 93984, SPE Europec/EAGE annual conference.*

Zimmerman, T., MacInnis, J., Hoppe, J., Pop, J., & Long, T. (1990). Application of emerging wireline formation testing technologies. In *OSEA 90105, 8th Offshore South East Asia conference.*

Subject Index

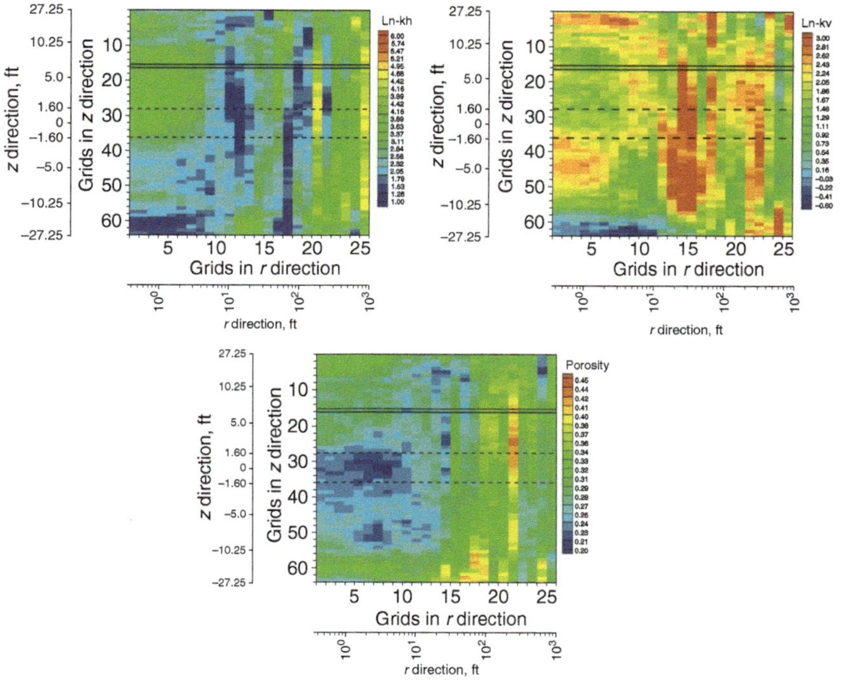

Plate 1 True log-horizontal permeability ($\ln k_h$) field (top left), true log-vertical permeability $\ln(k_v)$ field (top right), and true porosity (ϕ) field (bottom), Example 6. (see Figure 5.40, p. 289)

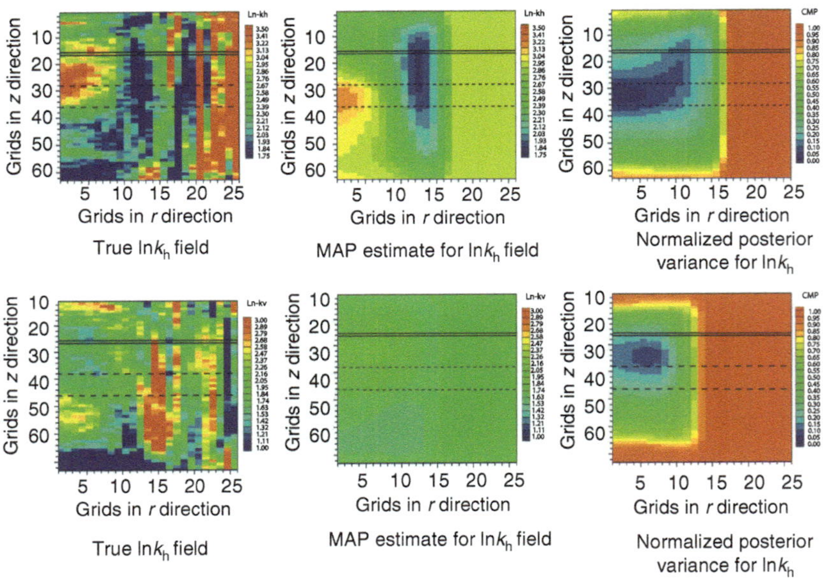

Plate 2 True, MAP estimates and normalized posterior variances for $\ln(k_h)$ and $\ln(k_v)$, packer and probe pressure matching simultaneously. (see Figure 5.44, p. 293)

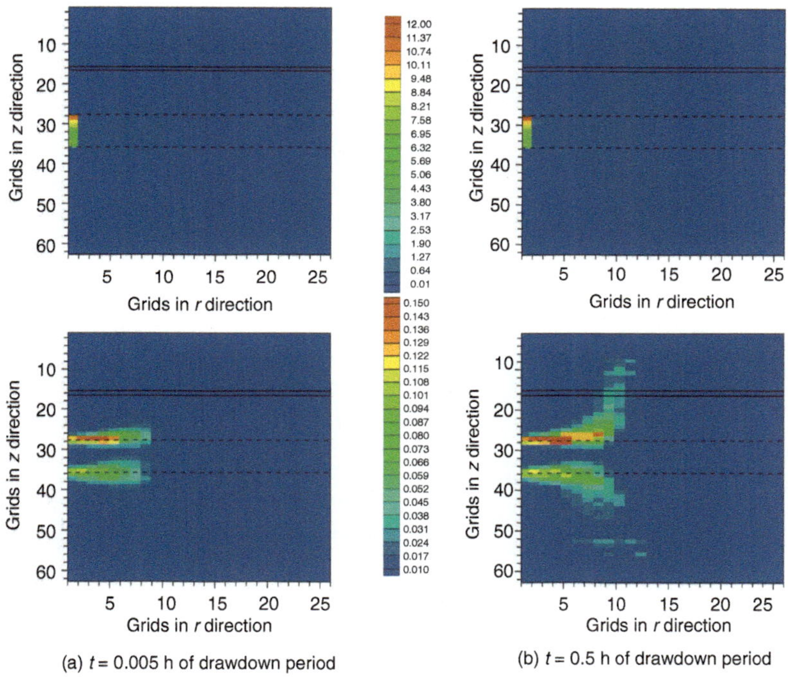

Plate 3 Sensitivity of packer pressure to $\ln(k_h)$ (top two figures) and $\ln(k_v)$ (bottom two figures). The figures on the left are for $t = 0.005$ hr, and the ones on the right are for $t = 0.5$ hr of drawdown period (see Figure 5.46, p. 295).

Source: Reproduced from Gok et al. (2005).

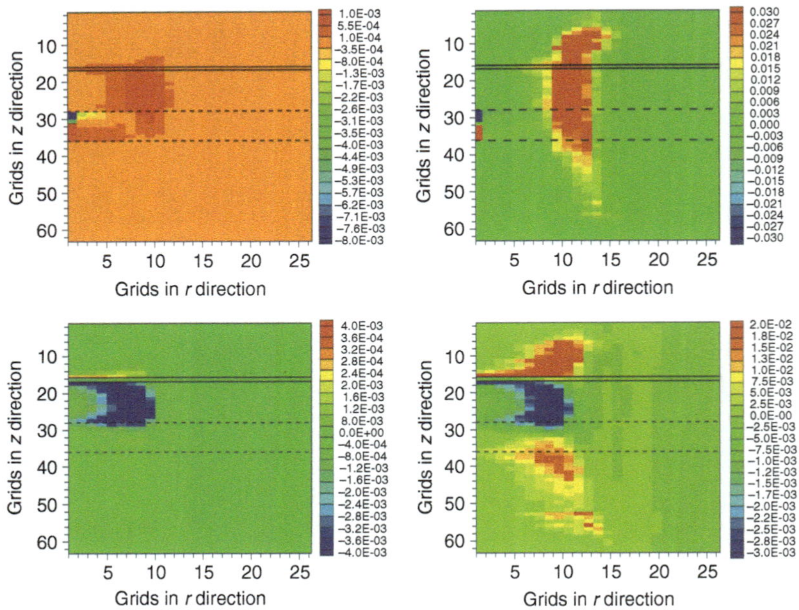

Plate 4 Sensitivity of probe pressure to $\ln(k_h)$ (top two figures) and $\ln(k_v)$ (bottom two figures). The figures on the left are for $t = 0.005$ hr, and the ones on the right are for $t = 0.5$ hr of drawdown period, reproduced from Gok et al. (2005). (see Figure 5.47, p. 296)

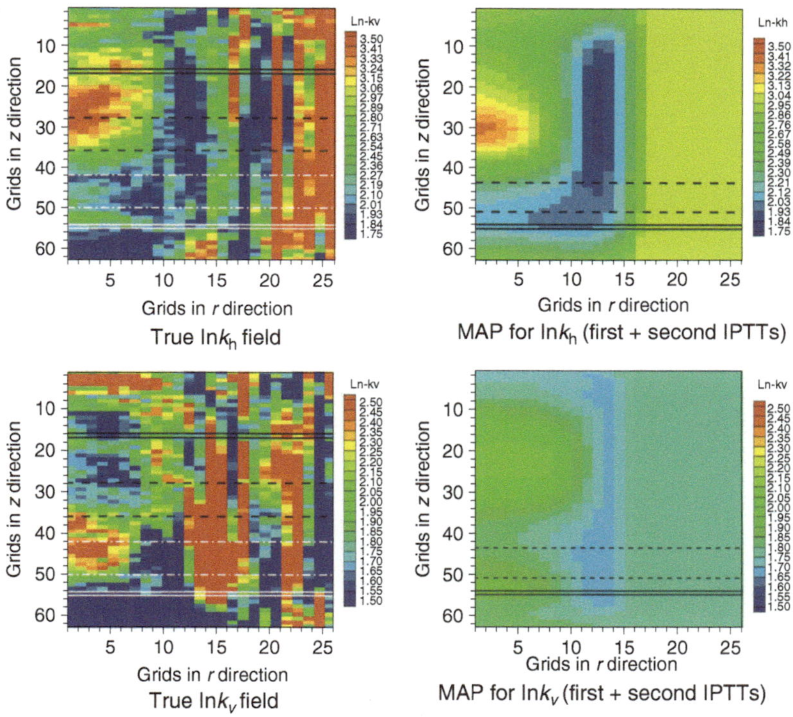

Plate 5 MAP estimates of log-horizontal and -vertical permeabilities conditioned to both packer and probe pressures of first and second IPTTs. (see Figure 5.48, p. 299)

FSC
www.fsc.org
MIX
Papier | Fördert
gute Waldnutzung
FSC® C083411

Kolibri 360 GmbH
Ferdinand-Jühlke-Straße 7
99095 Erfurt, Deutschland
produktsicherheit@kolibri360.de